Contents

Preface

This digest is the third of a series, the basic pattern of which was worked out in the production of No. 1—*Oils and oilseeds*. The digests are an attempt to present in a concise form basic data relating to the production and utilization of various groups of crops and their products, of economic importance to countries in the tropics and subtropics. For quick reference the data are arranged under standard headings and include particulars of growth requirements, planting and harvesting procedures, yield, products and their uses, processing techniques, trends in supply, demand and prices, and a bibliography.

The digests are in no way exhaustive or comprehensive. The aim is to provide a ready reference tool, for use particularly by non-specialists, and especially by practical workers in the developing countries concerned with advancing the rural economy. In addition, it is hoped that they may prove a useful starting point, or a framework to be expanded, for specialists and researchers, working within individual countries, or on individual crops, or their products.

Mrs. Kay has been helped in preparing this volume by a number of colleagues at the Institute, especially Drs. J. A. Cornelius, N. Macfarlane and H. C. Passam, who made helpful suggestions regarding the text, and Miss J. Church who provided the statistical data. Helpful advice and suggestions were also received from Mr. S. K. Karikari, Officer in Charge, Agricultural Research Station, Kade, University of Ghana, while he was undertaking a training assignment at the Institute. In addition, assistance and advice regarding the botanical nomenclature was generously given by Dr. R. M. Polhill, Royal Botanic Gardens, Kew, and Dr. C. L. A. Leakey, Tropical Agriculture Consultant, provided some pre-harvest information on the haricot bean. I wish to express my gratitude to them all for their efforts.

P. C. Spensley
Director

Tropical Products Institute
56/62 Gray's Inn Road
London WC1X 8LU
1979

Introduction

This digest, *Food legumes*, deals with those species of the plant family Leguminosae that may be consumed by human beings or domestic animals, commonly as mature dry seeds, ie the grain legumes, or pulses, but may also be consumed in certain instances as immature green seeds, or as green pods with the immature seeds enclosed. Species which provide only leaf or stem tissue, which is utilized as a foodstuff, have not been included. The two oleaginous legume crops, groundnut, *Arachis hypogaea* and soyabean, *Glycine max*, which are grown primarily for processing into edible oils and protein residues used largely for animal feeding, have already been dealt with in *Crop and product digest No.* 1—*Oils and oilseeds.* Similarly, the two minor legume crops, the African yam bean, *Sphenostylis stenocarpa* and the yam bean, *Pachyrrhizus erosus*, which are grown for their edible tubers, in addition to their seeds, have been dealt with in *Crop and product digest No.* 2—*Root crops.*

The food legumes, particularly the grain legumes, or pulses, are important foodstuffs in all tropical and subtropical countries, where they are second in importance only to cereals as a source of protein. In India, in fact, they provide the only high-protein component of the average diet and over 10 million tonnes are consumed annually.

The grain legumes have a high total protein content (average 20–26 per cent) and can be considered as a natural supplement to cereals, since, although they are usually deficient in the essential amino acids methionine and cystine, they contain adequate amounts of lysine, whereas cereals are deficient in lysine, but contain adequate amounts of methionine and cystine. Many of the grain legumes are world-wide in their use and most are reasonably palatable and acceptable under conditions of home cooking, although some may cause flatulence, and some also contain toxic principles which are usually removed by cooking.

World production of grain legumes was estimated to be of the order of 51 million tonnes in 1976, but some authorities consider that this tonnage could be underestimated by between 20 and 50 per cent, since in the tropics and subtropics many of the legume crops are grown and consumed at the rural level and do not enter official statistical reports.

In addition to their value as a foodstuff, the food legumes are also important in cropping systems because of their ability to produce nitrogen and so increase the overall fertility of the soil, thus partially replacing the use of expensive nitrogenous fertilizers. For example, a vigorous growing food legume such as the cowpea, *Vigna unguiculata*, can add as much as 40 lb/ac (45 kg/ha) of nitrogen to the soil, which is equivalent to 100 lb/ac (112 kg/ha) of urea, or 200 lb/ac (225 kg/ha) of ammonium sulphate.

The dominant food legumes vary from country to country, and from region to region, but most of them can be grown under a reasonably wide range of ecological conditions and many can be grown reasonably satisfactorily on poor soils without the application of artificial fertilizers.

This digest covers a total of 27 food legume crops, of these the following are of major importance: pea, *Pisum sativum*; haricot bean, *Phaseolus vulgaris*; chick pea, *Cicer arietinum*; broad bean, *Vicia faba*; pigeon pea, *Cajanus cajan*; cowpea, *Vigna unguiculata*; mung bean, *Vigna radiata* and lentil, *Lens culinaris*. Each crop in the digest is listed alphabetically under its first common name and other names are cross-referenced to this in an index of trivial names. The data, if available, are arranged under the following standard headings:

Common names
Widely used English names are given, the first being printed in capitals and being used for the alphabetical arrangement of the entries and for cross-referencing.

Botanical name
Family
Nomenclature closely follows the Kew Index and its supplements.

Other names
There is considerable confusion concerning the common names given to a number of food legumes. Some can apply to more than one species according to the locality, many are more correctly applicable to particular forms, types or cultivars within a species. There is also a wide variation in the spellings used. Many widely used common names which have been recorded in the literature are listed, with the country, or language to which they normally apply appended in parenthesis. Less common English names are given without the country being indicated.

Botany
Under this heading the annual or perennial nature of the plant, its approximate size, growth habit and form are described. Varietal differences and the systematics of the crop are also discussed, where appropriate.

Origin and distribution

Brief particulars of the origin and distribution of the crop are given.

Cultivation conditions

The main climatic regions in which it is possible to cultivate the crop are given in accordance with Van Royen and Bengtson*. The climates of the world are divisible into tropical, subtropical, intermediate or temperate and polar types.

Tropical climates have an average annual temperature of above 77°F (25°C), the average of all months exceeding 65°F (18°C). Subtropical climates have short mild winters and long growing seasons. There is a period of one to two months when freezing temperatures may occur, though the average temperature of the coldest month is above 43°F (6°C). The summer temperatures may be as high as those of the tropical climate. Intermediate climates ie those between subtropical and polar, have cold winters and warm to hot summers. They vary from areas where the winters are short to those where they are long and severe. All intermediate climates have a season of frost as well as a frost-free season.

The humid tropical climates are tropical rainforest, tropical monsoon, and tropical savanna. The tropical rainforest has no pronounced or prolonged dry season, an annual rainfall of 80–160 in. (2 000–4 000 mm) or more, a relative humidity of around 80 per cent, and a high and uniform temperature with annual means ranging from 77° to 80°F (25–27°C) with little seasonal variation. The tropical monsoon climate exhibits marked daily and seasonal temperature changes, has an annual rainfall of 40–80 in. (1 000–2 000 mm) with abundant rainfall during the wet season, alternating with a period of drought lasting from four to six months, or longer. The tropical savanna climate has a rainfall often exceeding 40 in. (1 000 mm) annually, evenly distributed over 120 to 190 days, with a prolonged drought often lasting six to seven months. The climate is hot with a moderate range of temperature. The dry tropical climates are subdivided into semi-arid or steppe type and arid or desert type. In the areas of tropical steppe climate the rainfall is occasional, though seasonal, and commonly averages 8–20 in. (200–500 mm) or more annually; the temperature being variable, but high at all seasons. The desert climate has a rainfall usually averaging less than 8 in. (200 mm) annually, and a daytime relative humidity commonly less than 50 per cent.

---

* Royen, W. van and Bengston, N. A. 1964. *Fundamentals of economic geography*. 5th ed. pp. 95–105. London: Constable and Company Ltd, 613 pp.

The subtropical climate is subdivided into dry subtropical or Mediterranean, and humid subtropical. The former has an average annual rainfall generally below 30 in. (750 mm), in some places below 20 in. (500 mm), with most of the rainfall occurring during the cool season. In some regions there is a moderate amount of summer rainfall, while in others there is practically no rain during this period. There are about six to eight months with an average temperature below 65°F (18°C). In the humid subtropical climate the rainfall averages about 30 in. (750 mm) per annum, with no pronounced dry season. There are generally four to six months with an average temperature below 65°F (18°C). In both types of subtropical climate frost may occur during the coldest period.

Humid intermediate climate has an annual rainfall which ranges from 20 in. (500 mm) in the drier parts to 80 in. (2 000 mm) in the more rainy sections. It may be divided into humid continental, east coast continental and west coast marine. Dry intermediate climates have an annual rainfall which is commonly less than 20 in. (500 mm). They may be subdivided into middle latitude steppe and middle latitude desert; the former having an annual rainfall from 6–20 in. (150–500 mm) and the latter less than 6 in. (150 mm) per annum.

Plant growth requirements are arranged under the main headings of temperature, rainfall and soil, with factors such as day-length and altitude mentioned where they are critical, the possibility of growth under irrigation is mentioned when describing rainfall requirements, and any positive evidence concerning the effects of fertilizers is included in the information on soil requirements.

Planting procedure
Information concerning the type or types of planting material is given. The usual methods of planting are discussed, together with details of field-spacing and, where applicable, seed-rate.

Pests and diseases
The most serious pests and diseases attacking the crop in various growing regions are mentioned.

Growth period
An approximate average, or a range, of time lengths from planting to harvesting is quoted.

Harvesting and handling
The commonest and best methods of harvesting and handling the primary product are indicated.

Primary product
The part of the plant for which the crop is primarily grown and the form in which it is commonly marketed is given. Normally, one form only has been selected and this is shown at the beginning of the heading.
This form is used as a basis for the quantitative data given in subsequent headings, unless otherwise stated under the heading.

Yield
A good average yield of the primary product is given. Yields obtained in different regions, or circumstances, may be quoted separately.

Main use
The main use, or uses, of the primary product is given.

Subsidiary uses
Additional uses of the primary product are entered under this heading.

Secondary and waste products
Useful by-products produced during the processing of the primary product, or prepared from other parts of the plant are listed, together with their relative uses, etc. Major waste products which result from primary or secondary product processing are given, with possible outlets where applicable.

Special features
Under this heading information is given on the chemical components of significance in the utility of the primary product. In addition, the presence of constituents that may be undesirable or call for special treatments, such as toxins, is indicated.

Processing
The processing operations through which the primary product may have to pass in order to produce a marketable commodity, are listed. In certain instances, similar information may also be given for secondary products.

Production and trade

Where information is available details are provided of: (i) the estimated average world production of the crop; (ii) the output of the major producing countries; (iii) shipments from the major exporting countries; (iv) shipments to the major importing countries. Averages are given for two recent five-year periods (normally 1965–69 and 1970–74), together with later annual figures, if these are available. An indication of price movements has also been included, wherever possible. Very fragmentary information has been included for certain crops, if this is all that is available.

Major influences

Any factors which might have a significant influence on the future supply of, and demand for, the commodity, are mentioned under this heading.

Bibliography

Textbooks, technical bulletins, research reports, papers given at recent symposia, articles in technical periodicals, and bibliographies are cited.
The literature covering the food legumes is extensive, particularly as regards the more important crops and the bibliographies for each individual crop are selective and by no means exhaustive. Emphasis has been given to material published between 1965 and 1976. The abbreviations used are those recommended in the *International list of periodical title word abbreviations*.
To avoid unnecessary repetition, the statistical references are not listed under each crop. The major sources of statistical information were: *Production yearbooks*, FAO, Rome; the trade returns of individual countries; *World pulse market survey* 1975, Department of Trade and Commerce, Agriculture, Fisheries and Food Products Branch, Canada.

Abbreviations

a	annum
ac	acre
av	average
°C	degrees Celsius
cm	centimetre
cm^2	square centimetre
cv.	cultivar
D	dextro
DNA	deoxyribonucleic acid
eg	for example
°F	degrees Farenheit
FAO	Food and Agricultural Organization of the United Nations
fob	free on board
ft	foot
g	gram
gal	Imperial gallon
h	hour
ha	hectare
ie	that is
in.	inch
$in.^2$	square inch
iod. val.	iodine value
iu	international unit
K	potassium
kg	kilogram
km	kilometre
L	laevo
l	litre
lb	pound
m	metre
m^3	cubic metre
mg	milligram
min	minute
mm	millimetre
mmhos/cm	millimhos per centimetre
MP	melting point
N	nitrogen
N_D	refractive index
nd	not dated
No.	number
O	oxygen
oz	ounce
P	phosphorus
p/a	per annum
pl	plants
Pol. No.	Polenske number
Pol. val.	Polenske value
ppm	parts per million
RM val.	Reichert-Meissl value
r/m	revolutions per minute
Repub.	Republic
S	sulphur
sap. val.	saponification value
SG	specific gravity
sp./spp.	species
ssp.	subspecies
syn.	synonym
T	long ton
t	tonne
val.	value
var.	variety
wk	week
wt	weight
yr	year

Geographic and language abbreviations used

Afg.	Afghanistan
Afr.	Africa
Alg.	Algeria
Am.	America
Ang.	Angola
Ant.	Antilles
Ar.	Arabic
Arg.	Argentina
As.	Asia
Aus.	Austria
Aust.	Australia
Bah.	Bahamas
Barb.	Barbados
Belg.	Belgium
Beng.	Bengal
Bom.	Bombay
Bots.	Botswana
Braz.	Brazil
Bulg.	Bulgaria
Burm.	Burma
C. Am.	Central America
Carib.	Caribbean
Col.	Colombia
C. Rica	Costa Rica
Cyp.	Cyprus
Czech.	Czechoslovakia
Dah.	Dahomey
Dut.	Dutch
E.	East
Egy.	Egypt
Eth.	Ethiopia
Fr.	France/French
Gab.	Gabon
Ger.	German
Gh.	Ghana
Gr.	Greece
Gren.	Grenada
Gu.	Guatemala
Guad.	Guadeloupe
Guy.	Guyana
Haw.	Hawaii
Hind.	Hindustani
Hun.	Hungary
Ind.	India
Indon.	Indonesia
Is.	Israel
It.	Italy
Iv. C.	Ivory Coast
Jam.	Jamaica
Jpn.	Japan
Ken.	Kenya
Kor.	Korea
Lat. Am.	Latin America
Maur.	Mauritius
Malag.	Malagasy Republic
Malays.	Malaysia
Mex.	Mexico
Mor.	Morocco
N.	North
Nic.	Nicaragua
P.R.	Puerto Rico
Pak.	Pakistan
Par.	Paraguay
Philipp.	Philippines
Pol.	Poland
Port.	Portugal/Portuguese
Pun.	Punjab
Qd.	Queensland
Re.	Reunion
Rhod.	Rhodesia
Rom.	Romania
S.	South
Salv.	El Salvador
Sen.	Senegal
Som.	Somali

Sp.	Spanish/Spain	UK	United Kingdom
Sri La.	Sri Lanka	USA	United States of America
St Luc.	Saint Lucia		
Sud.	Sudan	USSR	Union of Soviet Socialist Republics
Sur.	Surinam		
Sw.	Sweden	Venez.	Venezuela
Swah.	Swahili	Viet.	Vietnam
Tam.	Tamil	W.	West
Thai.	Thailand	W.I.	West Indies
Tog.	Togoland	Yug.	Yugoslavia
Trin.	Trinidad	Zam.	Zambia
Turk.	Turkey	Zan.	Zanzibar
Ug.	Uganda	Zar.	Zaire

Reference to trade names of agricultural chemicals, etc, implies no endorsement of the efficacy of these products or, any criticism of competing products not mentioned.

Common names	**ADZUKI BEAN, Adsuki, Atsuki Atzuki, or Azuki bean.**
Botanical name	*Vigna angularis* (Willd.) Ohwi & Ohashi, syn. *Phaseolus angularis* (Willd.) Wight.
Family	Leguminosae.
Other names	Adzukibohne (Ger.); Fagiolo adzuki (It.); Feijão adzuki (Ang.); Frijol adzuki (Sp.); Frijol diablito (Cuba); Haricot adzuki, H. à feuilles angulaires (Fr.); H. konde, Nirikia (Zar.); Poroto arroz (Arg.); Red bean[1] (Kor.).

Botany

An erect, bushy annual, usually 12–30 in. (30–75 cm) tall, depending upon the soil and cultivar; early-maturing types are usually determinate, late ones semi-determinate and much-branched. The leaves are pinnate, trifoliolate, somewhat hairy, the leaflets are ovate, sometimes slightly cleft with the odd leaflet supported on a long petiole. The leaves persist until the pods are fully mature. Usually there are between 6 and 12, bright-yellow flowers, borne on short pedicels. The pods are usually 3–4 in. (7.5–10 cm) long, thin-skinned, non-shattering, and when mature are normally straw coloured, but can be black or occasionally brown. They normally contain from 4 to 10 seeds, which are of various colours, but most commonly maroon. There are numerous cultivars which show considerable variation in the period of maturity, and the form and colour of the pods and seeds.

In Japan two main types are recognized, the common adzuki bean and the large-seeded adzuki bean, or Dainagon. The cultivar Chagara-wase, a common adzuki bean, has been long established. It has a short stem, light-brown pods and small seeds. It is early-maturing, but rather low yielding. In Hokkaido a leading cultivar of the common type of adzuki bean is the cultivar Takara-shozu, of medium height, 20–24 in. (50–60 cm), and also with light-brown pods and small seeds. Akatsuki-dainagon is a recently developed high-yielding cultivar of the large-seeded Dainagon type of adzuki bean.

---

[1] Also used for the rice bean, *Vigna umbellata*.

Adzuki bean

Origin and distribution

The adzuki bean is considered to have originated in China, or possibly India or Japan. It has been long established in the Far East, especially in Japan, Thailand, China, Korea and Taiwan. It has also been introduced successfully into the USA, S. America, Angola, Zaire, India and New Zealand.

Cultivation conditions

Temperature—a subtropical crop, requiring moderately high temperatures 59–86°F (15–30°C) throughout its growth period; for quick germination a soil temperature of above 60°F (16°C) is required. It does not tolerate frost, and cool conditions during the growing season have an adverse effect upon yield.

Rainfall—optimum yields are obtained with a moderate, evenly distributed rainfall, with a dry spell at harvest. The adzuki bean is more tolerant of heavy rainfall than many other pulse crops, such as the haricot bean; certain cultivars show a degree of tolerance to drought conditions.

Soil—the adzuki bean can be grown on a wide range of soil types from light to heavy clays, provided that they are reasonably fertile and well drained. It does not do well on very acidic soils and cannot stand waterlogging. In Japan when the adzuki bean is grown on volcanic soils it has shown a susceptibility to boron and nickel poisoning. In India the application of a basal dressing, 267 lb/ac (300 kg/ha) superphosphate, 89 lb/ac (100 kg/ha) potassium sulphate and 89 lb/ac (100 kg/ha) ammonium sulphate, followed by an additional application of ammonium sulphate 40 days after sowing, has been recommended to produce a good crop on soils of moderate fertility. In New Zealand the application of nitrogen 25 lb/ac (28 kg/ha) immediately before flowering has been recommended.

Altitude—the adzuki bean is better suited to higher elevations in the tropics. In the Philippines it is grown at altitudes above 1 400 ft (420 m).

Day-length—the adzuki bean was classified as a short-day plant by Allard and Zaumeyer (1944). However, there appears to be considerable variation in the sensitivity of cultivars to day-length. Recently Hartmann (1969) reported a day-neutral response in lines from Korea and Japan grown in Hawaii. In Japan research workers have reported longitudinal adaptation in the degree of sensitivity to day-length; early-maturing cultivars in the south being more sensitive to short days than those in the north.

Planting procedure

Material—seed, which retains its viability for about 2 years is used. Germination is hypogeal. Inoculation has been found to increase yields significantly in experimental trials in Australia.

Method—the seeds are usually sown with a drill, or by dibbing in rows, although in some areas they are broadcast. The seeds are normally sown at a depth of about 1 in. (2.5 cm). Weeding is necessary, usually about 20 and 40 days after sowing. In New Zealand effective weed control has been obtained with the pre-emergence application of dinoseb-amine and monolinuron. In Asia the adzuki bean is often sown at high rates directly into rice stubble, when weeding is not usually necessary.

Field-spacing—in India the recommended spacing is 12 to 20 in. (30–50 cm) between the rows and 4–5 in. (10–12.5 cm) between the plants. Other spacings reported to be used are 36 x 12–18 in. (90 x 30–45 cm) in the USA, 8 x 8 to 16 x 16 in. (20 x 20 to 40 x 40 cm) in Africa, and 4 x 10 in. (10 x 25 cm) in Taiwan.

Seed-rate—in India the seed-rate varies from 14 to 22 lb/ac (16–25 kg/ha); in Zaire it is reported to range from 7 to 27 lb/ac (8–30 kg/ha) and in the USA from 20 to 25 lb/ac (22–28 kg/ha).

Pests and diseases

In Japan over 20 insect pests have been listed as attacking the adzuki bean. The adzuki pod worm, *Matsumuraeses phaseoli*, is reported to be widespread, and particularly troublesome early in the season. The Japanese butter bean borer, *Ostrinia varialis*, is of economic importance in Hokkaido, but can be controlled effectively by spraying the crop three or four times during its growth cycle. Other pests affecting the crop in Japan include the cutworm, *Spodoptera litura*, the corn seed maggot, *Hylema platura* (*H. cilicrura*), the chafer, *Anomala rufocuprea*, the aphid, *Aphis craccivora*, and the soyabean cyst nematode, *Heterodera glycines*. In New Zealand the caterpillar of the small looper moth, *Plusia* spp., and eelworm nematodes have been reported, but are not serious.

The most destructive pest of stored adzuki beans is the weevil, *Callosobruchus chinensis*. Infestation begins soon after the sacked beans are placed in storage, and losses ranging from 5 to 50 per cent in the market value of summer stored beans have been reported in S. Korea.

Adzuki bean

In Japan the virus diseases, adzuki mosaic virus and cucumber mosaic virus are reported to be of considerable economic importance, and also the fungal disease *Sclerotinia sclerotiorum*. In addition, the following diseases have been reported in Japan, bacterial blight, *Xanthomonas phaseoli*, leaf blotch, *Clathrococcum nipponicum*, ascochyta leaf spot, *Ascochyta phaseolorum*, leaf spot, *Phyllosticta phaseolina*, rust, *Uromyces azukicola*, southern sclerotium blight, *Corticium rolfsii*, charcoal rot or ashy stem blight, *Macrophomina phaseoli*, leaf spot, *Cercospora cruenta*, anthracnose, *Colletotrichum phaseolorum* and powdery mildew, *Pshaerotheca fuliginea*. In the USA the adzuki bean is reported to be subject to wilt, *Fusarium* spp. and in Zaire to the root rot, *Sclerotium rolfsii*.

Growth period

Normally the adzuki bean reaches maturity in from 80 to 120 days depending upon the cultivar and the climatic and cultural conditions, although some early-maturing Japanese cultivars produce a crop in 60 days, and in New Zealand late-maturing types can take up to 190 days.

Harvesting and handling

About 80 per cent of the pods mature together, so that the crop can be harvested in a single operation. Sometimes crop losses can be high due to pod shattering. Mechanical harvesting has presented problems in New Zealand because of the short distance from the ground of the first pods, and experimental equipment, which severs the stems just below ground level, is being developed.

Primary product

Seed—the seeds, which constitute about 50 per cent of the pod, are oblong, approximately 0.3 x 0.15 in. (8 x 4 mm), sub-truncated at the ends. The following colours occur in order of frequency: maroon, straw to nearly white, grey (really black speckles on a greenish-yellow ground colour), maroon mottled with straw, black, brown, blue-black mottled with straw. The seeds have a long, narrow white hilum and in all cases the embryo is nearly white, and brittle in consistency. The seed weight of the common adzuki bean of Japan is approximately 0.4—0.5 oz (13–15 g)/100 seeds, and of the large-seeded Dainagon adzuki bean, 0.6—0.7 oz (18–20 g)/100 seeds.

Yield

In Japan the average yield during the period 1967–71 was 1 700 lb/ac (1 900 kg/ha) and in Taiwan 1 300 lb/ac (1 460 kg/ha), although in southern Taiwan the average commercial yield has recently been reported to be in excess of 2 130 lb/ac (2 400 kg/ha). In New Zealand yields of 1 200—2 000 lb/ac (1 340—2 240 kg/ha) of dressed seed have been obtained experimentally, and at the Simla Station in India, experimental yields of between 1 780 and 2 200 lb/ac (2 000—2 500 kg/ha) have been reported. Yields ranging from 80 to 600 lb/ac (90–670 kg/ha) have been reported from Oklahoma, USA.

Main use

Adzuki beans, particularly the maroon coloured types, are a popular foodstuff in Japan, where they are boiled or fried and often eaten with rice, or ground and used as a paste or a flour in the preparation of cakes, sweetmeats, etc. The whole beans are also sometimes 'popped' like maize, or candied.

Subsidiary uses

The beans may be used as an ingredient for soups or gruels, or split and used for dhal, or sprouted and eaten as a vegetable. They are sometimes used medicinally.

Secondary and waste products

The adzuki bean is sometimes grown as a green manure crop or for livestock feeding. It has also been used as a coffee substitute and in the manufacture of cosmetics. The possibility of quick-freezing the immature bean pods has been investigated, but their fibrous nature results in a product rather inferior to frozen French beans, *Phaseolus vulgaris*.

Special features

The approximate composition per 100 g of edible portion of the adzuki bean has been given as: moisture 13.0 g; protein 25.3 g; fat 0.6 g; total carbohydrate 57.1 g; fibre 5.7 g; ash 3.9 g; calcium 253 mg; iron 7.6 mg; vitamin A 15 iu; thiamine 0.57 mg; riboflavin 0.18 mg; niacin 3.2 mg. The amino acid content (mg/gN) has been reported as follows: isoleucine 280; leucine 490; lysine 440; phenylalanine 340; tyrosine 210; methionine 110; cystine 70; threonine 240; valine 340. The major oligosaccharide present in the seeds is stachyose; the presence of sucrose and raffinose has also been reported. The bean contains an oil with the following characteristics: SG $^{15°C}$ 0.9618; N_D $^{20°C}$ 1.4670; sap. val. 176; iod. val. 58; unsaponifiable matter 11 per cent. The fatty acids consist of 25 per cent solid acids, (stearic 50, palmitic 30, and carnaubic? 20 per cent) and 75 per cent liquid acids, mostly linoleic with oleic and a little linolenic. The reddish-black pigment present in the seeds of certain cultivars has been identified as delphinidin 3-monoglucoside. Pipecolic acid (0.65 g/kg) and a prosapogenin have also been isolated from the seeds.

Processing
Adzuki bean paste 'Ann', used for the preparation of confectionery products in Japan, is usually prepared by first soaking the beans in water, then boiling them until they are soft, followed by grinding and separation of the fibrous matter by centrifugation. The residue is next washed with water, pressed, and the resulting paste, either kneaded with sugar and used directly, or dried and ground to produce a powder.

Production and trade
Production—figures for the global production of adzuki beans are not available, since most countries do not show this bean as a separate item. Japan is the major producer and consumer; other important producers are Taiwan, Thailand, the People's Republic of China, N. and S. Korea and the far east of the USSR. Recently limited quantities have been grown for export in Colombia, Brazil and Canada. For the period 1965–69 production in Japan averaged 110 820 t/a and for 1970–74 123 000 t/a, with approximately 60 per cent of the total output being produced in Hokkaido. Japanese domestic production, however, fluctuates considerably from year to year according to weather conditions, since yields are adversely affected by cold weather in spring and early summer. Production in Taiwan, is reported to have averaged 2 880 t/a for the period 1965–69, 2 750 t/a for 1970 and 3 530 t/a for 1971, (later figures are not available), and to be mainly for domestic consumption, which is estimated to be some 3 000 t/a.

Trade—limited quantities of adzuki beans enter international trade. The People's Republic of China, Taiwan, N. and S. Korea, Colombia and Thailand are the chief exporting countries, although none of them show adzuki beans as a separate item in their trade returns. Japan is virtually the only importing country, the demand for imported supplies fluctuating considerably according to domestic production, which is never sufficient to meet demand.

Japan: Imports of adzuki beans
Quantity tonnes

	Annual average 1965–69	Annual average 1970–74	1975
Total	31 403	27 875	1 201
of which from:			
People's Repub. China	22 219[1]	20 801	671
S. Korea	1 474[1]	1 662	n/a
N. Korea	1 319[2]	n/a	n/a
Taiwan	29[1]	4 225	31
Colombia	94[3]	603	n/a
Thailand	251[1]	181	344
USA	14	69	155

Prices—small red adzuki beans fetch premium prices on the Tokyo Commodity Exchange and are often known as 'red diamonds'. Prices for average quality beans ranged between £110.00 and £150.00/t in the early 1970s. The value of Japanese imports averaged £152.40/t for the period 1970–74, compared with £78.40/t for 1965–69, and in 1975 reached £210.00/t.

Major influences

The adzuki bean, although at present of minor importance, could be more widely grown in the subtropics. It has a relatively high yield potential and is a nutritious foodstuff, without a pronounced bean flavour or antinutritional factors. It seems likely that imports into Japan will increase as the demand is rising and domestic production of food legumes is tending to fall. The possibility of obtaining supplies from countries other than the traditional producing countries in Asia has been receiving attention. Recent experimental trials in New Zealand have produced rather variable results, indicating that there is need for further agronomic research. Moreover, difficulties have been experienced with mechanized harvesting, essential in countries such as New Zealand with high labour costs. Although a valuable foodstuff the potential of the adzuki bean as an anti-erosion crop should not be overlooked and in fact this use has been recently receiving attention in India.

[1] Two year average.
[2] 1965 only.
[3] 1969 only.
n/a not available.

Bibliography

ABEELE, M. VAN DEN AND VANDENPUT, R. 1956. Le haricot adzuki. *Les principales cultures du Congo Belge*. 3rd ed. p. 858. Bruxelles: Direction de l'Agriculture, des Forêts et de l'Elevage, 932 pp.

ALLARD, H. A. AND ZAUMEYER, W. J. 1944. Responses of beans (*Phaseolus*) and other legumes to length of day. *US Dep. Agric., Tech. Bull.* 867, 24 pp.

AYKROYD, W. R. AND DOUGHTY, J. 1964. Legumes in human nutrition. *FAO Nutr. Stud.* 19, pp. 108; 115; 118. Rome: FAO, 138 pp.

BANG, Y. H. 1963. Laboratory evaluation of several chemical protectants against the southern cowpea weevil, *Callosobruchus chinensis*, on stored dried beans in Korea. *J. Econ. Entomol., 56*, 588–591.

CHADA, Y. R. (ed.). 1976. *V. angularis* (Willd.) Ohwi & Ohashi. *The wealth of India: Raw materials*. Vol. 10 (Sp–W). p. 475. New Delhi: Counc. Sci. Ind. Res., Publ. Inf. Dir., 591 pp.

CHENG, C. 1972. Current situation of food legume crops production in Taiwan, the Republic of China. Symp. food legumes. pp. 11–12. *Tokyo, Jpn., Minist. Agric. & For., Trop. Agric. Res. Cent., Trop. Agric. Res. Ser.* 6, 253 pp.

FOOD AND AGRICULTURAL ORGANIZATION OF THE UNITED NATIONS. 1959. *Tabulated information on tropical and subtropical grain legumes*. pp. 145–151. Rome: FAO, Plant Prod. Prot. Div., 367 pp.

HARTMANN, R. W. 1969. Photoperiod responses of *Phaseolus* plant introduction in Hawaii. *J. Am. Soc. Hortic. Sci., 94*, 437–440.

IIDA, W. 1972. Major diseases of leguminous crops in Japan. Symp. food legumes. pp. 101–107. *Tokyo, Jpn., Minist. Agric. & For. Trop. Agric. Res. Cent., Trop. Agric. Res. Ser.* 6, 253 pp.

KOBAYASHI, T., HASEGAWA, T. AND KEGASAWA, K. 1972. Major insect pests of leguminous crops in Japan. Symp. food legumes. pp. 109–126. *Tokyo, Jpn., Minist. Agric. & For., Trop. Agric. Res. Cent., Trop. Agric. Res. Ser.* 6, 253 pp.

MATLOCK, R. S. AND OSWALT, R. M. 1963. Adzuki bean, *Phaseolus angularis* (Willd.) W. R. Wight. *Okla. State Univ., Agric. Exp. Stn. Bull.* B–617, 4 pp.

MORI, T., WATANABE, K. AND FUJITA, I. 1966. [The influences of boron fertilizers, which are favourable to sugar beet and toxic to bean crops, on the Tokachi volcanic upland soils in Hokkaido]. *Hokkaido Natl. Exp. Stn., Res. Bull.* 90, pp. 61–74. (*Field Crop Abstr.*, *21*, 312).

MOTOMIYA, G. AND ITO, R. 1972. Domestic production, importation and utilization of food legumes and research organization in Japan. Symp. food legumes. pp. 23–32. *Tokyo, Jpn., Minist. Agric. & For., Trop. Agric. & For., Trop. Agric. Res. Cent., Trop. Agric. Res. Ser.* 6, 253 pp.

NARIKAWA, T. 1972. Kidney bean and azuki bean in Japan with reference to breeding in Hokkaido. Symp. food legumes. pp. 179–188. *Tokyo, Jpn., Minist. Agric. & For., Trop. Agric. Res. Cent., Trop. Agric. Res. Ser.* 6, 253 pp.

NOMURA, N. AND ASANUMA, K–I. 1970. [The effects of low temperature on the order and speed of flowering in the adzuki bean (*Phaseolus angularis* W. F. Wight)]. *Hokkaido Prefect. Agric. Exp. Stn., Bull.* 20, pp. 73–79. (*Field Crop Abstr.*, *24*, 2252).

PALMER, J. 1974. The growth habits and flowering of adzuki beans in New Zealand. *N.Z.J. Exp. Agric.*, *2*, 371–376.

PIPER, C. V. AND MORSE, W. J. 1914. Five oriental species of beans. pp. 3–12. *US Dep. Agric. Bull.* 119, 32 pp.

PURSEGLOVE, J. W. 1968. *Phaseolus angularis. Tropical crops: Dicotyledons.* Vol. 1. pp. 289–290. London: Longmans, Green and Co Ltd, 332 pp.

SACKS, F. M. 1977. A literature review of *Phaseolus angularis*—the adsuki bean. *Econ. Bot.*, *31*, 9–15.

SAIO, K. AND WATANABE, T. 1972. Advanced food technology of soyabean and other legumes in Japan. Symp. food legumes. pp. 209–216. *Tokyo, Jpn., Minist. Agric. & For., Trop. Agric. Res. Cent., Trop. Agric. Res. Ser.* 6, 253 pp.

SASANUMA, S., TAKEDA, K. AND HAYASHI, K. 1966. Black red pigment of 'adzuki bean': Studies on anthocyanins. *Bot. Mag.*, (*Tokyo*), *79*, 807–810. (*Field Crop Abstr.*, *21*, 313).

SATOH, H. 1974. [Responses of adzuki bean varieties to cultivating conditions]. *Hokkaido Prefect. Agric. Exp. Stn. Bull.* 29, pp. 61–71. (*Field Crop Abstr.*, *28*, 2060).

TANUSI, S., KASAI, T. AND KAWAMURA, S. 1972. [Determination of oligosaccharides in some edible legume seeds]. *J. Jpn. Soc. Food Nutr.*, *25*, 25–27. (*Field Crop Abstr.*, *26*, 4244).

Adzuki bean

TASAKI, J. 1965. [Studies of varietal differences on response to photoperiod, used for the classification of varieties of adzuki beans (*Phaseolus angularis* W. F. Wight)]. *Proc. Crop Sci. Soc., Jpn.*, *34*, 14–19. (*Field Crop Abstr.*, *20*, 1094).

TASAKI, J. AND HONMA, H. 1965. [Studies on the varietal difference in relation to the sensitivity to temperature on adzuki beans (*Phaseolus angularis* W. F. Wight)]. *Proc. Crop Sci. Soc., Jpn.*, *34*, 20–24. (*Field Crop Abstr.*, *20*, 1095).

THOMAS, T. A., PATEL, D. P. AND BHAGAT, N. R. 1974. Adzuki bean: a new promising pulse for the hills. *Indian Farming*, *23* (12), 29–30.

WHITE, D. 1972. Adzuki bean potential being assessed. *N.Z. J. Agric.*, *125* (3), 43–45.

Common names — **ASPARAGUS BEAN[1], Pea bean[2], Yard (long) bean.**

Botanical name — *Vigna unguiculata* ssp. *sesquipedalis*[3] (L.) Verdc., syn. *V. sesquipedalis* (L.) Fruhw.

Family — Leguminosae.

Other names — Banor (Philipp.); Bodi bean (Trin.); Boucouson (St Luc.); Cheong dau-kok, Ch'eung kong tau (China); Dolico gigante (It.); Dolique asperge, D. de Cuba[4], D. geánt (Fr.); Fagiolo asparagio (It.); Habichuela China (Cuba); Hamtak (Philipp.); Haricot kilomètre (Fr.); Increase pea (Barb.); Judia asparaga (Sp.); Juroku-sasage (Jpn.); Kachang bêlut, K. panjang[4], K. perut ayam (Malays.); Long bean (Aust.); Polon-mé (Sri La.); Rounceval pea (Barb.); Sitao (Philipp.); Six weeks bean, or pea (W.I.); Snake bean[5] (Aust.); Too-afuk yaou (Thai.); Tua kok[4] (Hong Kong).

Botany

A glabrous annual, often a vigorous climber, reaching 78–157 in. (2–4 m) in height and twining anti-clockwise, although prostrate forms also exist, and bush types are being developed in the Philippines. The leaves are trifoliolate, with rhomboid-ovate leaflets, 3–5 in. (7.5–12.5 cm) long, sometimes tinged with purple. The flowers are usually dirty-yellow or violet in colour, 0.75–1.0 in. (1.9–2.5 cm) long, with two or three on the end of a long peduncle; they open early in the morning, close flat by noon, and then fall very rapidly. The seed-pods vary in length and are fleshy, pendent, and contain 15–20 oblong or reniform seeds. The seeds are 0.35–0.5 in. (9–12 mm) long, usually brown or reddish-brown in colour, with a white hilum.

---

[1] Also frequently used for the winged bean, *Psophocarpus tetragonolobus*.
[2] Also used for small type haricot beans, *Phaseolus vulgaris*.
[3] For a detailed discussion of the systematics of the asparagus bean see cowpea, *Vigna unguiculata*.
[4] Also used for cowpea, *Vigna unguiculata*.
[5] Also sometimes used for Jack bean, *Canavalia ensiformis*.

Asparagus bean

Origin and distribution

The asparagus bean is thought to have originated in southern Asia, and is now widely grown throughout Asia, especially in the SE. Asian sub-continent,* Indonesia and the Philippines. It has also spread to the Caribbean area, but is less common in Africa.

Cultivation conditions

Temperature—for optimum yields a mean monthly temperature of between 68° and 86°F (20–30°C) is required. Although the asparagus bean can tolerate higher temperatures, it is susceptible to frost and cannot even tolerate temperatures around 40°F (4°C) for short periods.

Rainfall—a moderate, evenly distributed rainfall is required. In the low humid tropics yields tend to be reduced. In the Philippines the asparagus bean is grown successfully in areas with an average rainfall of about 75 in./a (1 880 mm/a).

Soil—the soil requirements of the asparagus bean are similar to those of the cowpea, *Vigna unguiculata*, and it is grown on many types of soils from sandy loams to clay. It cannot tolerate waterlogging and prefers a slightly acid to neutral soil, pH 5.5 to 6.0. Little is known of its fertilizer requirements, but the application, just prior to planting, of a 1:1 mixture, by weight, of a NPK fertilizer 14:14:14 and urea, at the rate of 220 lb/ac (250 kg/ha), or NPK fertilizer, 12:24:12 or 6:10:4 at the rate of 356–620 lb/ac (400–700 kg/ha), has been recommended.

Altitude—it is usually grown at low or medium elevations in the tropics, up to about 3 000 ft (900 m).

Day-length—some cultivars are sensitive to day-length, others are day-neutral.

Planting procedure

Material—seed is used, germination is epigeal and the percentage of germination is normally high; the seed can retain its viability for about 3 years.

Method—the asparagus bean is frequently grown in small market gardens mixed with other crops such as sweet potatoes, taros and yams. The seed is usually planted 1–2 in. (2.5–5 cm) deep in rows, or about 4 to 6 in hills.

* Throughout this digest the term SE. Asia sub-continent refers to India, Pakistan and Bangladesh.

Climbing types require support and normally the seed is planted at the base of bamboo canes, which are usually tied to form a trellis. The seed-bed should be cultivated to a fine tilth and free of weeds. Weeding is usually necessary, particularly during the early stages of growth; thinning and irrigation may also be required. In Trinidad, experimental trials with the pre-emergence use of the herbicides prometryne, ametryne, diphenamid and nitrofen, has resulted in increased yields.

Field-spacing—the average spacing used is 12–20 x 40 in. (30–50 x 100 cm).

Seed-rate—in the Philippines the average seed-rate is 27–31 lb/ac (30–35 kg/ha), elsewhere it can range from about 22 to 45 lb/ac (25–50 kg/ha).

Pests and diseases
The asparagus bean is susceptible to most pests and diseases which attack the cowpea, *Vigna unguiculata*. The virus disease, yellow mosaic, is of considerable economic importance, particularly where infestation from aphids, *Aphis* spp. is a problem. In the Philippines cutworms, *Agrotis* spp., and bean fly, *Ophiomyia* (*Melanagromyza*) *phaseoli*, are reported to be troublesome, although crop losses are minimized by spraying with Sevin, Meptox, Folidol or Imidan.

Growth period
The green pods are normally ready for harvesting about 50 to 65 days after sowing, depending upon the cultivar. The seed can be harvested 60 to 80 days after sowing.

Harvesting and handling
The green pods are picked by hand every 2 to 4 days, usually when they are about one-half or two-thirds mature, otherwise they become tough and fibrous. If the mature seeds are required the pods are left to ripen and turn yellow. They are then usually picked by hand, threshed and handled similarly to cowpeas.

Primary product
Pod—the asparagus bean is grown principally for its nutritious, immature pod. The pods vary in length from about 12 to 36 in. (30–90 cm) and are about 0.5 in. (1.25 cm) broad. They are fleshy, becoming more or less

inflated and flabby as they mature, and finally shrinking about the widely spaced seeds as they ripen. The pods can be white, green or reddish-purple, according to the cultivar.

Yield

In the Philippines the yield of immature pods normally ranges from 1.6 to 2.2 T/ac (4–5.5t/ha), but may reach 4T/ac (10t/ha).

Main use

The immature pods are a nutritious foodstuff, sometimes known in the Philippines as 'poor man's meat'. They are eaten raw in salads, when very immature and tender, or if more mature, boiled as a vegetable.

Subsidiary uses

The mature seeds may be eaten as a pulse, similarly to cowpeas.

Secondary and waste products

In Asia the green leaves are sometimes boiled and eaten as a vegetable like spinach; their protein content ranges from 2 to 5.3 per cent. The asparagus bean is also sometimes grown as a green manure crop. The juice from the leaves is reported to be used in traditional medicine in Malaysia and Indonesia.

Special features

The pod consists of approximately 91 per cent of edible material and has the following approximate composition: moisture 88.2–90.6 per cent; protein 2.0–4.3 per cent; fat 0.15–0.2 per cent; carbohydrate 4.1–8.2 per cent; crude fibre 1.4–1.5 per cent; ash 0.7 per cent; calcium 42.0 mg/100 g; phosphorus 46.0 mg/100 g; iron 0.9/100 g; sodium 5.0 mg/100 g; potassium 230 mg/100 g; vitamin A 570 iu/100 g; thiamine 0.12 mg/100 g; riboflavin 0.13 mg/100 g; niacin 1.20 mg/100 g; ascorbic acid 22 mg/100 g. In the Philippines green-podded cultivars have an average protein content of about 2.8 per cent, white-podded ones 2.1 per cent and red 3.3 per cent. The amino acids present per 100 g of edible material are tryptophan 24–26 mg; methionine 18–22 mg; lysine 129–136 mg. The protein content of mature pods can be as high as 9 per cent and for ripe mature seeds 24 per cent.

Processing

The asparagus bean may be quick-frozen, but the product is unlikely to achieve widespread commercial acceptance because it is so very similar to the

French or snap bean, *Phaseolus vulgaris*, in flavour and appearance, although the product tends to be a little tougher and more fibrous. Moreover, the pods wilt very rapidly after picking and must be processed with the minimum of delay in order to achieve an acceptable product.

Production and trade

There are no statistical data relating to the asparagus bean, in most countries it is grouped with the cowpea, *Vigna unguiculata*.

Major influences

The asparagus bean is a nutritious leguminous crop which could be more widely cultivated in many tropical areas as a substitute for French or snap beans.

Bibliography

BAILEY, L. H. 1971. *Manual of cultivated plants*. p. 576. New York: Macmillan Co, 1116 pp.

CARANDANG, E. C. 1968. Sitao. *Culture of vegetables*. pp. 70–72. Manila, Repub. Philipp., Dep. Agric. Nat. Resource, Bur. Plant Ind., 150 pp.

CHUNG, H. L. AND RIPPERTON, J. C. 1929. Utilization and composition of oriental vegetables in Hawaii. pp. 44–45; 60. *US Dep. Agric., Hawaii Agric. Exp. Stn. Bull.* 60, 64 pp.

DEANON, J. R. (JR.) AND SORIANO, J. M. 1967. The legumes. *Vegetable production in South East Asia*. (Knott, J. E. and Deanon, J. R. eds.). pp. 66–96. Los Baños, Laguna: Univ. Philipp. Coll. Agric., 366 pp.

FOOD AND AGRICULTURE ORGANIZATION. 1959. *Tabulated information on tropical and subtropical grain legumes*. p. 306. Rome: FAO, Plant Prod. Prot. Div., 367 pp.

HERKLOTS, G. A. C. 1972. Cowpea. *Vegetables in South-east Asia*. pp. 260–267. London: George Allen and Unwin Ltd, 525 pp.

LOPEZ, H., NAVIA, J. M., CLEMENT, D. AND HARRIS, R. S. 1963. Nutrient composition of Cuban foods: (iii) Foods of vegetable origin. *J. Food Sci.*, *28*, 607.

MOODY, K. 1973. Weed control in tropical grain legumes. *Proc. 1st Int. Inst. Trop. Agric., Grain legume improvement workshop*. pp. 162–183. Ibadan, Nigeria, Int. Inst. Trop. Agric., 325 pp.

Asparagus bean

PALO, A. V. 1972. Production of food legumes in the Philippines with special reference to leguminous vegetables. Symp. food legumes. pp. 189–195. *Tokyo, Jpn., Minist. Agric. & For., Trop. Agric. Res. Cent., Trop. Res. Ser.* 6, 253 pp.

SHEPHERD, A. D. AND NEUMANN, H. J. 1958. New processed vegetables may diversify agriculture and diet. *Chemurg. Dig.*, *17* (11), 10.

SUMMERFIELD, R. J., HUXLEY, P. A. AND STEELE, W. 1974. Cowpea (*Vigna unguiculata* (L.) Walp.). *Field Crop Abstr.*, *27*, 301–312.

TERRA, G. J. A. 1966. Tropical vegetables. p. 82. *Amsterdam, Neth., Dep. Agric. Res., R. Trop. Inst., Commun.* 54e, 107 pp.

Common names — **BAMBAR(R)A GROUNDNUT, Earth pea, Ground bean[1], Kaffir pea[2], Madagascar groundnut.**

Botanical name — *Voandzeia subterranea* (L.) Thou.

Family — Leguminosae.

Other names — Aboboi, Akyii (Gh.); Congo goober, Djokomaie (Zar.); Epi roro (Nig.); Ful abungawi (Sud.); Gertere (Ar.); Gobbe (Sur); Guerte (Ar.); Guijiya, Gujuya (Nig.); Haricot de Behanzin (Tog.); H. pistache (Fr.); Hluba bean, Igiuhluba (S. Afr.); Intoyo, Juga/o bean (Zam.); Kachang bogor (Indon.); K. menila, K. poi, K. tanah (Malays.); Mandubi d'Angola (Braz.); Manila bean[3]; Nela-kadalai (Malays.); Njama (Malawi); Njugo bean (C. Afr.); Njugu mawe (Zan.); Ntoyo (Zam.); Nzama (Malawi); Nzumbil (Zar.); Okboli ede (Nig.); Pararu[1] (W. Afr.); Pistache malagache (Malag.); Pois arachide, P. bambarra, P. de terre, P. souterrain (Fr.); Stone groundnut; Vanzon, Voandzou (Fr.); Voanjobory (Malag.).

Botany

A bunched, herbaceous, annual legume with almost submerged, trailing stems, approximately 4–6 in. (10–15 cm) long, but showing a wide range in growth habit. There are many different types of bambara groundnuts, which are sometimes classified into open or spreading types, compact or bunched types, and intermediate or semi-bunch types. The plant has a compact, well-developed tap-root which produces numerous downward growing lateral roots on its lower part. Both the main and lateral roots form an association with *Rhizobium* spp. and produce small rounded or lobed nodules. Branched, hairy stems, usually 10 to 120 in number, arise from the surface crown and each is made up of about 12 internodes from which leaf and flower buds

---

1 Also used for Kersting's groundnut, *Kerstingiella geocarpa*.
2 Also sometimes used for the cowpea, *Vigna unguiculata*.
3 Also used for the winged bean, *Psophocarpus tetragonolobus*.

arise alternately. The leaves are trifoliolate, the petiole is thickened and the base may be pink, purple or bluish-green in colour, according to the type. The leaflets are oblong or lanceolate 1–3 in. (2.5–7.5 cm) long, 0.3–1.2 in. (0.8–3 cm) wide. The flowers are typically papilionaceous and are borne in a raceme on long, hairy peduncles which arise from the nodes on the stem. They have yellowish-white, deep-yellow or reddish-pink petals, sometimes with reddish-brown striations, a few have light-pinkish petals. The bunch types are usually self-pollinated, the spreading are cross-pollinated by ants. After flowering the gynophore elongates and pushes the ovary into the soil, where the round, or oval, hard, wrinkled pods, 0.5–1 in. (1.25–2.5 cm) diameter, containing one, or more, hard seeds, develop. Both wild and cultivated forms exist; according to Hepper (1963) the former is *V. subterranea* var. *spontanea* and the latter *V. subterranea* var. *subterranea*.

Origin and distribution

The bambara groundnut is indigenous to tropical Africa, but is now found in Asia, parts of N. Australia, S. and C. America.

Cultivation conditions

Temperature—although a very adaptable crop, the bambara groundnut prefers a sunny, hot climate with average day temperatures between 68° and 82°F (20–28°C) and a frost-free period of 100 to 120 days.

Rainfall—the bambara groundnut requires an evenly distributed moderate rainfall from sowing until flowering. In the Malagassy Republic optimum yields are obtained with an annual rainfall of 36–48 in. (900–1 200 mm), but the crop can be grown successfully in dry savanna areas with a rainfall of approximately 24 to 30 in. (600–750 mm). It is also tolerant of excessive rain, except at maturity.

Soil—although adapted to a wide range of soils, the bambara groundnut thrives on light sandy, well-drained loams, with a pH of 5.0–6.5. However, it can be grown on poor soils unsuitable for groundnuts and is one of the best crops for savanna ochrosols. When planted in a new area inoculation with soil from a field where the crop has been previously grown has been recommended, but is not necessary if *Rhizobia* of the cowpea group are present. Calcareous soils are reported to be detrimental to the crop. Fertilizer trials are somewhat inconclusive, but when grown on well-fertilized soils, eg after maize, the application of 100–150 lb/ac (112–170 kg/ha) of superphosphate (19 per cent P_2O_5) has been recommended. On soils of low fertility the application of 200–400 lb/ac (225–450 kg/ha) of a general fertilizer, followed by 36–53 lb/ac (40–60 kg/ha) of ammonium sulphate 3 weeks after planting is reported to have a significant effect on yields.

Altitude—the bambara groundnut is normally grown at elevations up to 5 200 ft. (1 600 m).

Planting procedure

Material—the use of shelled seed which has been treated with thiram is recommended, but whole pods are sometimes used.

Method—the bambara groundnut may be planted as a pure stand often as the first crop in a rotation and followed by cassava, but is more usually grown as an intercrop with cereals (usually pearl millet), root crops or other legumes. The soil should be ploughed 7 in. (17.5 cm) deep and worked to a fine tilth, and 1–4 seeds are sown in a hole about 2–3 in. (5–7.5 cm) deep; in many parts of Africa flat-topped ridges are preferred. North of the equator sowing takes place August-September; south of the equator September-December. The time of sowing is critical for optimum yields. In Malawi for example, January-sown seed yields 20 per cent less than December-sown. Weeding is essential and is usually carried out 15 days after germination and as the clusters join. Ridging is often carried out to encourage the development of the pods underground.

Field-spacing—usually 4–6 in. (10–15 cm) apart in single rows 18 in. (45 cm) apart on the flat, or in double rows 8 in. (20 cm) apart on flat-topped ridges 36 in. (90 cm) apart, is used.

Seed-rate—the following average seed-rates have been reported: Tanzania 31 lb/ac (35 kg/ha), Kenya 22–40 lb/ac (25–45 kg/ha), S. Africa 45–53 lb/ac (50–60 kg/ha), E. Nigeria 53–62 lb/ac (60–70 kg/ha). With close spacing in pure stands, the seed-rate may be as high as 170 lb/ac (190 kg/ha).

Pests and diseases

The bambara groundnut is reported to have few serious pests. Leafhoppers, *Empoasca facialis* and *Hilda patruelis*, and the larvae of *Diacrisia maculosa* and *Lamprosema indicata* sometimes attack the crop, but are not usually troublesome. Root-knot nematodes, *Meloidogyne javanica*, can reduce yields considerably if the crop is grown in infested soils. The seed is attacked by a number of storage pests including the bruchid beetles, *Callosobruchus maculatus, C. subinnotatus*, and in addition, in W. Africa, *Piezotrachelus ugandanus*, and in Sierra Leone *Ctenocampa hilda*.

The bambara groundnut is generally considered to be comparatively free from serious diseases during growth. However, in N. Nigeria an unidentified leaf virus is reported to affect yields adversely and in NW. Tanzania the crop is reported to be frequently affected by a wilt disease which is often fatal. This disease has been identified recently as *Fusarium oxysporum* forma specialis *voandzeiae*, and also affects groundnuts. In addition, the following fungal diseases have been reported as attacking the crop, but none appears to reduce yields seriously: *Sphaerotheca voandzeiae, Cercospora voandzeiae* and *C. canescens, Ascochyta phaseolorum, Phaeolus manihotis, Meliola vignae-gracilis, Colletotrichum capsici, Corticium spp., Elsinoe spp., Erysiphe polygoni, Leptosphaerulina trifolii, Phyllosticta voandzeiae* and *Synchytrium dolichi.*

Growth period
Bunch types normally mature in 90–120 days, and the spreading types in 120–150 days.

Harvesting and handling
With small-scale cultivation the pods are often removed as they mature to minimize losses due to rotting and premature germination; with large areas harvesting is carried out when the plants begin to yellow and drop their leaves. The plants are uprooted by hand, or are ploughed out, but since a high percentage of the pods can become detached, (up to 50–60 per cent), gleaning is very important for maximum yields. The pods are usually sun-dried on the ground or mats for several days, before being stored in the shell in baskets or sealed pots, or shelled, and then stored in sacks or plastic bags. Traditionally the crop is shelled by hand with flails followed by winnowing, but modified groundnut shellers are also used successfully. Shelling percentage is about 75 per cent by weight. The shelled nuts are very susceptible to insect infestation and are sometimes fumigated with phosphine, carbon tetrachloride or carbon disulphide. Seed stored in jute sacks in Upper Volta has been found to be severely attacked (up to 39 per cent) by insects, whereas preliminary tests showed that, stored in plastic bags, it was relatively immune from attack.

Primary product
Seed—the bambara groundnut is a hard, smooth, edible seed, usually round, and varying in size, up to about 0.6 in. (1.5 cm) in diameter. It can vary in

colour from white, cream or ivory, to dark-brown, red, or black, and may be speckled or patterned with a combination of these colours. With most cultivars there is a marked white hilum, which in light-coloured seeds is sometimes surrounded by a black or brown eye. One hundred seeds weigh approximately 1.7–2.6 oz (50–75 g).

Yield

Average yields of seed in Africa are reported to be 580–760 lb/ac (650–850 kg/ha), although yields as low as 50–100 lb/ac (56–112 kg/ha) have been reported from Zambia, and as high as 3 200 lb/ac (3 580 kg/ha) in Rhodesia.

Main use

The bambara groundnut is a complete food and is eaten in various forms either immature or fully ripe. The fresh semi-ripe seeds are more palatable than the hard mature ones. The immature seeds are normally eaten fresh, or boiled, or grilled, while the mature ones are often pounded into a flour and then mixed with oil or butter to form a porridge, or sometimes they are roasted in oil. Occasionally the young whole pods are washed and boiled, or used in soups. The commercial canning of the bambara groundnut has been successfully developed in Rhodesia and Ghana.

Subsidiary uses

Flour is sometimes prepared from roasted or unroasted seeds and they can also be used for a livestock food after soaking them in water.

Secondary and waste products

The roasted ground meal has been used occasionally as a coffee substitute. The leaves and haulms may be utilized as an animal feedingstuff. An analysis of the dried leaves has been given as: dry matter 90.2 per cent; crude protein 15.9 per cent; crude fibre 31.7 per cent; ash 7.5 per cent; fat 1.8 per cent; N-free extract 43.1 per cent; all expressed as percentage of dry matter.

Special features

The approximate composition of the bambara groundnut is: moisture 11.0 per cent; total carbohydrate 61.7 per cent; fat 6.3 per cent; protein 17.7 per cent; fibre 4.9 per cent; ash 3.3 per cent; thiamine 0.28 mg/100 g; riboflavin 0.12 mg/100 g; niacin 2.1 mg/100 g; vitamin A 30 iu/100 g; ascorbic acid 1.0 mg/100 g; calcium 73 mg/100 g; iron 7.6 mg/100 g; phosphorus 0.38 mg/100 g. The amino-acid content (mg/g N) has been reported as: leucine

494–510; lysine 400–430; valine 331–340; phenylalanine 219–360; isoleucine 275–280; threonine 219–240; methionine 113–120; cystine 70–180. The protein is of high biological value because of its relatively high lysine content. The oil contains the following fatty acids: palmitic 19.4 per cent; stearic 11.8 per cent; oleic 24.4 per cent; linoleic 34.2 per cent; arachidic 5.3 per cent; behenic 4.9 per cent.

The presence in the raw bambara groundnut of a trypsin inhibitor, which is inactivated by autoclaving, has been reported.

Production and trade

Bambara groundnuts are cultivated extensively in Africa at a subsistence level. They are reported to be the third most important leguminous crop, south of the Sahara, being superseded only by cowpeas and groundnuts. However, there is little reliable information regarding production in many African countries. Total production in Africa has been estimated to be about 330 000 t/a and the more important producers are: Nigeria (100 000 t/a); Upper Volta (65 000 t/a); Niger (30 000 t/a); Ghana (20 000 t/a); Togo (8 000 t/a); the Ivory Coast (7 000 t/a).

Most of the production is consumed domestically and bambara groundnuts do not usually enter international trade, although there have been several unsuccessful attempts to develop exports to Europe for use as an animal feedingstuff.

Major influences

The bambara groundnut has lost importance in many parts of Africa because of the expansion of groundnut production, but in recent years there has been renewed interest in the crop for cultivation in the arid savanna zones, because of its resistance to drought conditions and ability to yield a reasonable crop when grown on poor soils. Little work has been done on the agronomy of the crop or its improvement, and there is an urgent need to develop improved cultivars.

Bibliography

Armstrong, G. M., Armstrong, J. K. and Billington, R. V. 1975. *Fusarium oxysporum* forma specialis *Voandzeiae*, a new form species causing wilt of bambara groundnut. *Mycologia*, *67*, 709–714.

Aykroyd, W. R. and Doughty, J. 1964. Legumes in human nutrition. *FAO Nutr. Stud.* 19, pp. 116; 118. Rome: FAO, 138 pp.

BUSSON, F. 1965. *Voandzeia* Thouars. *Plantes alimentaires de l'ouest Africain: Étude botanique, biologique et chimique*. pp. 250–255. Marseille: Leconte, 568 pp.

COBLEY, L. S. 1956. Bambara groundnut, Earthnut—*Voandzeia subterranea. An introduction to the botany of tropical crops*. pp. 162–163. London: Longmans, Green and Co Ltd, 357 pp.

COMMONWEALTH BUREAU OF PASTURES AND FIELD CROPS. nd. Bambara groundnuts (*Voandzeia subterranea*). Annotated bibliography 1075, 1959–67. *Commonw. Agric. Bur., Bur. Pastures & Field Crops, Hurley, Maidenhead, England*, 2 pp.

DILHAC, P., BUSSON, F. AND GEORGIN, A. 1959. Note sur l'étude chimique des graines de *Voandzeia subterranea* (Thouars). [Note on the chemical study of the seeds of *Voandzeia subterranea* (Thouars)]. *Riz Rizic., 5*, 120.

DOKU, E. V. 1968. Flowering, pollination and pod formation in bambara groundnut (*Voandzeia subterranea*) in Ghana. *Exp. Agric., 4*, 41–48.

DOKU, E. V. 1969. Growth habit and pod production in bambara groundnut (*Voandzeia subterranea*). *Ghana J. Agric. Sci., 2*, 91–95.

DOKU, E. V. and KARIKARI, S. K. 1970. Flowering and pod production of bambara groundnut *Voandzeia subterranea* (Thouars) in Ghana. *Ghana J. Agric. Sci., 3*, 17–26.

DOKU, E. V. AND KARIKARI, S. K. 1971. Bambara groundnut. *Econ. Bot., 25*, 255–262.

DOKU, E. V. AND KARIKARI, S. K. 1971. Operational selection in wild bambara groundnuts. *Ghana J. Sci., 11*, 47–56.

DUFOURNET, R. 1957. Note sur le voanjobory (*Voandzeia subterranea* Dup. Thouars) cultivé à Madagascar. [Note on the bambara groundnut (*Voandzeia subterranea* Dup. Thouars) grown in Madagascar]. *Riz Rizic., 3*, 169–172.

EBBELS, D. L. AND BILLINGTON, R. V. 1972. Fusarium wilt of *Voandzeia subterranea* in Tanzania. *Trans. Br. Mycol. Soc., 58*, 336–338.

FERRÃO, J. E. M. AND XABREGAS, J. 1957/59. O valor alimentar da *Voandzeia subterranea* Thouars. [Nutritional value of *Voandzeia subterranea*]. *Agron. Angolana*, (11), pp. 1–23.

GÖHL, B. 1975. *Voandzeia subterranea* Thouars. *Tropical feeds: Feeds information summaries and nutritive values*. p. 229. Rome: FAO, 661 pp.

Bambar(r)a groundnut

HEPPER, F. N. 1963. The bambara groundnut (*Voandzeia subterranea*) and Kersting's groundnut (*Kerstingiella geocarpa*) wild in West Africa. *Kew Bull.*, *16*, 395–403.

HEPPER, F. N. 1970. Bambara groundnut (*Voandzeia subterranea*). *Field Crop Abstr.*, *23*, 1–6.

HERKLOTS, G. A. C. 1972. Bambara groundnut. *Vegetables in South-east Asia*. pp. 268–270. London: George Allen and Unwin Ltd, 525 pp.

JACQUES-FELIX, H. 1950. Pour une enquête sur le voandzou (*Voandzeia subterranea* Thou.). [An enquiry on the bambara groundnut (*Voandzeia subterranea* Thou.)]. *Agron. Trop*, *5*, 62–73.

JOHNSON, D. T. 1968. The bambara groundnut: A review. *Rhod. Agric. J.*, *65*, 1–4.

JOHNSON, R. M. 1954. Bambara groundnuts. *Colon. Plant Anim. Prod.*, *4*, 76.

KARIKARI, S. K. 1971. Economic importance of bambara groundnut. *World Crops*, *23*, 195–196.

KARIKARI, S. K. 1972. Correlation studies between yield and some agronomic characters in bambara groundnut (*Voandzeia subterranea* Thouars). *Ghana J. Agric. Sci.*, *5*, 79–83.

KINYAWA, P. L. 1969. Bambara groundnut research at Ukiriguru. *Tanzania, Ukiriguru research notes*, (24), pp. 3–4. (*Trop. Abstr.*, *26*, u 2373).

LEPIGRE, M. 1965. Étude sur les possibilités d'amélioration de la conservation des haricots au Togo en milieu rural. [Study on the possibilities of improving bean storage under farming conditions in Togo]. *Agron. Trop.*, *20*, 388–430.

OKIGBO, B. N. 1973. Grain legumes in the farming systems of the humid lowland tropics. *Proc. 1st Int. Inst. Trop. Agric., Grain legume improvement workshop*. pp. 211–223. Ibadan, Nigeria, Int. Inst. Trop. Agric., 325 pp.

OWUSU-DOMFEH, K. 1972. Trypsin inhibitor activity of cowpeas (*Vigna unguiculata*) and bambara beans (*Voandzeia subterranea*). *Ghana J. Agric. Sci.*, *5*, 99–102.

OYENUGA, V. A. 1968. *Voandzeia subterranea* Thouars (Bambara groundnut). *Nigeria's foods and feedingstuffs: Their chemistry and nutritive value*. pp. 52–53. Ibadan: University Press, 99 pp.

POINTEL, J. G. 1968. Contribution à la conservation du niébé du voandzou, du maïs, des arachides et du sorgho. [Contribution to the storage of cowpea, bambara groundnut, maize, groundnuts and sorghum]. *Agron. Trop., 23*, 982–986.

PURSEGLOVE, J. W. 1968. *Voandzeia* Thou. *Tropical crops: Dicotyledons.* Vol. 1. pp. 329–332. London: Longmans, Green and Co Ltd, 332 pp.

RACHIE, K. O. AND ROBERTS, L. M. 1974. Grain legumes of the lowland tropics. *Adv. Agron. 26*, 1–132.

RACHIE, K. O. AND SILVESTRE, P. 1977. Grain legumes. *Food crops of the lowland tropics.* (Leakey, C.L.A. and Wills, J. B. eds.). pp. 41–74. Oxford: Oxford University Press, 345 pp.

RASSEL, A. 1960. Le voandzou (*Voandzeia subterranea* Thouars) et sa culture au Kwango. [The bambara groundnut (*Voandzeia subterranea* Thouars) and its cultivation at Kwango]. *Bull. Agric. Congo Belge, 51*, 1–26.

SMARTT, J. 1976. *Voandzeia* Thou. *Tropical pulses.* pp. 78–81. London: Longman Group Ltd, 347 pp.

STANTON, W. R., DOUGHTY, J., ORRALA-TETEH, R. AND STEELE, W. 1966. *Voandzeia subterranea. Grain legumes in Africa.* pp. 128–132. Rome: FAO, 183 pp.

UCCIAI, E. AND BUSSON, F. 1963. Contribution à l'étude des corps gras de *Voandzeia subterranea* Thouars. [A contribution to the study of the fatty material of *Voandzeia subterranea*]. *Oléagineux, 18*, 45–48.

Common names — **BROAD BEAN, Faba bean[1], Field bean[2], Horse bean[3], Pigeon bean[1], Tick bean[1], Windsor bean[4].**

Botanical name — *Vicia faba* L.

Family — Leguminosae.

Other names — Ackerbohne (Ger.); Ater-bahari, Ater-bar-ativeri, Bagila, Bakela (Eth.); Bakla (Pun.); Baldenga, Baldunga (Eth.); Bob (Yug., Pol.); Bobik (Pol.); Boby kormouvje (USSR); Boerboon (S. Afr.); Bondböna (Sw.); Double bean[5] (Ind.); Duiveboon (Dut.); English bean; European bean; Faveira (Port.); Feldbohne (Ger.); Fève (Fr.); F. des marais (Zar.); Féverole (Fr.); Févette (Mor.); Ful (Ar.); Ful masri (Sud.); Groote boon (Dut.); Grosse fève (Fr.); Haba (Arg.); H. caballar, H. comun (Sp.); Katjang babi (Indon.); Kuru bakla (Turk.); Kyamos (Gr.); Laba major (Sp.); Labboon (Belg.); Lóbab (Hun.); Ontjet (Indon.); Paardeboon (Belg.); Pacae (Peru); Pè-lêt-ma (Burm.); Pferdbohne (Ger.); Pol hagina (Is.); Puffebohne (Ger.); Pul (Is.); San-du-si, San-to-pè (Burm.); Saubohne (Ger.); Sora-mame (Jpn.); Tayôk-pè (Burm.); Tuinbonen (Dut.); Vicos (Gr.); Vika bob (Czech.); Waalse boon, Wierboon (Dut.); Yeshil bakla (Turk.).

Botany

An erect, robust, leafy, glabrous, simple-stemmed annual, normally ranging from 2 to 6 ft (60–180 cm) in height, or in dwarf forms, from 1 to 1.5 ft (30–45 cm), and with one or more basal branches. Unlike most other members

---

1 Used for small-seeded types of broad bean.
2 Also used for the haricot bean, *Phaseolus vulgaris* and in India for the hyacinth bean, *Lablab purpureus*.
3 Also sometimes used for the sword or jack bean, *Canavalia* spp., in Jamaica.
4 Used for large-seeded types of broad bean.
5 Also used in India for the lima bean, *Phaseolus lunatus*.

of the genus *Vicia* the broad bean is without tendrils. It has a well-developed, sometimes branched, tap-root, with extensive lateral roots, which bear clusters of small, lobed nodules, and grow out horizontally before turning sharply downwards. The erect, stiff, stout stems are hollow. The leaves are alternate, pinnate, and consist of between two and six leaflets, borne on a fairly long petiole. The leaflets are entire, oval-shaped, 2–3 in. (5–7.5 cm) in length, attached to the grooved petiole by small pulvini. They are often alternate, and terminate in a small point or rudimentary tendril. The fragrant flowers, (1–6), are borne in the axils of the leaves and are 1 to 1.5 in. (2.5–4 cm) long, dull-white in colour, with a purplish blotch on the wings. Flowering progresses from the lower part to the top part of the stem and takes from 14 to 20 days. Cross-fertilization varies from 3 to 50 per cent depending upon the cultivar, climate and insect activity. The pod is fleshy, green, 2 to 3 in. (5–7.5 cm) long in field cultivars and up to 12 in. (30 cm) in horticultural cultivars. The pod has a pointed beak with a tendency for the calyx to persist at its base. When immature it has a white, velvety interior, which on maturity becomes tough, hard, and constricted between the seeds. The seeds usually number two to six, and vary in shape, size and colour, according to the cultivar.

There are numerous different types of broad beans which show considerable variation in their growth period, yield, size and colour of the seeds. The systematics of the species are very confused. At one time some authorities recognized two distinct botanical varieties, *hortensis*, the culinary or garden broad bean, and *equina*, the field or horse bean. Muratova (1931) using the size of the seed recognized two subspecies: (i) *paucijuga* and (ii) *faba*, and subdivided the latter into: (a) var. *minor*, (more correctly var. *minuta*) which has small, rounded seeds, about 0.4 in. (1 cm) long, typified by tick beans; (b) var. *equina*, with medium-sized seeds, average length 0.6 in. (1.5 cm), typified by horse beans; (c) var. *major*, (ie var. *faba*) the culinary, or horticultural, broad bean, with large, broad, flat seeds, average length 1 in. (2.5 cm) and longer pods. However, Hanelt (1972) has suggested that *paucijuga* is only a geographical race of the subspecies *minor* and recognized subspecies *faba* (vars. *faba*, *equina*) and subspecies *minor* var. *minuta*. Recently, however, Cubero (1974) has suggested that there are four subspecies (*minor*, *equina*, *faba* and *paucijuga*).

A large number of cultivars of broad beans have been developed to suit local conditions in the producing countries and the purpose for which the crop is to be utilized. In the UK for example, the horticultural, white-cultivar

Broad bean

Threefold White is widely grown for processing, and Blue Rock and Maris Bead are high-yielding tick beans grown mainly for pigeon and poultry food. In Canada the improved small-seeded cultivars Ackerperle Diana and Erfordia have been developed recently for animal feed. In recent years plant breeders in the UK and France have been attempting to develop self-pollinating, self-fertilizing 'hybrid' varieties, or pure lines, in an attempt to improve and stabilize yields.

Origin and distribution

Broad beans are one of the most ancient of food crops and originated in the Near East, quickly spreading to Europe, N. Africa, along the Nile to Ethiopia, and from Iraq and Syria to India, and from there to China. They were well established in Europe, including the UK, by the Iron Age, and were the only bean known in Europe in the pre-Colombian era. They are now widely cultivated throughout the temperate and subtropical regions and are frequently grown as a winter crop on the edge of the tropics, eg in the Sudan and Burma, or at high altitudes in the tropics, eg in Uganda.

Cultivation conditions

Temperature—essentially a subtropical or temperate crop, for optimum yields broad beans require an average temperature of between 65° and 80°F (18–27°C), during the growth period. Higher temperatures, particularly at flowering, cause blossom dropping and failure to set seed, in addition to aggravating disease problems. When grown in the humid tropics they usually make good growth, produce flowers, but fail to set seed. In temperate regions such as the UK both winter and spring types are grown. The optimum conditions for the winter types are fairly mild winters, average temperature approximately 35°F (2°C), without prolonged or severe frost, followed by a warm dry spring and summer. Spring types require a warm spring with a moderate rainfall, followed by a warm summer to hasten maturity.

Rainfall—broad beans require a moderate, fairly evenly distributed rainfall of approximately 25 to 40 in./a (650–1 000 mm/a). They are generally considered to be the least drought resistant of the grain legume crops. Experiments in Rhodesia indicate that their peak moisture requirements are 9 to 12 weeks after establishment. In the northern Sudan, where broad beans are grown entirely under irrigation, water stress at flowering has been found to be very detrimental to yield.

Soil—deep, fertile, well-drained soils are essential for high yields; clays, silts or heavy loams are all satisfactory, provided the pH is 6–7, preferably 6.5, and there are adequate reserves of organic matter. Below pH 5.5 growth is liable to suffer and acid soils should be treated with ground limestone, or chalk, well in advance of sowing. Broad beans cannot tolerate waterlogging or saline soils. Nutrient requirements vary considerably according to soil types, but there must be an adequate supply of potash, otherwise the crop is liable to become heavily infested with the fungal disease chocolate spot, *Botrytis* spp. In the UK, the application of a dressing of farmyard manure (FYM) 10T/ac (25t/ha), followed by 220–340 lb/ac (250–380 kg/ha) superphosphate, and 112 lb/ac (125 kg/ha) potassium muriate has been recommended, or where FYM is not available, 340–450 lb/ac (380–500 kg/ha) of superphosphate and 220 lb/ac (250 kg/ha) of potassium muriate. If the soil is very fertile these amounts may be halved. Band placement, 1–2 in. (2.5–5 cm) beside the seed has been suggested for optimum results. In Canada the general application of 20–30 lb/ac (22–34 kg/ha) of phosphoric acid has been recommended. In addition, on sandy, or sandy loam soils the application of 15–30 lb/ac (17–34 kg/ha) of potash has been suggested. In tropical areas the application of superphosphate at rates of 45–89 lb/ac (50–100 kg/ha) of phosphoric acid has been recommended, and if potassium is deficient 22–44 lb/ac (25–50 kg/ha) of potash. In Egypt, where the soils are lacking in nitrogen and phosphorus, but are generally adequate in potassium, the suggested economical fertilizer treatment is 32 lb/ac (36 kg/ha) of nitrogen, plus 64 lb/ac (72 kg/ha) of phosphoric acid.

Altitude—broad beans are sometimes grown in the tropics at elevations above 4 000 ft (1 200 m), but results are very variable. In Uganda experimental plots at elevations of between 6 000 and 8 000 ft (1 800–2 400 m) have given satisfactory results. In Ethiopia, broad beans are grown between approximately 6 000 and 10 000 ft (1 800–3 000 m), especially between 6 700 and 8 300 ft (2 000–2 500 m), in the so-called cereal-pulse zone.

Planting procedure

Material—sound, plump seed with a germination rate of at least 80 per cent should be used. Germination is hypogeal. In the UK effective strains of the bacteria *Rhizobium leguminosarum* are present in the soils so that inoculation is not necessary, but in some countries, eg Canada and Egypt, inoculation is recommended. The pre-sowing treatment of the seed with an insecticide/fungicide protectorant is widely practised; thiram or captan are recommended as effective fungicides.

Broad bean

Method—the seed should be drilled into a moist, firm, weed-free seed-bed. In Middle East countries the seed is sometimes broadcast, although this practice is not recommended. In the UK, cap-feed drills, usually associated with shoe coulters, are often used, although disc coulter drills are also satisfactory, provided that the seed-bed is even. In the Netherlands, modified bulb planters are often used. The seed should be planted 3–4 in. (7.5–10 cm) deep to ensure good moisture conditions for germination and seedling establishment. Broad beans are poor competitors with weeds, particularly during the early stages of growth, efficient weed control is therefore essential. In the UK the use of the pre-emergence herbicides chlorpropham plus diuron or fenuron, or simazine, is reported to give effective control over most weeds. Dinoseb-acetate has also been recommended for post-emergence use. Wild oats, *Avena* spp., which are a serious problem, are usually checked by the pre-drilling application of tri-allate or the post-emergence application of barban. Parasitism by broomrape, *Orobanche crenata*, is common in the Middle East. In Egypt it causes serious crop losses, estimated to range from 30 to 100 per cent. Several herbicides have been tested against broomrape and the use of eptam applied as a post-emergence spray has been shown to be reasonably effective. Soil fumigation by dibromochloropropane is also reported to provide satisfactory control, and the herbicide ozak (terbutol), if this is incorporated deeply into the soil before sowing. In addition, pre-sowing flood, followed by irrigation every 30 days is reported to give reasonably effective control.

Field-spacing—the essential factor for obtaining a high yield of broad beans is to ensure that a large number are established within a given area; the actual arrangement is relatively unimportant. In the UK row width is therefore determined according to the convenience of growing the crop and can range from 6 to 7 in. (15–17.5 cm), narrow, to 18 to 24 in. (45–60 cm), wide, with the seeds placed 5–6 in. (12.5–15 cm) apart. In other countries such as the Sudan and Morocco wider spacings are often used, usually the rows are 30–40 in. (75–100 cm) apart with 6–10 in. (15–25 cm) between the plants.

Seed-rate—although there is evidence that tillering occurs at low seeding rates, this is not sufficient to equal the increased yields obtained with higher plant densities. Seed-rates vary considerably, but in many countries are considered to be sub-optimal. The following have been reported: Egypt 180 lb/ac (200 kg/ha); Peru 100–110 lb/ac (110–125 kg/ha); Greece 130 lb/ac (150 kg/ha); Sudan 70 lb/ac (78 kg/ha) and Canada 120–160 lb/ac (130–180 kg/ha). In the UK, it has been demonstrated that for large-seeded types of broad beans, a seed-rate of 400 lb/ac (450 kg/ha) will produce the maximum yield, but 200–300 lb/ac (225–340 kg/ha) is considered to be the economic optimal range and for some small-seeded types, eg with tick beans, 170 lb/ac (190 kg/ha) is often satisfactory.

Pests and diseases

Pests—broad beans are attacked by several pests. The black bean aphid, or black fly, *Aphis fabae*, can be devastating unless controlled effectively. It causes damage through penetration of the plant tissue and is also a vector of several virus diseases. In addition, it encourages the development of sooty moulds, *Cladosporium* spp., and recently it has been suggested that its presence may also increase the susceptibility of broad beans to chocolate spot, *Botrytis* spp. The use of systemic insecticides, such as menazon, prior to flowering has been suggested as an effective preventative measure. The use of granular compounds, such as disulfoton, is also reported to be effective and has the advantage that it is not harmful to bees, which are of importance in the pollination of the flowers. When the aphids are infesting the plants, then, provided precautions are taken to protect bees and other pollinating insects, the following insecticides are recommended in order to obtain a speedy kill: dimethoate, formothion, malathion, or phosphamidon. In addition to the black bean aphid, the pea aphid, *Acyrthosiphum pisum*, and the vetch aphid, *Megoura viciae*, can sometimes be troublesome, and are also capable of transmitting virus diseases. Pea and bean weevils, *Sitona* spp., can also cause considerable damage on broad bean crops. The adult form feeds on the foliage and the larvae on the root nodules. Satisfactory control is obtained by spraying with DDT or the use of gamma-BHC dust. In France the adult form of *Otiorhynchus ligustici* and the broad bean weevil, *Bruchus rufimanus*, are reported to be troublesome. The latter pest is also of ecomonic importance in Greece and Egypt. In the latter country the leaf-miner, *Liriomyza congesta*, has become a serious pest and is responsible for considerable yield losses. In the Sudan, *Heliothrips* spp., can sometimes cause serious damage. Stem eelworm, *Ditylenchus dipsaci*, has recently become a serious pest of crops in the UK. Rotation plus disinfection of the soil can reduce the problem. Since the seed is a source of infection fumigation with methyl bromide has been suggested as an effective control measure. Like most grain legumes, broad beans are susceptible to infestation from various beetles in the field before harvest, which persists and can cause serious damage when the beans are stored.

Broad bean

In the UK, birds, especially wood pigeons and rooks, often cause serious damage to the winter sown crop, especially where there is uneven coverage of the seed or shallow planting. In Egypt, sparrows, particularly the local variety *Passer domesticus* var. *niloticus*, have become a serious pest.

Diseases—there are a number of fungal diseases which can cause extensive damage to broad beans, especially when they are grown in tropical or subtropical countries. Chocolate spot, caused by the fungi *Botrytis fabae* and *B. cinerea*, can cause very serious crop losses because of the reduction of leaf area growth. No fungicidal treatment is completely effective, and there is correlation between humidity and its severity and the level of potassium in the soil. The removal of all the haulms of the previous crop, good ploughing, and the use of fertilizers to prevent potassium deficiency are probably the best control methods. Leaf spot, *Ascochyta fabae*, can cause serious losses in humid conditions accompanied by high temperatures. The disease is seedborne so that it is essential to use clean seed. Stem rots, mainly caused by the fungus, *Sclerotinia trifoliorum* var. *fabae*, are sometimes troublesome, especially in wet soils. Root rots caused by *Rhizoctonia solani* and *Fusarium* spp. have been reported to be a problem in certain tropical and subtropical countries, eg Egypt, but treating the seed with captan or pentachloronitrobenzene is effective generally. In the Sudan, powdery mildews, *Leveillula taurica*, *Erysiphe umbelliferarum* and *E. polygoni*, are a problem and the crop is usually sprayed with lime sulphur two or three times a season. In several countries, eg Peru and Egypt, crop losses are also caused by broad bean rust, *Uromyces viciae-fabae*. Two virus diseases, mosaic and leaf-roll virus, attack broad beans, causing stunted growth or complete collapse. As has been mentioned they are spread by aphids, so that efficient control of aphids is essential.

Growth period

The maturing period for broad beans is from 90–220 days, depending upon the cultivar and climatic conditions. In the UK it is 150–210 days for winter beans and 90–120 days for the spring sown crop. In the Sudan, Egypt, Greece and Cyprus it is from 135–150 days, in Canada 90–120 days and in Peru about 210 days.

Harvesting and handling

When grown for dry seed, broad beans should be harvested fully mature, when the leaves have either fallen, or become yellow, or even blackened, and the lower pods have begun to turn black and become dry. Field drying on the standing plant is to be recommended, but unfortunately many cultivars tend to lose seed by shattering as the pods dry, and must be harvested when the lower pods begin to blacken. With small-scale cultivation the plants are cut 2–3 in. (5–7.5 cm) below the soil surface with special knives attached to tool frames and set at an angle of about 150° to the row. After being allowed to dry on windrows for a few days the crop is threshed, special care being taken to minimize seed breakage, since broad beans are very susceptible to mechanical damage. Combine-harvesters can be used, provided that there is sufficient clearance between the drum and the concave of the combine, and the drum is operated at a slow speed. Speeds in the range of 300 to 500 r/m have been recommended to minimize splitting and cracking of the beans. In the UK and most other European countries, the beans must be dried immediately to a moisture content of 14–15 per cent before storage. As excessive or sudden heat causes the beans to split, it is recommended that this is accomplished by passing unheated air, or air with a 7–10°F (4–5.5°C) temperature lift, through the beans at uniform depth. Recently a few farmers in the UK have been treating moist stockfeed beans with propionic acid, (about 1 per cent), to inhibit fungal and bacterial growth, with the object of storage without drying, but the process is still at a development stage.

Broad beans are very susceptible to insect infestation during storage, particularly *Bruchus rufimanus* and *B. pisorum*. This is usually controlled by fumigation with methyl bromide, aluminium phosphide, or for small quantities, a mixture of 3 parts ethylene chloride and 1 part carbon tetrachloride.
In the Sudan, the beans are frequently stored in the open, in heaps, or in sacks on wooden slats, after fumigation under a tarpaulin, using methyl bromide.

When grown for consumption as a vegetable the broad bean pods are harvested at the green or snap stage, and the plants are frequently picked over two or three times by hand. Increasing quantities of broad beans are being grown in the UK for canning and quick-freezing and these are harvested mechanically and vined in the field, similarly to peas. Investigations have shown that for canning the optimum maturity is at a tenderometer reading of 136, using a 5 oz (142 g) sample of raw beans, but most processors specify a reading of 120–150. After vining the beans are placed in large 'tanks' or 'bins' for transport either directly to the factory or more usually to a farm cleaning and cooling line. After cleaning and cooling they are returned to the bins, which are lined with black polythene, which can be tied at the top to seal them. In this way discolouration during transport to the factory is minimized, particularly if ice is added to keep the beans chilled. Broad beans quickly deteriorate and are usually cooled to 41°F (5°C) before being transported to the processing unit. Even in the pod broad beans quickly lose their dark, bright-green appearance and become yellow, or if damaged, grey or black, with waterlogging. They can be stored at 32°F (0°C) for a maximum period of four to five weeks, but there is a loss of culinary quality. If grown for forage, broad beans are normally harvested a little before the end of flowering and when the first pods are well formed.

Primary product

Seed—which shows great variation according to the cultivar; some are strongly compressed, angled and about 1 in. (2.5 cm) across, while others are nearly globular and about 0.24 in. (6 mm) in diameter. Their external colour may be white, buff, brown, green, purple, or black; all have a prominent hilum, which often exceeds 0.24 in. (6 mm) in length, and may be white, buff or black. One hundred seeds weigh approximately 1.4–6.3 oz (40–180 g), according to the cultivar.

Yield

Yields of the dry beans vary considerably, according to the cultivar, climate and cultural practices, particularly the seed-rate used. The following average yields have been reported: UK, winter beans, 1.25–1.76 T/ac (3.1–4.37 t/ha); spring beans, 1.0–1.5 T/ac (2.5–3.75 t/ha); Morocco, 0.44–0.64 T/ac (1.1–1.6 t/ha); Peru 0.6–0.9 T/ac (1.5–2.25 t/ha); Denmark 1.4 T/ac (3.5 t/ha); Egypt 0.68 T/ac (1.7 t/ha). Yields of fresh green broad beans for use as a vegetable average between 4.5 and 5.0 T/ac (11.25–12.5 t/ha), although some growers in Kent are reported to obtain about 10 T/ac (25 t/ha).

Main use

The dry mature seeds are frequently utilized as a nutritious foodstuff, particularly in the Middle East, India and other parts of Asia. In the Middle East they are often consumed after baking (Foul midamis). In Ethiopia they are also utilized ground into a porridge. In northern Europe and Canada, small and medium-sized broad beans (faba, tick and horse beans) are used mainly for feeding livestock or poultry, up to 30 per cent for dairy cattle, 15 per cent for pigs, but for poultry feeding, supplementation with the amino acid methionine is advisable. In the UK tick beans are very popular for feeding pigeons.

The green immature beans are a popular vegetable when eaten boiled and considerable quantities are produced for canning and quick-freezing, especially in the UK.

Subsidiary uses
Recently broad beans have attracted interest as a source of vegetable protein suitable for use as a meat extender, or substitute, in products such as sausages and meat pies. In addition, the possibility of utilizing the beans as a skim milk substitute for feeding calves has been investigated. Broad beans are sometimes dehulled and ground to produce flour. Trials in Canada indicate that beans with a low moisture content (about 9 per cent) are the most satisfactory for milling. In India the beans are sometimes roasted and eaten like groundnuts.

Secondary and waste products
Broad beans are sometimes grown as a green manure crop or for silage. The haulms left after harvesting may also be used as an animal feedingstuff. Their approximate composition is: moisture 85.0 per cent; crude protein 2.5 per cent; fat 0.4 per cent; crude fibre 4.9 per cent; N-free extract 5.4 per cent; ash 1.8 per cent; calcium 0.22 per cent; phosphorus 0.04 per cent; digestible crude protein 2.0 per cent. Recent investigations indicate that the beans could be utilized, similarly to soyabeans for the production of fermented food products such as tempeh and miso. A process for the extraction of starch and protein from the seed-coat is reported to have been developed in Belgium.

Special features
The approximate composition of broad beans has been given as: moisture 11 per cent; crude protein 26–33 per cent; total carbohydrate 51–66 per cent; fat 2 per cent; crude fibre 8 per cent; ash 4 per cent; calcium 90 mg/100 g; vitamin A 100 iu/100 g; niacin 2.3 mg/100 g; thiamine 0.54 mg/100 g; riboflavin 0.29 mg/100 g; ascorbic acid 4 mg/100 g. The crude fibre is concentrated mainly in the seed-coat which constitutes approximately 14–15 per cent of the seed. The protein consists of globulins, albumins and glutelins. Two globulins, legumin and vicilin, and one albumin, legumelin, have been identified. Legumin is the major protein; the albumin level is relatively low. The average amino acid content (mg/gN) has been reported as: isoleucine 250; leucine 443; lysine 404; phenylalanine 270; tyrosine 200;

methionine 46; cystine 50; threonine 210; valine 275; arginine 556; histidine 148; alanine 259; aspartic acid 702; glutamic acid 942; glycine 258; proline 249; serine 280. Experiments have shown that the application of sulphur-containing fertilizers can increase the methionine and cystine content. Baking the beans to prepare Foul midamis has been found to reduce their amino acid content by 8 to 25 per cent.

There is considerable variation in the composition of broad beans, according to the cultivar, climate, time of sowing, etc. In the UK for example, spring sown beans have an average crude protein content of 31.4 per cent, compared with 26.5 per cent for winter sown beans. Winter sown broad beans have also been found to provide more available carbohydrate (46–48 per cent), (that is starch and taka-diastase digestible polysaccharides), than spring sown types (30–42 per cent).

The starch content ranges from 30 to 42.3 per cent and is reported to be inversely related to the protein content. It consists of about 64 per cent amylopectin and 36 per cent amylose. It gelatinizes at a lower temperature than wheat starch, (61° to 70°C), and sets to a slightly higher cold paste viscosity. In a recent examination of the carbohydrate content of broad beans from different origins, the furfural generator content, expressed in terms of xylose, ranged from 4.6 to 6.9 per cent; value for total ethanol-soluble sugars ranged from 4.9 to 7.2 per cent and the sucrose content varied from 0.55 to 2.2 per cent. The major constituents of the ethanol-soluble sugars were verbascose, stachyose and higher-molecular-weight alpha-galactosides, which appear to be important in animal feeding.

Broad beans contain a brownish-yellow oil with the following characteristics: SG $^{18°C}$ 9.385; N_D $^{20°C}$ 1.484; acid val. 18.2; sap. val. 190.2; iod. val. 118.5; RM val. 0.57; unsaponifiable matter (chiefly sitosterol) 1.98 per cent. The fatty acid composition of the oil has been reported as: unsaturated, oleic 45.8 per cent; linoleic 30.0 per cent; linolenic 12.8 per cent; saturated 11.4 per cent (stearic 8.2 per cent). In addition, the presence of cholesterol (0.04 per cent) and the enzyme lipoxygenase has been reported.

Two heat-labile trypsin inhibitors, thought to be glycoproteins, are also present in the beans. Heating for 40 minutes at 230°F (110°C) is stated to inactivate them. The presence of auxins, growth regulating substances, has been detected in the roots and pipecolic acid has been isolated from the germinating beans. In addition, broad beans also contain toxic principles which can cause favism, a disease characterized by haemolytic anaemia, and which occurs principally in countries of the Mediterranean littoral, where it is an important nutritional problem. Recent evidence suggests that the substances responsible for favism are pyrimidines that occur naturally as beta-glycosides. These substances known as divicine and *iso*-uramil are the aglycone moieties of vicine and convicine, respectively. In addition, the presence of L-dopa, (3–(3, 4–dihydroxyphenyl)–L–alanine), has been reported.

The seed-coat of many cultivars contains leucodelphinidin and leucocyanidin, which cause discolouration during cooking.

Processing
Broad beans are a popular processed vegetable.

Canning—white-seeded types, such as Threefold White, are used for canning since green coloured beans are liable to turn greenish-brown or grey when processed. The beans are picked at an immature stage and are usually depodded in the field using specially adapted viners. The beans are blanched for 2 minutes at 170°F (77°C), no longer, or they may split. The beans are next thoroughly washed, inspected, and then filled into vegetable lacquered cans, either by hand, or with modified pea or bean fillers. Hot brine (1.6 per cent salt and 1.4 per cent sugar) is added and the cans exhausted to a centre temperature of 170°F (77°C), before being processed for 25 minutes at 240°F (116°C) for A1 (211 x 400) or A2 (307 x 408). After processing the cans should be cooled as rapidly as possible.

Quick-freezing—considerable quantities of broad beans are also quick-frozen. As with the canned product only white-seeded types produce a satisfactory green coloured final product. The beans are normally blanched for about 1 minute in boiling water, prior to freezing and packing.

Production and trade
Production—broad beans are the fourth most important of the food legume crops. Annual production averaged 5 214 400 t/a for the years 1970–74, compared with 4 931 000 t/a for 1965–69, an increase of 6 per cent. In 1975 world production was estimated to have amounted to 6 201 000 t and in 1976 to have been 6 187 000 t.

Broad bean

Broad beans: Major producing countries

Quantity tonnes

	Annual average 1965–69	Annual average 1970–74	1975	1976
People's Repub. China	2 460 000	3 410 000	4 350 000	4 458 000
Italy	394 000	317 000	252 000	220 000
Egypt	300 000	280 000	234 000	237 000
Morocco	129 000	218 000	213 000	230 000
Ethiopia	127 000	148 000	200 000	200 000
UK	166 000	173 000	91 000	110 000

Trade—considerable quantities of broad beans enter international trade for use as a foodstuff and for livestock feeding. Statistical data are fragmentary, since many countries do not separate broad beans from other dry beans in their trade returns. However, from the available data, it is estimated that world trade averaged at least 225 000 t/a for the period 1970–74, compared with 180 000 t/a for 1965–69, an increase of 25 per cent.

Broad beans: Major exporting countries

Quantity tonnes

	Annual average 1965–69	Annual average 1970–74	1975	1976
UK	n/a	73 514[1]	93 937	21 558
Ethiopia	23 269[2]	21 383	22 113	n/a
Morocco	90 578[3]	38 760	5 215	n/a
People's Repub. China	11 217[4]	15 112	n/a	n/a

1 Three year average.
2 Four year average.
3 Two year average.
4 1969 only.
n/a Not available.

Broad beans: Major importing countries

Quantity tonnes

	Annual average 1965–69	Annual average 1970–74	1975	1976
Italy	73 266	92 007	40 773	39 398
Netherlands	28 522	38 937	7 632	2 405
France	11 407	36 606	8 211	n/a
Ger. Fed. Repub.	36 021[1,2]	24 653[2]	4 203	2 274
Japan	22 383	21 433	20 774	n/a

Prices—producer prices in Italy have shown a steady upward trend since 1969; annual average 1966–69 £42.52/t, 1970–74 £70.38/t. The average fob value of Ethiopian exports has also shown a steady upward trend in recent years, from an annual average value of £30.75/t for the years 1966–69 to £58.00/t for 1970–74. Information for the UK is available only since 1972 and the average fob value of exports has been: 1972 £39.00/t, 1973 £56.00/t, 1974 £69.00/t, 1975 £74.00/t, 1976 £122.00/t; five year average, 1972–76 £72.00/t.

Major influences

Broad beans are grown widely throughout the temperate and subtropical areas of the world and have potential in areas, such as western Canada, where the growing season of 90–120 days is inadequate for soyabean cultivation. In the tropics they are a possible crop for cultivation at higher elevations. In most countries they are grown for human consumption as dry beans, although there is a considerable demand for the green immature beans for use as a vegetable, in countries such as the UK. In recent years there has been increased interest in the crop for livestock feeding and as a source of protein isolates. The commercial development of textured protein meat substitutes could result in an increased demand. In certain countries, eg the UK and Denmark, despite disease problems, they are a popular break-crop in intensive cereal production. However, currently, yields are very variable, often because of pollination difficulties, and there is need to develop self-fertile types.

---

1 Four year average.
2 Includes some dry beans, other than broad beans.
n/a Not available.

Broad bean

From a nutritional point of view, although broad beans have a relatively high protein content, they are deficient in the essential amino acids, methionine and cystine, and there is also a need to breed improved high-yielding types with higher methionine and cystine contents. In addition, the problem of favism needs further investigation, especially in those countries where it is endemic and food products prepared from broad beans are likely to be used for infant feeding.

Bibliography

ANON. 1965. Favism in Britain. *Br. Med. J.*, *2*, 1140.

ANON. 1970. The horsebean—Denmark's answer to the soyabean. *Oil Mill Gaz.*, *74* (11), 40.

ANON. 1974. Protein beano from beans. *New Sci.*, *62* (902), 681.

ARNON, I. 1972. Pulses or grain legumes. *Crop production in dry regions.* Vol. 2. pp. 217–260. London: Leonard Hill Books, 683 pp.

ARTHEY, V. D. and WEBB, C. 1969. The relationship between maturity and quality of canned broad beans (*Vicia faba* L.). *J. Food Technol.*, *4*, 61–74.

ASSEM, M. A. 1974. Insect pests of food legumes. *Proc. 1st FAO/SIDA Semin., Improvement and production of field food crops for plant scientists from Africa and the Near East.* pp. 570–584. Rome: FAO, 684 pp.

AYKROYD, W. R. and DOUGHTY, J. 1964. Legumes in human nutrition. *FAO Nutr. Stud.* 19, pp. 115–117. Rome: FAO, 138 pp.

BHATTY, R. S. 1974. Chemical composition of some faba bean cultivars. *Can. J. Plant Sci.*, *54*, 413–421.

BLAND, B. F. 1971. Horse or field beans. *Crop production: Cereals and legumes.* pp. 269–303. London/New York: Academic Press, 466 pp.

BOND, D. A. 1976. Field bean. *Evolution of crop plants.* (Simmonds, N.W. ed.). pp. 179–182. London: Longman Group Ltd, 339 pp.

BOND, D. A. and TOYNBEE-CLARKE, G. 1968. Protein content of spring and winter varieties of field beans (*Vicia faba* L.) sown and harvested on the same dates. *J. Agric. Sci.*, (*Camb.*), *70*, 403–404.

BRIEN, R. M., CHAMBERLAIN, E. E., COTTIER, W., CRUICKSHANK, I. A. M., DYE, D. W., JACKS, H. and REID, W. D. 1955. Diseases and pests of peas and beans in New Zealand and their control. pp. 69–74. *N.Z. Dep. Sci. Ind. Res., Bull.* 114, 91 pp.

BURNETT, D., AUDUS, L. J. and ZINSMEISTER, H. D. 1965. Growth substances in the roots of *Vicia faba*. *Phytochemistry*, *4*, 891–904.

CACKETT, H. E. and METELERKAMP, H. R. R. 1963. The relationship between evapotranspiration and the development of the field bean crop. *Rhod. J. Agric. Res.*, *1*, 18–21.

CASTELL, A. G. 1976. Comparison of faba beans (*Vicia faba*) with soyabean meal or field peas (*Pisum sativum*) as protein supplements in barley diets for growing-finishing pigs. *Can. J. Anim. Sci.*, *56*, 425–432.

CERNING, J., SAPOSNIK, A. and GUILBOT, A. 1975. Carbohydrate composition of horsebeans (*Vicia faba*) of different origins. *Cereal Chem.*, *52*, 125–138.

CHADA, Y. R. (ed.). 1976. *V. faba* Linn. *The wealth of India: Raw materials.* Vol. 10 (Sp-W). pp. 460–465. New Delhi: Counc. Sci. Ind. Res., Publ. Inf. Dir., 591 pp.

COMMONWEALTH BUREAU OF PASTURES AND FIELD CROPS. nd. Field beans (*Vicia faba*) agronomy. Annotated bibliography 1137, 1956–70. *Commonw. Bur. Pastures & Field Crops, Hurley, Maidenhead, Berks., England*, 16 pp.

CRAMP, K. V., EWAN, J. W., HUME, W. G., FINCH, C. G. and SHEARD, G. F. 1962. Beans. pp. 2–4. *Lond., Minist. Agric. Fish. Food Bull.* 87, 24 pp.

CUBERO, J. I. 1973. Evolutionary trends in *Vicia faba*. *Theor. Appl. Genet.*, *43*, 59–65.

DESHMUKH, A. D. and SOHONIE, K. 1961. Distribution of pipecolic acid in germinating legumes and their growing parts. *J. Sci. Ind. Res.* (*India*), *20C*, 329–330.

DICKINSON, D., KNIGHT, M. and REES, D. I. 1957. Varieties of broad beans suitable for canning. *Chem. Ind.*, (46), pp. 1503.

DRAPER, S. R. 1976. Changes in the amount of protein and lignocellulosic material accompanying the development of the field bean. *J. Sci. Food Agric.*, *27*, 23–27.

DUTHIE, I. F. 1976. Fabalac—a new dried skim milk substitute. *Flour Anim. Feed Milling*, *158*, 15–16; 18.

EDEN, A. 1968. A survey of the analytical composition of field beans (*Vicia faba* L.). *J. Agric. Sci.* (*Camb.*), *70*, 299–301.

Broad bean

EDEN, A. 1968. Field beans in stock-feeding. *Agriculture, (UK), 75*, 209–212.

EL-BEHEIDI, M. A. and SALEM, A. H. 1971. The effect of different cultural methods on the yield of field beans. *Beitr. Trop. Subtrop. Landwirthsch. Tropenveterinärmed., 9*, 247–251.

EL-NADI, A. H. 1969. Water relations of beans: (i) Effect of water stress on growth and flowering. *Exp. Agric., 5*, 195–207.

EL-NADI, A. H. 1970. Water relations of beans: (ii) Effects of differential irrigation on yield and seed size of broad beans. *Exp. Agric., 6*, 107–111.

EL NAHRY, F., DARWISH, N. M. and NADA, E. 1977. Effect of local methods of preparation on the amino acid constituents of broad beans. *Qual. Plant., Plant Food Human Nutr., 27*, 151–160.

EL-SAEED, E. A. K. 1968. Agronomic aspects of broad beans (*Vicia faba* L.) grown in the Sudan. *Exp. Agric., 4*, 151–159.

EPPENDORFER, W. H. 1971. Effects of S, N and P on amino acid composition of field beans (*Vicia faba*) and responses of the biological value of the seed protein to S-amino acid content. *J. Sci. Food Agric., 22*, 501–505.

ESKIN, N. A. M. and HENDERSON, H. M. 1974. A study of lipoxygenase from small faba beans. *Madrid, 4th Int. Congr. Food Sci. & Technol., Work. Doc. 1b, Chem. and biochem. of food deterioration, Pap.* 17, pp. 49–51.

EVANS, L. E., ROGALSKY, J. R., HODGSON, G. C., CAMBELL, L. D., DEVLIN, T. J., INGALLS, J. R., STOTHERS, S. C. and SEALE, M. E. 1975. Growing and using fababeans. *Can. Dep. Agric., Publ.* 1540, 21 pp.

EVANS, L. E., SEITZER, J. F. and BUSHUK, W. 1972. Horse beans—a protein crop for western Canada. *Can. J. Plant Sci., 52*, 657–659.

FLINK, J. and CHRISTIANSEN, I. 1973. The production of a protein isolate from *Vicia faba. Lebensm.-Wiss. Technol., 6*, 102–106.

FOOD AND AGRICULTURE ORGANIZATION OF THE UNITED NATIONS. 1970. Amino acid content of foods and biological data on proteins. *FAO Nutr. Stud.* 24, pp. 50–51. Rome: FAO, 286 pp.

GABR, M. 1973. Legumes and green leafy vegetables in infant and young child nutrition. *FAO/WHO/UNICEF, Protein Advis. Group, PAG Bull.* 3 (2), 46–49.

GANE, A. J., KING, J. M., GENT, G. P., BIDDLE, A. J., HANDLEY, R. D. and BINGHAM, R. J. B. 1975. *Pea and bean growing handbook*. Vol. 2. *Beans*. Peterborough, England, Processors and Growers Res. Organ., 7 sections, (loose leaf).

GENT, G. P. 1968. Broad beans for processing. *Agriculture*, (*UK*), *75*, 278–280.

GINDRAT, D. 1969. Les principaux champignons parasites de la féverole. [The principal fungal diseases of the broad bean]. *Rev. Suisse Agric.*, *1*, 111–115.

GOULDEN, D. S. 1974. Prospects for field beans in New Zealand. *N.Z. J. Agric.*, *129*, 37–38.

HAGBERG, A. and SJÖDIN, J. 1975. Broad bean. *Food protein sources*. (Pirie, N. W. and Swaminathan, M. S. eds.). pp. 117–119. London: Cambridge University Press, 260 pp.

HAKAM, M. M. 1974. Legume field crops improvement in the Arab Republic of Egypt. *Proc. 1st FAO/SIDA Semin., Improvement and production of field food crops for plant scientists from Africa and the Near East*. pp. 77–81. Rome: FAO, 684 pp.

HAKAM, M. M. and IBRAHIM, A. A. 1974. Cultural practices of grain legumes in the Arab Republic of Egypt. *Proc. 1st FAO/SIDA Semin., Improvement and production of field food crops for plant scientists from Africa and the Near East*. pp. 457–465. Rome: FAO, 684 pp.

HAMISSA, M. R. 1974. Fertilizer requirements for broad beans and lentils. *Proc. 1st FAO/SIDA Semin., Improvement and production of field food crops for plant scientists from Africa and the Near East*. pp. 410–416. Rome: FAO, 684 pp.

HANELT, P. 1972. Die infraspezifische Variabilität von *Vicia faba* und ihre Gliederung. [The infraspecific variability of *Vicia faba* L. and its subdivision]. *Kult'pflanz*, *20*, 75–128. (*Plant Breed. Abstr.*, *43*, 5554).

HAWTIN, G. C., BOND, D. A., LANE, M., KAMBAL, A. E. and KHIDIR, M. O. (eds.). 1974. Broad beans. *Guide for field crops in the tropics and the sub-tropics*. (Litzenberger, S. C. ed.). pp. 129–137. Washington, Off. Agric. Tech. Assist. Bur., Agency Int. Dev., 321 pp.

Broad bean

HERBERT, D. A. and FELMINGHAM, J. D. 1969. Canning and freezing. *Food industries manual*. 20th ed. (Woollen, A. H. ed.). pp. 16; 46. London: Leonard Hill Books, 509 pp.

HEWETT, P. D. 1973. The field behaviour of seed-borne *Ascochyta fabae* and disease control in field beans. *Ann. Appl. Biol.*, *74*, 287–295.

HIGAZI, M. I. and READ, W. W. C. 1974. A method for determining vicine in plant material and in blood. *J. Agric. Food Chem.*, *22*, 570–571.

HODSON, G. L. and BLACKMAN, G. E. 1956. An analysis of the influence of plant density on the growth of *Vicia faba*: (i) The influence of density on the pattern of development. *J. Exp. Bot.*, *7*, 147–165.

IBRAHIM, G. and HUSSEIN, M. M. 1974. A new record of root rot of broad bean (*Vicia faba*) from the Sudan. *J. Agric. Sci.* (*Camb.*), *83*, 381–383.

KAMEL, A. H. 1967. A study of the possibility of eradicating the broad bean weevil *Bruchus rufimanus* Boh. (*Coleoptera, Bruchidae*) by fumigation of seeds. *J. Stored Prod. Res.*, *3*, 365–369.

KAO, C. 1974. Fermented foods from chick pea, horse bean and soybean. *Diss. Abstr. Int.*, B35 (5), 2250.

KOTHARY, K. and SOHONIE, K. 1960. Proteins of double bean (*Faba vulgaris* Moench.): (i) Isolation, fractionation and amino acid composition of the proteins; (ii) Studies in *in vitro* digestion. *J. Sci. Ind. Res.*, (*India*), *19C*, 14–16; 16–18.

LAST, F. T. and NOUR, M. A. 1961. Cultivation of *Vicia faba* L. in northern Sudan. *Emp. J. Exp. Agric.*, *29*, 60–72.

LAWES, D. A. 1974. Field beans: improving yield and reliability. *Span*, *17*, 21–23.

LIENER, I. 1975. Antitryptic and other antinutritional factors in legumes. *Nutritional improvement of food legumes by breeding*. (Milner, M. ed.). pp. 239–258. New York/London: John Wiley and Sons, 399 pp.

LINEBACK, D. R. and KE., C. H. 1975. Starches and low-molecular weight carbohydrates from chick pea and horse bean flours. *Cereal Chem.*, *52*, 334–347.

LOCK, A. 1969. Canning of vegetables—broad beans. *Practical canning*. pp. 222–224. London: Food Trade Press Ltd, 415 pp.

MARQUARDT, R. R. and CAMPBELL, L. D. 1975. Performance of chicks fed faba bean (*Vicia faba*) diets supplemented with methionine, sulfate and cystine. *Can. J. Anim. Sci.*, *55*, 213–218.

MCEWEN, T. J., MCDONALD, B. and BUSHUK, W. 1974. Faba beans (*Vicia faba minor*) physical, chemical and nutritional properties. *Madrid, 4th Int. Congr. Food Sci. & Technol., Work Doc. 8a, New food sources of key nutrients*) *Pap*. 10, pp. 24–26.

MCFARLANE, J. A. 1970. Pulses. *Food storage manual*. pp. 335–346. Rome: FAO, World food programme, 799 pp., plus indexes, (loose leaf).

MUNCK, L., HOLMBERG, E., KARLSSON, K. E., LÖFQUIST, B. and SJÖDIN, J. 1975. Breeding for protein quantity and quality in field bean (*Vicia faba*) and barley (*Hordeum vulgare*). *Nutritional improvement of food legumes by breeding*. (Milner, M. ed.). pp. 197–204. New York/London: John Wiley and Sons, 399 pp.

MURATOVA, V. 1931. Common beans (*Vicia faba*). *Bull. Appl. Bot. Genet. Plant Breed., Suppl.*, *50*, 285 pp.

NORTH, J. J. 1968. Growing field beans. *Agriculture* (*UK*), *75*, 14–17.

NOUR, M. A. 1957. Control of powdery mildew diseases in the Sudan with special reference to broad beans. *Emp. J. Exp. Agric.*, *25*, 119–131.

PALMER, R. and THOMPSON, R. 1975. A comparison of the protein nutritive value and composition of four cultivars of faba beans (*Vicia faba* L.) grown and harvested under controlled conditions. *J. Sci. Food Agric.*, *26*, 1577–1583.

PATWARDHAN, V. N. and WHITE, J. W. (Jr.). 1973. Problems associated with particular foods. *Toxicants occurring naturally in foods*. pp. 477–507. Washington, Natn. Acad. Sci., 624 pp.

PICARD, J. 1963. La coloration des téguments du grain chez la féverole (*Vicia faba* L.) étude de l'hérédité des différentes colorations. [The colouration of the seed-coat of broad bean seeds a study of the heredity of different colours]. *Ann. Amél. Plant.*, *13*, 97–117.

PICARD, J. and BERTHELEM, P. 1957. La féverole. (The field bean). *Ann. Amél. Plant., Ser. B*, *7*, 287–311.

POWELL, D. F. 1974. Fumigation of field beans against *Ditylenchus dipsaci*. *Plant Pathol.*, *23*, 110–113.

Broad bean

Purseglove, J. W. 1968. *Vicia faba* L. *Tropical crops: Dicotyledons*. Vol. 1. pp. 319–321. London: Longmans, Green and Co Ltd, 332 pp.

Rizk, Z. 1974. Diseases of food legumes. *Proc. 1st FAO/SIDA Semin., Improvement and production of field food crops for plant scientists from Africa and the Near East*. pp. 557–559. Rome: FAO, 684 pp.

Salem, S. A. 1975. Changes in carbohydrates and amino acids during baking of *Vicia faba*. *J. Sci. Food Agric.*, *26*, 251–253.

Salih, H. S. A., Ishag, H. M. and Siddig, S. A. 1973. Effect of sowing date on incidence of Sudanese broad bean mosaic in, and yield of, *Vicia faba*. *Ann. App. Biol.*, *74*, 371–378.

Siegel, A. and Fawcett, B. 1976. *Food legume processing and utilization (with special emphasis on application in developing countries)*. Ottawa, Int. Dev. Res. Cent., IDRC–TS 1, 88 pp.

Sjödin, J. 1974. Breeding for improved nutritional quality in field beans (*Vicia faba* L.). *Proc 1st FAO/SIDA Semin., Improvement and production of field food crops for plant scientists from Africa and the Near East*. pp. 295–301. Rome: FAO, 684 pp.

Soper, M. H. R. 1953. The influence of farm practice on the productivity of field beans. *Emp. J. Exp. Agric.*, *21*, 22–32.

Soper, M. H. R. 1956. Field beans in Great Britain. *Field Crop Abstr.*, *9*, 65–70.

Soper, M. H. R. 1958. Field beans. *Lond., Min. Agric. Fish Food*, 15 pp.

Smartt, J. 1976. *Vicia faba* L. *Tropical pulses*. pp. 49–50. London: Longman Group Ltd, 348 pp.

Tomkins, R. G. 1965. The storage of broad beans. *UK, Agric. Res. Counc., Ditton and Covent Gard. Lab., Annu. Rep.* 1964/65, p. 24.

Warsy, A. S. and Stein, M. 1973. Trypsin inhibitors of broad bean (*Vicia faba* L.). *Qual. Plant, Plant Food Hum. Nutr.*, *23*, 157–169.

Watson, J. W., McEwen, T. J. and Bushuk, W. 1975. Note on the dry milling of fababeans. *Cereal Chem.*, *52*, 272–273.

Westphal, E. 1974. *Vicia faba* L. Pulses in Ethiopia, their taxonomy and agricultural significance. pp. 205–213. *Wageningen, PUDOC, Cent. Agric. Publ. Doc., Agric. Res. Rep.* 815, 278 pp.

WHITTEMORE, C. T. and TAYLOR, A. G. 1973. Digestibility and nitrogen retention in pigs fed diets containing dried and undried field beans treated with propionic acid. *J. Sci. Food Agric.*, *24*, 1133–1136.

WILLIAMS, P. F. 1975. Growth of broad beans infected by *Botrytis fabae*. *J. Hortic. Sci.*, *50*, 415–424.

WILSON, B. J., MCNAB, J. M. and BENTLEY, H. 1972. Trypsin inhibitor activity in the field bean (*Vicia faba* L.). *J. Sci. Food Agric.*, *23*, 679–684.

WOOLLEN, A. 1973. Tastes good. *Aust. Food Manuf.*, *42* (8), 4.

ZAHRAN, M. K. 1974. Improvement of food crop production through weed control in Egypt. *Proc. 1st FAO/SIDA Semin., Improvement and production of field food crops for plant scientists from Africa and the Near East.* pp. 605–613. Rome: FAO, 684 pp.

Common names CHICK PEA, Gram, Bengal gram.

Botanical name *Cicer arietinum* L.

Family Leguminosae.

Other names Adas[1] (Ind.); Adungare[2] (Eth.); Agrâo de bico (Port.); Alhamos (Ar.); Ater cajeh, Atir saho (Eth.); Bagoly borsó (Hun.); Beiqa, Blabi (Ar.); Boot (kaley), But (mah), Buthalai (Ind.); Café français (Fr.); Cece (bianco), Ceci, Cecio, Cesari, Cesco (It.); Ceseron (Fr.); Chahna, Chala, Chania, Chan(n)a, Chela (Ind.); Chemps (Ar.); Chicaro (Port.); Chicher (Ger.); Chimbera (Eth.); Chimtza (tarbutit) (Is.); Chola, Chono, Chota but, Chunna (Ind.); Cicererbis (Ger.); Cicérolé (Fr.); Cicerewt (Dut.); Ciche (Fr.); Coffee pea; Csicseri borsó (Hun.); Djelbane (N. Afr.); Dwergertjie (S. Afr.); Echte Kicher (Ger.); Egyptian pea; Erebinthos, Erevinthos (Gr.); Ervanços (Port.); Fontanellerbse (Ger.); Gairance, Gairoutte (Fr.); Garavance, Garbanza (Carib.); Garbanzo[3] (Lat. Am.); Garvance (Fr.); Gewöhnliche Kichererbse (Aus.); Gráo (de bico) (Port.); Gravancos (Sp., Lat. Am.); Hagoly-borsó (Hun.); Hamaz (N. Afr.); Hamiça (Ar.); Hammes, Ham(m)o(u)s (Ar.); Harimandha-kam (Ind.); Himmos (akhdar), Hommes, Hom(m)os malana (Ar.); Hummous (Sud.); Hyokko-mame (Jpn.); Ikiker (N. Afr.); Jumez (Ar.); Kabkaza (Sud.); Kachang kuda (Malays.); Kadala(i), Kadale, Kadli (Ind.); Kala-pè (Burm.); Karbantos (Is.); Karikadale (Ind.); Kebkabeik (Sud.); Keker (Dut.); Kicher (en),

---

[1] Also used for the lentil, *Lens culinaris*.
[2] Also frequently used as a general term for beans.
[3] Sometimes used in Venezuela for the grass pea, *Lathyrus sativus*.

Kichererbse (Ger.); Kiker (N. Afr.); Konda kadala (Sri La.); Korkadala (Ind.); Kulapia (Burm.); Lablabi, Leblebi[4] (Ar.); Mdengu (Ken.); Mukhudo (USSR); Nachius (Turk.); Nachunt (USSR); Nakhut (Afg., Bul.); Nakud (Iran); Nakut (USSR); Nohot (Rom.); Nohud, (Turk.); Nokhut (Iran); Nut (Pol., USSR); Omnos (Ar.); Pisello cece, P. cornuto (It.); Pois bécu, P. blanc, P. breton, P. café, P. chabot, P. chiche, P. citron, P. cornu, P. de brebis, P. gris, P. pointu, P. tête de belier (Fr.); Rebinthia, Revithia (Gr.); Sangalu (Ind.); Shembera, Shihu, Shimbera, Shimbrah, Shumbra (Eth.); Sigró (Sp.); Sin-gaung-kala-pè (Burm.); Sisser erwt (Dut.); Slanutok (Bulg.); Spizole (It.); Stragaliais[4] (Gr.); Sundal kadalai (Ind.); Yellow gram (E. Afr.); Ziesererbsen, Ziserbohne (Ger.).

Botany

A small, herbaceous, annual shrub, showing considerable variation in form. Some types are semi-erect with a main stem and only a few branches, while others are semi-spreading with profuse branching. Normally the plants grow to a height of 18–24 in. (45–60 cm) and are frequently bluish-green in colour and covered with glandular hairs. The tap-root is well developed, and can reach 12 in. (30 cm) or more in length; it generally has four rows of lateral roots covered with nodules. The main stem is rounded and the branches are usually quadrangular and ribbed. The leaves are pinnate or odd-pinnate, rigid, alternate and compound, each having from 11 to 18 leaflets. These are oval, normally about 0.24 in. (6 mm) in length and 0.15 in. (4 mm) in breadth, with serrated edges. The flowers are usually solitary and formed in the axils of the leaves on a jointed peduncle, 1–1.6 in. (2.5–4 cm) long. They are normally pinkish, purplish, red fading into blue, or white, or occasionally greenish-white or blue, and are generally self-fertilized (90–95 per cent). The seed-pods are oblong about 1 in. (2.5 cm) long and 0.4 in. (1 cm) broad, generally containing one or two relatively large seeds, which vary in colour.

[4] Usually used to denote roasted seeds.

Chick pea

Distinct geographical forms of chick peas exist and four races have been recognized, namely *orientale, asiaticum, mediterraneum* and *eurasiaticum*; in addition, five subraces within the race *orientale* and four within the race *asiaticum*, are also recognized. In the races *mediterraneum* and *eurasiaticum* the seeds are large, or very large, with a white seed-coat, in the other races the seeds are smaller, and the seed-coats can vary in colour. In India growers commonly recognize two types of chick pea: *Deshi* gram which has small seeds, brown or yellowish in colour, usually wrinkled with a beak at the end, and *Kabul* or *Kabuli* gram, which has white, larger smooth seeds. A number of improved cultivars of chick peas have been developed, notably in India, where a number of cultivars with a high yield potential have become available in recent years. These include: C–235, C–214, Early S–3, BGS–1, BGS–2, Chaffa, L–550, L–144, G–130 and BR–78.

Origin and distribution

The origin of the chick pea is a little obscure. Some authorities consider that the species did not exist in the wild, but was developed artificially by man, and that there were several centres of origin in Asia and the eastern Mediterranean. In ancient times cultivation quickly spread throughout the Mediterranean region and the SE. Asian sub-continent, and gradually extended to the drier parts of Africa, notably Ethiopia. Chick peas were introduced successfully into the New World, and have become an important crop in Mexico, Argentina and Chile. More recently they have been introduced into Australia, although cultivation there is not yet of commercial importance.

Cultivation conditions

Temperature—essentially a subtropical crop, chick peas grow best as a post-monsoon, cool season crop in the SE. Asian sub-continent, during the dry season in Ethiopia and during spring and summer in the Mediterranean. They are fairly tolerant of high temperatures during the later stages of growth, but resistance to frost varies according to cultivar, some are resistant, others are very susceptible to damage. For optimum growth day/night temperatures within the range of 79–64°F (26–18°C) and 84–70°F (29–21°C) are required. There is also a considerable variation according to the cultivar in the soil temperature required for germination, but generally it should exceed 41°F (5°C), and preferably be above 59°F (15°C). For high yields bright sunshine is required; dull, cloudy weather, particularly if accompanied by high humidity, reduces flowering and seed-setting.

Rainfall—chick peas are a drought resistant crop and are normally grown in areas with an annual rainfall of about 25–30 in. (650–750 mm), although they can be grown successfully in areas with a rainfall of about 40 in./a (1 000 mm/a). In India they are often grown as a winter crop on residual moisture in the soil, sometimes supplemented by irrigation. Excessive humidity, heavy rain or hailstorms, are known to have a very detrimental effect on seed-setting. Optimum seed-set is reported to occur when the minimum relative humidity varies between 21 and 41 per cent. Chick peas are sometimes grown under irrigation, when they normally receive 8–10 ac/in. (822–1 208 m^3) of water in two or three irrigations.

Soil—chick peas can be grown on a wide range of soil types provided that the drainage is good, as they cannot withstand waterlogging. For optimum results clay loams are required, but in India chick peas are grown successfully on the moderately heavy, grey and brown alluvial soils of the upper Gangetic basin and on the black cotton soils of the Deccan plateau. Lack of aeration can result in poor nodulation and a low rhizobial activity and cultural practices, such as bunding, have been found to improve productivity significantly. A pH of 6–9 is favourable, acid soils, pH 4.6, encourage the development of fusarium wilt. Chick peas are fairly tolerant of alkalinity and salinity and will grow quite well up to a conductivity value of 1.2 mmhos/cm, and up to a sodium absorption rate of 4.5.

The precise manurial requirements of chick peas require detailed study and in many areas the crop is not fertilized. In India recently the application of phosphoric acid 36 lb/ac (40 kg/ha) and 14 lb/ac (16 kg/ha) of nitrogen has been recommended. If the soil is deficient in zinc, as in parts of India, the application of 22 lb/ac (25 kg/ha) of zinc sulphate has been found to be beneficial. Liming is sometimes recommended on acid soils, but requires care since it can result in seeds with a tough seed-coat.

Altitude—chick peas have been grown successfully at elevations of between 1 000 and 4 000 ft (300–1 200 m) in Kenya.

Day-length—chick peas have been found to react favourably to long photoperiods of 16 hours. Although the behaviour of cultivars varies, in general chick peas are moderately sensitive to photoperiod and the vegetative period is shortened in long days, but short days (9 hours) do not prevent flowering.

Chick pea

Planting procedure

Material—seed of good germination capacity (above 75 per cent) is used. The germination capacity varies considerably according to the cultivar and storage conditions. The seed of many cultivars retains its viability for two or three years under normal storage conditions, but some large white-seeded Mediterranean cultivars are reported to lose their viability after one year, or even less. Germination is hypogeal. Inoculation of the seed has been found to increase yields significantly (24 to 62 per cent).

Method—the seed is normally sown 2–4 in. (5–10 cm) deep, either broadcast, or in drills, in soil which has been levelled and worked to a rough tilth. A fine tilth has been found to decrease germination. The depth of ploughing normally varies from 6 to 10 in. (15–25 cm). Deep tillage is essential for compact or badly aerated soils.

Time of sowing is of considerable importance and varies according to the cultivar and the area. The recommended sowing dates are as follows:

India and Pakistan	—late September until mid-December, with October the optimum time;
Bangladesh	—mid-September until mid-March;
Egypt and Israel	—mid-October until mid-December;
Mediterranean area	—February until April;
Mexico	—mid-October until the end of December;
Ethiopia	—September until January in most areas;
Sudan	—end of October until the end of November.

Chick peas may be grown as a pure crop, but in many areas are grown as a mixed crop, often with cereals such as wheat or barley. When grown as a pure crop, a single intercultivation is often given about three weeks after planting to suppress weed growth. In India experimental weeding trials using bar harrows and rollers gave a fourfold increase in yield compared with the non-weeded control. In certain countries attempts have been made to evaluate the use of herbicides. In the USSR chlorazine has been recommended as a pre-sowing and pre-emergence treatment, and in Bulgaria the use of prometryne and simazine is considered to be economical under mechanized cultivation. Other herbicides which are reported to give effective weed control include: noruron, benfluralin, trifluralin, linuron and alachlor. When the plants begin to branch the shoot tips are sometimes clipped, a practice said to induce better branching.

Field-spacing—in the SE. Asian sub-continent chick peas are planted in rows 10–12 in. (25–30 cm) apart, with 4–12 in. (10–30 cm) between the plants, but in the USSR and Bulgaria the inter-row distance is frequently reduced to 6 in. (15 cm). In Mexico field-spacings up to 56 x 4 in. (140 x 10 cm) are reported to be used for some cultivars on alluvial soils, and in the Mediterranean and Middle East distances between the rows usually range from 12 to 32 in. (30–80 cm).

Seed-rate—in India the seed-rate varies considerably from 24 to 32 lb/ac (27–36 kg/ha) in the Punjab, 40 to 50 lb/ac (45–56 kg/ha) in Maharashtra, and up to 60 to 80 lb/ac (67–90 kg/ha) in central India. Where large-seeded types such as the cultivars L–104 and L–75 are grown, the seed-rate averages from 80 to 100 lb/ac (90–112 kg/ha). In Mexico the optimum seed-rate is reported to be 40 lb/ac (45 kg/ha). In Greece the average seed-rate is high, 107 lb/ac (120 kg/ha), while in Morocco it is reported to range between 45 and 71 lb/ac (50–80 kg/ha).

Pests and diseases

Pests—chick peas are relatively free from insect attack in the field, probably because of their glandular secretions. The gram caterpillar, gram pod borer, or American cotton bollworm, *Heliothis armigera*, is the most widespread and serious pest in the field, but damage varies considerably from year to year and season to season. In Iran for example, crop losses of 90 per cent have been recorded in certain seasons. In parts of India losses of 20 per cent are fairly common. Spraying with insecticides such as DDT, malathion, endrin, mevinphos or BHC, can control this pest, but the breeding of resistant cultivars is likely to be a more effective solution. Other pests which may also attack chick peas in India, include the red gram plume moth, *Exelastis atomosa*, and the cutworms, *Agrotis ipsilon* and *Ochropleura flammatra.* Effective control is reported to be obtained by the use of insecticides such as DDT. The semi-looper, *Plusia orichalcea*, is troublesome occasionally on chick pea crops in India. The caterpillars can defoliate plants completely unless effectively controlled by the use of DDT or endrin. Miner flies, *Liriomyza* spp. can also be troublesome in certain areas, eg Spain, Israel and the USSR. In Israel, *L. cicerini* has recently become a serious pest, and in Tadzhikistan (USSR) crop losses ranging from 10 to 40 per cent have been reported. Spraying with DDT, Thiofos, parathion or carbaryl is reported to give effective control. The parasite *Opius cicerini* has been suggested as a possible biological control. In parts of India recently root-knot nematodes, *Meloidogyne arenaria*, have been causing considerable damage.

Chick peas are very susceptible to infestation during storage by the bruchid beetles, *Callosobruchus chinensis*, *C. maculatus* and *C. analis*. Infestation commences in the field and spraying with DDT, BHC, dieldrin or rotenone is often recommended to control the incipient bruchid population at this stage. Research is also in progress on the development of resistant cultivars.

Diseases—chick peas are attacked by relatively few serious diseases. Gram blight caused by *Ascochyta rabiei* (*Mycosphaerella rabiei*) is widespread, it frequently occurs in epidemic form in Punjab, Bihar and Jammu, in India. The disease is seed-borne and its development is favoured by wet weather and low temperatures. The use of disease-free seed, burning of all plant debris, and mixed cropping, have been recommended as control measures, in addition to growing resistant cultivars. Wilt, *Fusarium oxysporum* var. *ciceri* is another widespread disease of chick peas of economic importance. In northern India and Pakistan, 15 per cent of the crop is reported to be infected annually and some years this can reach between 30 and 70 per cent. The disease can also result in serious crop losses in Spain and Mexico. Temperatures between 77° and 95°F (25–35°C) are optimum for growth; lack of sufficient depth of soil is also reported to increase its incidence. Sowing date, sowing method, and the time of irrigation, have also been found to influence the incidence of wilt, and there is need to develop cultivars with a long-term resistance. Other diseases which may affect chick peas include the stem rot, *Sclerotinia sclerotiorum*, foot rot, *Operculella padwickii*, verticillium wilt disease, *Verticillium albo-atrum*, rust, *Uromyces ciceris-arietini*, and a seedling blight or root-rot, *Phytophthora cryptogaea*. Recently *Kabuli* gram in India has been reported to be infected by *Colletotrichum dematium*.

Chick peas are usually relatively free from virus diseases, although a yellowing disease, probably the bean yellow mosaic virus, can be troublesome, and in Iran several serious cases of virus attack have been reported. In addition, chick peas may occasionally suffer from phosphorus or manganese deficiency diseases, developing a yellowing or browning of the leaves, resulting in stunted growth and low seed yields, and in severe cases the death of the plant.

Growth period

In India the crop normally takes from 90 to 180 days to reach maturity, although most of the recently developed improved cultivars mature in approximately 115 to 125 days. An exception is the new *Kabuli* cultivar L–550 which takes 160–165 days. In Egypt the growth period is normally 140–150 days, in Greece 180 to 225 days and in Morocco 90 to 155 days.

Harvesting and handling

For dry seed production chick peas are normally harvested when the leaves become yellow or reddish-brown. In many countries harvesting is carried out by hand. The plants are cut just above the soil surface or uprooted, stacked in heaps and left to dry for about one week, before being threshed and winnowed. In some countries eg Spain, the USSR and Bulgaria, combine-harvesters may be used, although there is sometimes a reduction in yield as the lowest pods on the plants may be lost. In addition, the distances between the rows must be adapted to suit the harvesters. Erect cultivars such as Plovdiv–19, have been developed in the USSR for mechanized cultivation. Traditionally threshing was carried out by bullocks trampling on the pods, or by beating them with sticks, but the use of threshing machines which also sieve and clean the seeds is becoming more widespread. In some countries, eg Spain, Morocco and Mexico, the seeds are size-graded. In Mexico for example, the following export grades per oz (28 g) weight, are used: Extra, 36–38 seeds, Fino, 38–40 seeds, Sublime 40–44 seeds.

Chick peas should be dried to a moisture content of between 8 and 15 per cent, before being stored, preferably in airtight bins or clean sacks. Small producers frequently store the seeds in closely woven baskets or earthenware pots, covered with a layer of sand. As has already been mentioned, chick peas are very susceptible to insect infestation, particularly bruchids, during storage. Regular dusting with pyrethrum, derris, BHC, DDT or lindane, or fumigation with methyl bromide are suggested control methods. Infestation by *Callosobruchus chinensis* has been found to result in a loss of seed viability, a reduction of the thiamine and tryptophan content, and an increase in free fatty acid content, in addition to affecting the flavour adversely.

Primary product

Seed—chick pea seeds are generally small, rounded or angular, approximately 0.3 to 0.4 in. (7–10 mm) long and 0.2 to 0.3 in. (5–8 mm) broad, with pointed anterior ends and smooth, or wrinkled, seed-coats. Their colour may vary from white, cream, yellowish with an orange tinge, brown, to dark-green or black. Sometimes minute black dots are present, or there is a mosaic pattern. The hilum is deep, greyish with a coloured margin. In many areas white-seeded types of chick peas are preferred because they are bigger and considered to be more nutritious. One hundred seeds normally weigh between 0.4 and 2.9 oz (12–78 g).

Yield

In the SE. Asian sub-continent the overall average yield of seed is about 530 lb/ac (600 kg/ha), but varies considerably from region to region, the standard of crop management, and from season to season. Brown-seeded types tend to yield higher than green-seeded types. Using improved cultivars and efficient crop management yields of 1 000 lb/ac (1 120 kg/ha) or more, are attainable. The recently released improved Indian cultivar C–235 for example, is reported to have a yield potential of about 2 670 lb/ac (3 000 kg/ha). In many other countries average yields tend to be higher, for example: Egypt 1 500 lb/ac (1 680 kg/ha), Turkey 1 090 (1 220 kg/ha), Bulgaria 1 230 lb/ac (1 380 kg/ha), the Sudan 950 lb/ac (1 060 kg/ha), Mexico and Israel 890 lb/ac (1 000 kg/ha). Experimentally yields in excess of 3 560 lb/ac (4 000 kg/ha) have been obtained, which indicates the potential productivity of chick peas.

Main use

In the principal producing areas chick peas are used in various forms, mainly for human consumption. In India about 75 per cent of production is consumed as dhal. Elsewhere the whole mature seeds are used in a wide range of dishes, often after they have been soaked and boiled. In addition, the seeds may be parched, or roasted in hot pans and eaten similarly to roasted groundnuts, or used as an ingredient of various sweetmeats.

Subsidiary uses

In Asia and the Middle East the green seeds are sometimes eaten raw and relished as a snack, and in addition, are shelled and sold as a substitute for fresh peas, when these are not available. In some countries, eg in Mexico, chick peas are used mainly for livestock feed. The seeds are sometimes ground into a flour or meal, which is frequently mixed with wheat flour to produce unleavened bread or used in the preparation of various confectionery products. In recent years the flour has been utilized as a constituent of low-cost protein enriched foods, particularly for infant feeding. Chick peas may also be used to produce fermented food products.

Secondary and waste products

Chick pea seeds are sometimes sprouted similarly to mung beans and used as a vitamin C rich vegetable. Broken seeds and the residue from dhal production may be used for livestock feeding. Starch suitable for textile sizing is occasionally extracted from chick peas, the yield is approximately 21 per cent. The fresh young plant shoots are sometimes used in Asia as a vegetable, similar to spinach. An approximate analysis has been given as follows: moisture 60.6 per cent; protein 8.2 per cent; fat 0.5 per cent; carbohydrate 27.2 per cent; ash 3.5 per cent; calcium 0.31 per cent; phosphorus 0.21 per cent. The immature pods are also used occasionally as a vegetable.

The straw left after the seeds have been harvested is a valuable forage crop, its average composition (dry weight basis) is: crude protein 12.9 per cent; fibre 36.3 per cent; N-free extract 38.1 per cent; fat 1.5 per cent; ash 11.2 per cent; calcium 2.2 per cent; phosphorus 0.5 per cent; magnesium 0.5 per cent; sodium 0.3 per cent; potassium 3.0 per cent. In Indonesia chick peas are reported to be grown as a fodder crop producing approximately 36 T/ac (90 t/ha) of green matter per year.

The vegetative parts of the chick pea are covered with glandular hairs which exude an acidic liquid which contains about 94.2 per cent malic acid, 5.6 per cent oxalic acid, and 0.2 per cent acetic acid. The concentration varies between 0.36 and 1.30 per cent and reaches its maximum at flowering, falling to about 1 per cent at maturity. At one time in India the plant exudations were collected and widely used medicinally, or as vinegar.

Special features

Whole mature chick peas have the following approximate composition: moisture 10 per cent; protein 17 per cent; fat 5 per cent; fibre 4 per cent; carbohydrate 61 per cent; ash 3 per cent; calcium 0.19 per cent; phosphorus 0.24 per cent; vitamin A 316 iu/100 g; vitamin B 100 mg/100 g; vitamin C 5 mg/100 g. The seed-coat constitutes about 14 per cent of the seed and has a high fibre and calcium content. The major constituents of chick peas vary according to the maturity of the seed and recently it has been shown that on a fresh weight basis the nutritive value of the whole seed increases with ripening, although on a moisture-free basis, the immature seeds are nutritionally superior. In addition, the composition of mature chick peas has been found to vary greatly according to soil and climatic conditions, and the cultivar. Crude protein contents ranging from 12.3 to 31.6 per cent have been reported, the lower values being obtained from seed from plants where there has been little or no nodulation. In general *Kabuli* type chick peas have a higher protein content than *Deshi* types. The main constituent of the protein is globulin, 88 per cent of total nitrogen. The proximate amino acid content, (mg/gN), has been given as: isoleucine 277; leucine 468; lysine 428; methionine 65; cystine 74; phenylalanine 358; tyrosine 183; threonine 235; valine 284; arginine 588; histidine 165; alanine 271; aspartic acid 725; glutamic acid 991; glycine 251; proline 263; serine 318. Chick peas are deficient in the essential amino acid tryptophan, an average value of 174 mg/100 g of edible portion has been reported.

The carbohydrate content consists mainly of starch, in addition to a number of sugars, of which sucrose is predominant, although the presence of glucose, fructose, raffinose, stachyose and xylose, has also been reported. The starch grains range from large oval-shaped (17 to 29μ) to small spherical (6 to 7μ) granules. The amylose content is reported to range from 28.3 to 33.5 per cent and the gelatininzation temperature from 62.5° to 68°C.

The B-complex vitamins in chick peas are: thiamine 0.4 mg/100 g of edible portion; riboflavin 0.3 mg/100 g; niacin 2.5 mg/100 g; total folic acid 3.47 μ g/g (dry wt). Maleic acid accounts for 62 per cent of the total organic acids present in the seed, citric acid a further 25 per cent, and malonic acid 7.0 per cent.

Chick peas normally have an oil content of approximately 2 to 5 per cent. The oil from *Deshi* type seeds has the following characteristics: SG $^{40°C}$ 0.9356; $N_D^{38°C}$ 1.4845; iod. val. (Wijs) 111.7; acid val. 2.4; sap. val. 184.6; RM val. 0.61; unsaponifiable matter 3.4 per cent. Oil from *Kabuli* type seeds: SG$^{40°C}$ 0.9301; $N_D^{38°C}$ 1.4825; iod val. (Wijs) 113.2; acid val. 2.6; sap. val. 185.4; RM val. 0·60; unsaponifiable matter 4.0 per cent. The component fatty acids of the oils are: *Deshi*, oleic 52.10 per cent; linoleic 38.00 per cent; myristic 2.74 per cent; palmitic 5.11 per cent and stearic 2.05 per cent. *Kabuli*, oleic 50.30 per cent; linoleic 40.00 per cent; myristic 2.28 per cent; palmitic 5.74 per cent; stearic 1.61 per cent and arachidic 0.07 per cent.

In addition, chick peas are a relatively rich source of lecithin. The presence of heat-labile trypsin inhibitors has also been reported and a number of flavanoids. The effects of the inclusion of chick peas in diets on the lowering of blood cholesterol levels has received attention recently, but their possible therapeutic value requires further investigation. The presence of aflatoxin on chick peas has been reported from the Sudan.

Processing

Dhal—is prepared traditionally by first cleaning the seeds, then sprinkling them with water and leaving them in piles overnight to loosen the seed-coat. After which they are dehusked and split (milling), either by hand, or in power operated abrasion type mills. Sometimes a suspension of turmeric powder in water is mixed with the dry dhal as it leaves the mill to give it an attractive yellow colour. This traditional milling method is laborious and time-consuming and the yield and quality of the dhal are sub-optimal, usually averaging 75 per cent, compared with the maximum theoretical yield of 88 per cent. The Central Food Technological Research Institute, (CFTRI), Mysore, has recently developed an improved milling method which gives a yield of 84 per cent. In this process the chick peas are exposed to heated air for a predetermined time in specially designed units to loosen the seed-coats which are then removed in improved abrasion type milling machines.

Puffing—puffed chick peas are popular in the SE. Asian sub-continent. For puffing, the seeds are soaked in water and then mixed with sand, heated to 428°F (250°C) and toasted for 15–25 seconds. After removing the sand by sieving, the seed-coat is removed by passing the seeds between a hot plate and a rough roller. Recently the effect of moisture conditioning and the use of hardening agents to give good puffing has been investigated by CFTRI.

Canning—chick peas are sometimes canned either in brine or tomato sauce, but the final product is liable to become discoloured and mucilaginous. The following processing method has been suggested for canning in brine: the cleaned chick peas are first soaked in three parts water for approximately 18 hours, after which they are drained, rinsed, blanched in boiling water for 5 to 10 minutes and then immediately rinsed in cold water. They are next filled into lacquered cans and hot brine containing 1 per cent citric acid is added, and 10 oz (280 g) cans are processed at a pressure of 10 lb/in^2 (0.703 kg/cm^2) for 60 minutes. Some canners use brine containing 1.5 per cent sodium chloride, 3.0 per cent sucrose and 0.1 per cent monosodium glutamate.

Chick pea

Dehydration—the use of green whole chick peas is generally restricted to a short period during the harvest season, but attempts have been made to develop the production of a dehydrated product, which retains the original green colour and can be served as a substitute. Experiments have shown that a satisfactory product may be obtained by blanching the green seeds in an aqueous solution containing 0.4 per cent magnesium oxide and 0.025–0.5 per cent sodium bicarbonate, followed by drying for 3 hours in a cross-flow drier at 149–158°F (65–70°C), with an air velocity of approximately 560–600 ft/min (170m/min), and final drying completed after equilibration of moisture overnight.

Production and trade

Production—chick peas are the third most important food legume crop, constituting about 15 per cent of the world's reported total output of grain legumes. Estimated world production for the years 1970–74 averaged 6 494 000 t/a, compared with 6 317 000 t/a for 1965–69, an increase of almost 3 per cent, but somewhat below the record level of 8 300 000 t achieved in 1960. Provisional figures for 1976 indicate that world production was approximately 7 466 000 t, compared with 5 538 000 t in 1975.

Chick peas: Major producing countries

Quantity tonnes

	Annual average 1965–69	Annual average 1970–74	1975	1976
India	4 781 000	4 874 000	4 015 000	5 933 000[1]
Pakistan	571 200	529 000	550 000	601 000[1]
Mexico	147 600	231 000	195 000	190 000[1]
Ethiopia	174 000	194 000	148 000	109 000[1]
Turkey	97 600	161 000	172 000	165 000[2]

1 FAO provisional estimate.
2 Unofficial estimate.

Trade—there is a large inter-state trade in chick peas in India. The states of Bihar and West Bengal are normally deficient in supplies and United Provinces and Haryana usually have supplies surplus to their own requirements. Although grown primarily for local consumption considerable quantities of chick peas also enter international trade. Unfortunately many countries do not show this grain legume as a separate item in their trade returns so that it is difficult to estimate the total tonnage entering world trade. However, from the available statistical data, world trade probably averaged some 108 700 t/a for the years 1970–74, compared with some 54 800 t/a for 1965–69, an increase of 100 per cent.

Chick peas: Major exporting countries

Quantity tonnes

	Annual average 1965–69	Annual average 1970–74	1975	1976
Morocco	23 187	31 889	22 586	n/a
Mexico	3 355	30 880	n/a	n/a
Turkey	9 811	25 141[1]	n/a	n/a
Ethiopia	7 986	7 088	945	n/a
Lebanon	2 379	2 602[1]	n/a	n/a
Portugal	5 638	1 889[1]	604	n/a

1 Four year average.
n/a not available.

Chick peas: Major importing countries

Quantity tonnes

	Annual average 1965–69	Annual average 1970–74	1975	1976
Spain	10 480[1]	28 282	n/a	31 928
Lebanon	3 335[2]	5 712[3]	n/a	n/a
USA	2 802	4 074	4 560	3 490
Portugal	374	1 563	469	n/a
Venezuela	1 769	1 547[2]	n/a	n/a

1 Three year average.
2 1969 only.
3 Four year average.
n/a not available.

Chick pea

Prices—chick pea prices have shown an upward trend in recent years as demand has tended to exceed supplies in several countries. Spain, annual average producer price, 1965–69 £94.95/t, 1970–74 £160.49/t. India, annual average wholesale price, 1965–69 £46.03/t, 1970–74 £54.44/t. Pakistan, annual average wholesale price, 1965–69 £46.63/t, 1970–74 £70.68/t. The average fob value of chick peas exported from Turkey was £105.75/t for the five year period 1970–74, compared with £45.00/t for 1965–69. The average value of Moroccan exports was £146.00/t in 1975 compared with £106.40/t for the period 1970–74, and an average value of £54.00/t during the late 1960s.

Major influences

Chick peas are an important food legume widely consumed in Asia, the Middle East and several Mediterranean countries. They are the leading legume crop of the SE. Asian sub-continent, where they rank fifth in importance as a food crop after rice, wheat, millets and sorghum. Until recently very little research had been carried out on the crop and average yields are low and tend to fluctuate according to seasonal weather conditions, particularly rain at the time of sowing, and the incidence of diseases. Improved high-yielding disease resistant cultivars are being developed and experimental results indicate that the crop could have a high level of productivity.

Chick peas are a nutritious foodstuff and because of their bland flavour are suitable for incorporation in protein-rich food mixtures suitable for infant feeding. However, they are deficient in the sulphur containing essential amino acids, and tryptophan. Traditionally chick peas have been regarded as a poor man's food and are not popular in some areas because of their bland, insipid flavour. During the late 1960s and early 1970s many farmers in the SE. Asian sub-continent tended to replace chick peas by the new high-yielding Mexican wheat cultivars which has resulted in a downward trend in production and increased demand.

Bibliography

ABU-SHAKRA, S., MIRZA, S. AND TANNOUS, R. 1970. Chemical composition and amino acid composition of chick pea seeds at different stages of development. *J. Sci. Food Agric.*, *21*, 91–93.

ADAMS, J. M. 1977. A bibliography on post-harvest losses in cereals and pulses with particular reference to tropical and subtropical countries. *Rep. Trop. Prod. Inst.*, G. 110, iv + 23 pp.

AGEEB, O. A. A. AND AYOUB, A. T. 1977. Effect of sowing date and soil type on plant survival and grain yield of chick peas (*Cicer arietinum* L.). *J. Agric. Sci., (Camb.)*, *88*, 521–527.

ARAULLO, E. V. 1974. Processing and utilization of cowpea, chick pea, pigeon pea and mung bean. *Proc. symp. interaction of agriculture with food science, Singapore*, 1974. (MacIntyre, R. ed.). pp. 131–142. Ottawa, Int. Dev. Res. Cent., IDRC–033e, 166 pp.

ARGIKAR, G. P. 1958. Parching quality of some types of gram (*Cicer arietinum* L.). *Poona Agric. Coll. Mag.*, *48* (4), 8–11.

ARGIKAR, G. P. 1970. Gram *Cicer arietinum* Linn. *Pulse crops of India.* (Kachroo, P. and Arif, M. eds.). pp. 54–135. New Delhi: Indian Counc. Agric. Res., 334 pp.

ARNON, I. 1972. Pulses or grain legumes. *Crop production in dry regions.* Vol. 2. pp. 217–260. London: Leonard Hill Books, 683 pp.

AZIZ, M. A., KHAN, M. A. AND SHAH, S. 1960. Causes of low setting of seed in gram (*Cicer arietinum*). *Agric. Pak.*, *11*, 37–48.

BALASUBRAMANIAN, V. AND SINHA, S. K. 1976. Nodulation and nitrogen fixation in chick pea (*Cicer arietinum* L.) under salt stress. *J. Agric. Sci. (Camb.)*, *87*, 465–466.

BARJA, I., MUÑOZ, P., SOLIMANO, G., VALLEJOS, E., RADRIGÁN, M. E. AND TAGLE, M. A. 1974. Infant feeding formula based on chick pea (*Cicer arietinum*): Its use as the sole food in healthy infants. *Indian J. Nutr. Diet.*, *11*, 335–341.

BEZUNEH, T. 1975. Status of chick pea production and research in Ethiopia. *International workshop on grain legumes.* pp. 95–101. Hyderabad, India, Int. Crops Res. Inst. Semi-arid Trop., 350 pp.

BHARGAVA, R. N. AND SUBBA RAO, M. S. 1972. Grow BR–78 green seeded gram for more profit. *Indian Farming*, *22* (1), 35.

CHANDRA, S. AND ARORA, S. K. 1968. An estimation of protein, ascorbic acid and mineral matter content in some indigenous and exotic varieties of gram (*Cicer arietinum* L.). *Curr. Sci.*, *37*, 237–238.

CHENA, G. R., CRISPIN, A. M. AND LARREA, E. R. 1967. El garbanzo, un cultivo importante en Mexico. [The chick pea an important crop in Mexico]. *Mexico, Inst. Nac. Invest. Agric., Foll. Misc.* 16, pp. 1–45.

Chick pea

CHOPRA, K. AND SWAMY, G. 1975. *Pulses: An analysis of demand and supply in India. Institute for social and economic change; Monograph 2.* New Delhi: Sterling Publishers, PVT, Ltd, 132 pp.

ÇÓLAKOĞLU, M. AND BILGIR, B. 1974. The studies of leblebi (roasted chick peas) processing and their composition. *Madrid, 4th Int. Congr. Food Sci. & Technol., Work Doc. 5a, New preservation methods, Pap.* 29, pp. 77–79.

COMMONWEALTH BUREAU OF PASTURES AND FIELD CROPS. nd. Chick pea. Annotated bibliography 1228, 1951–70. *Commonw. Agric. Bur., Bur. Pastures & Field Crops, Hurley, Maidenhead, Berks, England,* 12 pp.

CORBIN, E. J. 1975. Present status of chick pea research in Australia. *International workshop on grain legumes.* pp. 87–94. Hyderabad, India, Int. Crops Res. Inst. Semi-arid Trop., 350 pp.

CUBERO, J. I. 1975. The research on the chick pea (*Cicer arietinum*) in Spain. *International workshop on grain legumes.* pp. 117–122. Hyderabad, India, Int. Crops Res. Inst. Semi-arid Trop., 350 pp.

DART, P. J., ISLAM, R. AND EAGLESHAM, A. 1975. The root nodule symbiosis of chick pea and pigeon pea. *International workshop on grain legumes.* pp. 63–83. Hyderabad, India, Int. Crops Res. Inst. Semi-arid Trop., 350 pp.

DAVIES, J. C. AND LATEEF, S. S. 1975. Insect pests of pigeon pea and chick pea in India and prospects for control. *International workshop on grain legumes.* pp. 319–331. Hyderabad, India, Int. Crop Res. Inst. Semi-arid Trop., 350 pp.

DE, H. N., KHAN, S. A., MULLICK, N. I. AND ISLAM, A. 1969. Nutritive value of Bengal gram (chick pea): Assessment of the nutritive value of formulated foods based on Bengal gram (chick pea) and fish protein concentrate and egg powder by determination of their protein efficiency. *Bangladesh, Counc. Sci. Ind. Res. Lab., Dacca, Sci. Res., 6,* 148–155.

DHAWAN, C. L. AND MADAN, M. L. 1951. The effect of certain soil factors on the yield of major crops in the Punjab: (iii) Gram. *Indian J. Agric. Sci., 21,* 45–62.

DIXIT, P. D. 1932. Studies in Indian pulses: A note on the cytology of Kabuli and Desi gram types. *Indian J. Agric. Sci., 2,* 385–390.

EL BARDI, T. A. 1977. Pulses 2: Chickpeas. *Abstr. Trop. Agric., 3*(3), 9–18.

ESER, D. 1975. The situation of research of chick pea agriculture in Turkey. *International workshop on grain legumes.* pp. 123–128. Hyderabad, India, Int. Crops Res. Inst. Semi-arid Trop., 350 pp.

ESHEL, Y. 1967. Effect of sowing date on growth and seed yield components of chick pea (*Cicer arietinum* L.). *Israel J. Agric. Res.*, *17*, 193–197.

FOOD AND AGRICULTURE ORGANIZATION OF THE UNITED NATIONS. 1970. Amino acid content of foods and biological data on proteins. *FAO Nutr. Stud.* 24, pp. 50–51. Rome: FAO, 285 pp.

GHIRARDI, P., MARZO, A. AND FERRARI, G. 1974. Lipid classes and total fatty acids pattern of *Cicer arietinum. Phytochemistry*, *13*, 755–756.

GÖHL, B. 1975. *Cicer arietinum* L. *Tropical feeds: Feeds information summaries and nutritive values.* pp. 173; 511; 527. Rome: FAO, 661 pp.

HABISH, H. A. 1972. Aflatoxin in haricot and other pulses. *Exp. Agric.*, *8*, 135–137.

HAFIZ, A. AND ASHRAF, M. 1957. Association of morphological characters in gram plants with blight reaction. *Agric. Pak.*, *8*, 1–4.

HAKAM, M. M. 1974. Legume field crops improvement in the Arab Republic of Egypt. *Proc. 1st FAO/SIDA Semin., Improvement and production of field food crops for plant scientists from Africa and the Near East.* pp. 77–81. Rome: FAO, 684 pp.

HAWARE, M. P. AND NENE, Y. L. 1976. Some uncommon but potentially serious diseases of chickpea. *Trop. Grain Legume Bull.*, (5), pp. 26–30.

HAWTIN, G. C. 1975. The status of chick pea research in the Middle East. *International workshop on grain legumes.* pp. 109–116. Hyderabad, India, Int. Crops Res. Inst. Semi-arid Trop., 350 pp.

HULSE, J. H. 1975. Problems of nutritional quality of pigeon pea and chick pea and prospects of research. *International workshop on grain legumes.* pp. 189–207. Hyderabad, India, Int. Crops Res. Inst. Semi-arid Trop., 350 pp.

JADHAV, P. S., JAIN, T. C. AND PRASANNALAKSHMI, S. 1975. *Sorghum, millets and peas: A bibliography of Indian literature*, 1969–73. pp. 62–70. Hyderabad, India, Int. Crop Res. Inst. Semi-arid Trop., 116 pp.

JAFFARI, J. 1975. The status of chick peas (*Cicer arietinum*) in Iran. *International workshop on grain legumes.* pp. 103–107. Hyderabad, India, Int. Crops Res. Inst. Semi-arid Trop., 350 pp.

Chick pea

Jeswani, L. M. 1975. Varietal improvement of seed legumes in India. *Food protein sources.* (Pirie, N. W. and Swaminathan, M. S. eds.). pp. 9–18. London: Cambridge University Press, 260 pp.

Joshi, S. N. 1972. Variability and association of some yield components in gram (*Cicer arietinum* L.). *Indian J. Agric. Sci.*, *42*, 397–399.

Kandaswamy, T. K. and Natarajan, C. 1974. A note on phyllody disease on Bengal gram (*Cicer arietinum* L.). *Madras Agric. J.*, *61*, 1019–1020.

Kandé, J. 1967. Valeur nutritionnelle de deux graines de légumineuses: le pois chiche (*Cicer arietinum*) et la lentille (*Lens esculenta*). [Nutritional value of two grain legumes: chick pea (*Cicer arietinum*) and lentil (*Lens esculenta*)]. *Ann. Nutr. Aliment.*, *21* (2), 45–67.

Kanwar, J. S. 1974. Improvement of crops and their relationship to nutrition and food science technology in the semi-arid tropics. *Proc. symp. interaction of agriculture with food science, Singapore*, 1974. (MacIntyre, R. ed.). pp. 53–64. Ottawa, Int. Dev. Res. Cent., IDRC–033e, 166 pp.

Kanwar, J. S. and Singh, K. B. (ed.). 1974. Chick peas (*Cicer arietinum*). *Guide for field crops in the tropics and the subtropics.* (Litzenberger, S. C. ed.). pp. 115–121. Washington, Off. Agric. Tech., Assist. Bur. Agency Int. Dev., 321 pp.

Kao, C. 1974. Fermented foods from chick pea, horsebean and soybean. *Diss. Abstr. Int.*, B35 (5), 2250.

Khan, M. A., Almas, K., Abid, A. R. and Yaqoob, M. 1976. The effect of gram flour on the quality of wheat protein. *Pak. J. Agric. Sci.*, *13*, 167–172.

Khan, M. A. and Chowdhury, M. A. 1976. Effect of growth regulators on flowering and yield of gram (*Cicer arietinum* L.). *Pak. J. Agric. Sci.*, *13*, 105–112.

Kheswalla, K. F. 1941. Foot-rot of gram (*Cicer arietinum* L.) caused by *Operculella padwickii* Nov. gen., Nov. spec. *Indian J. Agric. Sci.*, *11*, 316–318.

Kowale, B. N. and Misra, U. K. 1975. Effect of feeding Bengal gram (*Cicer arietinum*) at different protein levels on plasma lipids of rats. *Agric. Biol. Chem.*, *39*, 901–903.

Krishna Murti, C. R. 1975. Biochemical studies on Bengal gram. *J. Sci. Ind. Res.*, (*India*), *34*, 266–281.

KURIEN, P. P., DESIKACHAR, H. S. R. AND PARDIA, H. A. B. 1972. Processing and utilization of grain legumes in India. Symp. food legumes. pp. 225–236. *Tokyo, Jpn., Minist. Agric. For. & Trop. Agric., Trop. Agric. Res. Cent., Trop. Agric. Res. Ser.* 6, 253 pp.

LAL, B. M., PRAKASH, V. AND VERMA, S. C. 1963. The distribution of nutrients in the seed parts of Bengal gram. *Experientia, 19*, 154–155.

LAL, B. M., ROHEWAL, S. S., VERMA, S. C. AND PARKASH, V. 1963. Chemical composition of some pure strains of Bengal gram (*Cicer arietinum* L.). *Ann. Biochem. Exp. Med., 23*, 543–548.

LAL, S. 1976. Relationship between grain and biological yields in chick pea (*Cicer arietinum* L.). *Trop. Grain Legume Bull.*, (6), pp. 29–31.

LINK, K. C., LUH, B. S. AND SCHWEIGERT, B. S. 1975. Folic acid content of canned garbanzo beans. *J. Food Sci., 40*, 562–565.

LINEBACK, D. R. AND KE, C. H. 1975. Starches and low-molecular-weight carbohydrates from chick pea and horse bean flours. *Cereal Chem., 52*, 334–347.

LUTHRA, J. C. AND BEDI, K. S. 1932. Some preliminary studies on gram-blight with reference to its cause and mode of perennation. *Indian J. Agric. Sci., 2*, 499–515.

LUTHRA, R. C., GILL, A. S. AND SINGH, K. B. 1973. C–214 an ideal gram for Punjab. *Indian Farming, 23* (6), 29.

MAESEN, L. J. G. VAN DER. 1972. *Cicer* L., a monograph of the genus with special reference to the chick pea (*Cicer arietinum* L.) its ecology and cultivation. *Wageningen, Meded, Landbouwhogesch.*, 72–10, 342 pp.

MATHUR, B. N., HANDA, D. K. AND SINGH, H. G. 1969. Note on the occurrence of *Meloidogyne arenaria* as a serious pest of *Cicer arietinum. Madras Agric. J., 56*, 744.

MISHRA, R. P., SHARMA, N. D. AND JOSHI, L. K. 1975. A new disease of gram (*Cicer arietinum* L.). *Curr. Sci., 44*, 621–622.

MOOLANI, M. K. 1966. Effect of N–P fertilization on growth and yield of gram (*Cicer arietinum* L.). *Ann. Arid Zone, 5*, 127–133.

MURTY, B. R. 1975. Biology of adaption in chick pea. *International workshop on grain legumes.* pp. 239–251. Hyderabad, India, Int. Crops Res. Inst. Semi-arid Trop., 350 pp.

Chick pea

NAIK, M. S. AND NARAYANA, N. 1959. Relation of maturity to the composition of seeds of gram (*Cicer arietinum*). *Indian J. Appl. Chem.*, *22*, 239–242.

NARAYANA RAO, M., RAJAGOPALAN, R., SWAMINATHAN, M. AND SUBRAHAMANYAN, V. 1959. The chemical composition and nutritive value of Bengal gram (*Cicer arietinum*). *Food Sci.*, *8*, 391–394.

NENE, Y. L. AND REDDY, M. V. 1976. Preliminary information on chick pea stunt. *Trop. Grain Legume Bull.*, (5), pp. 31–32.

PAK, N., BERNIER, L., DUFFAU, G., MACAYA, J., SORIANO, H., MUÑOZ, P., HIDALGO, R., SOTO, A., ZERENE, N. AND TAGLE, M. A. 1975. Infant-feeding formula based on chick pea (*Cicer arietinum*), its use in the treatment of acute diarrhoea with dehydration. Preliminary communication. *Indian J. Nutr. Diet.*, *12*, 42–46.

PALIWAL, K. V. AND ANJANEYULU, B. S. R. 1967. Growth of wheat and gram under saline-alkali field conditions. *Madras Agric. J.*, *54*, 169–175.

PARPIA, H. A. B. 1975. Utilization problems in food legumes. *Nutritional improvement of food legumes by breeding*. (Milner, M. ed.). pp. 281–295. New York/London: John Wiley and Sons, 399 pp.

PINGALE, S. V., KADKOL, S. B. AND SWAMINATHAN, M. 1956. Effect of insect infestation on stored Bengal gram (*Cicer arietinum* L.) and green gram (*Phaseolus radiatus* L.). *Food Sci.*, *5*, 211–213.

PURSEGLOVE, J. W. 1968. *Cicer arietinum* L. *Tropical crops: Dicotyledons*. Vol. 1. pp. 246–250. London: Longmans, Green and Co Ltd, 332 pp.

PUSHPAMMA, P. 1975. Evaluation of nutritional value, cooking quality and consumer preferences of grain legumes. *International workshop on grain legumes*. pp. 213–220. Hyderabad, India, Int. Crops Res. Inst. Semi-arid Trop., 350 pp.

RAMANATHAN, L. A. AND BHATIA, B. S. 1970. Dehydrated green Bengal gram (*Cicer arietinum*). *J. Food. Sci. Technol.*, *7*, 208–209.

RAMANUJAM, S. 1976. Chick pea. *Evolution of crop plants*. (Simmonds, N. W. ed.). pp. 157–159. London: Longman Group Ltd, 339 pp.

RAO, P. N. AND RAO, D. R. 1974. The nature of major protein components of the new varieties of *Cicer arietinum*. *J. Food Sci. Technol.*, *11*, 139–140.

ROYES, W. V. AND FINCHAM, A. G. 1975. Grain quality in *Cajanus* and *Cicer*. *International workshop on grain legumes*. pp. 209–211. Hyderabad, India, Int. Crops Res. Inst. Semi-arid Trop., 350 pp.

SANDHU, T. S., SINGH, K. B. AND SINGH, H. 1975. L–550: A new kabuli gram. *Indian Farming*, *24* (12), 21; 30.

SASTRI, B. N. (ed.). 1950. *Cicer* Linn. (Leguminosae). *The wealth of India: Raw materials*. Vol. 2 (C). pp. 154–161. Delhi: Indian Counc. Sci. Ind. Res., 427 pp.

SATTAR, A., ARIF, A. G. AND MOHY-UD-DIN, M. 1953. Effect of soil temperature and moisture on the incidence of gram wilt. *Pak. J. Sci. Res.*, *5*, 16–21.

SATTAR, A. AND HAFIZ, A. 1952. Disease of gram (*Cicer arietinum* L.). Researches on plant diseases of the Punjab. Sci. Monogr. 1. pp. 43–68. *Lahore, Pak. Assoc. Adv. Sci.*, 158 pp.

SAXENA, M. C. AND YADAV, D. S. 1975. Some agronomic considerations of pigeon peas and chick peas. *International workshop on grain legumes*. pp. 31–61. Hyderabad, India, Int. Crops Res. Inst. Semi-arid Trop., 350 pp.

SEN, S. AND MUKHERJEE, D. 1961. Preliminary studies on defective seed setting in gram. *Sci. Cult.*, *27*, 185–188.

SEN GUPTA, P. K. 1974. Diseases of major pulse crops in India. *PANS*, *20*, 409–415.

SENEWIRATNE, S. T. AND APPADURAI, R. R. 1966. Bengal gram. *Field crops of Ceylon*. pp. 173–175. Colombo: Lake House Investments Ltd, 376 pp.

SHARMA, B. M. AND YADAV, J. S. P. 1976. Availability of phosphorus to gram as influenced by phosphate and irrigation regime. *Indian J. Agric. Sci.*, *46*, 205–210.

SHAW, F. J. F. AND KHAN, A. R. 1931. Studies in Indian pulses: (ii) Some varieties of Indian gram *Cicer arietinum* L. *India, Mem. Dep. Agric., Bot. Ser.*, *19*, 27–47.

SHEHNAZ, A. AND THEOPHILUS, F. 1975. Effect of insect infestation on the chemical and nutritive value of Bengal gram, (*Cicer arietinum*) and field bean (*Dolichos lablab*). *J. Food Sci. Technol.*, *12*, 299–302.

SIDDAPPA, G. S. 1959. Canning of dried Bengal gram (*Cicer arietinum*). *Indian J. Hortic.*, *16*, 170–174.

SIEGEL, A. AND FAWCETT, B. 1976. *Food legume processing and utilization* (*with special emphasis on application in developing countries*). Ottawa, Int. Dev. Res. Cent., IDRC–TS1, 88 pp.

Chick pea

Singh, K. B. and Auckland, A. K. 1975. Chick pea breeding at ICRISAT. *International workshop on grain legumes*, pp. 3–17. Hyderabad, India, Int. Crop Res. Inst. Semi-arid Trop., 350 pp.

Singh, K. B. and Singh, H. 1973. Improved gram for increased yield. *Indian Farming*, *22* (12), 19–20; 22.

Singh, R. G. 1971. Response of gram (*Cicer arietinum* L.) to the application of nitrogen and phosphate. *Indian J. Agric. Sci.*, *41*, 101–106.

Singh, R. S. 1970. Effect of number of cultivations and increasing levels of N and P on yield and quality of gram (*Cicer arietinum* L.). *Madras Agric. J.*, *57*, 267–270.

Smartt, J. 1976. *Cicer* L. *Tropical pulses*. pp. 43–45. London: Longman Group Ltd, 348 pp.

Sohoo, M. S. and Singh, K. B. 1972. G–130 a new Bengal gram variety for Punjab. *Indian Farming*, *22* (1), 27–28.

Srikantia, S. G. 1975. Chick pea and pigeon pea: Some nutritional aspects. *International workshop on grain legumes*. pp. 221–223. Hyderabad, India, Int. Crops Res. Inst. Semi-arid Trop., 350 pp.

Subramanian, A., Raj, S. M. and Veakatachalam, C. 1974. A note on effect of graded doses of phosphate and spacing on the yield of Bengal gram. *Madras Agric. J.*, *61*, 791–792.

Sur, S. C., Gupta, K. S. and Sen, S. 1966. Optimum time of sowing gram (*Cicer arietinum* L.) in West Bengal. *Indian J. Agric. Sci.*, *36*, 173–179.

Swaminathan, M. S. and Jain, H. K. 1975. Food legumes in Indian agriculture. *Nutritional improvement of food legumes by breeding*. (Milner, M. ed.). pp. 69–82. New York/London: John Wiley and Sons, 399 pp.

Tripathi, R. S., Dubey, C. S., Khan, A. W. and Agrawal, K. B. 1975. Effect of application of rhizobium inoculum on the yield of gram (*Cicer arietinum* L.) varieties in Chambal commanded area of Rajasthan. *Sci. Cult.*, *41*, 266–269.

Venkataraman, L. V. and Jaya, T. V. 1976. Influence of germinated green gram and chick pea on growth of broilers. *J. Food Sci. Technol.*, *13*, 13–16.

Verma, S. C. and Lal, B. M. 1966. Physiology of Bengal gram seed: Changes in phosphorus compounds during ripening of the seed. *J. Sci. Food Agric.*, *17*, 43–46.

VERMA, S. C., LAL, B. M. AND PARKASH, V. 1964. Changes in the chemical composition of the seed parts during ripening of Bengal gram (*Cicer arietinum* L.) seed. *J. Sci. Food Agric.*, *15*, 25–31.

WESTPHAL, E. 1974. *Cicer arietinum* L. Pulses in Ethiopia, their taxonomy and agricultural significance. pp. 84–90. *Wageningen, PUDOC, Cent. Agric. Publ. Doc., Agric. Res. Rep.* 815, 278 pp.

YEGNA NARAYAN AIYER, A. K. 1966. Bengal gram (*Cicer arietinum*). *Field crops of India.* 6th ed. pp. 110–115. Bangalore City: Bangalore Printing and Publishing Co Ltd, 564 pp.

Common names CLUSTER BEAN, Guar.

Botanical name *Cyamopsis tetragonoloba* (L.) Taub.

Family Leguminosae.

Other names Aconite bean; Bakuchi (Ind.); Cyamopse â quatre ailes (Fr.); Dridhabija, Gavar, Gawar, Gorakshaphalini, Gorani, Gorchikuda, Gori kayi, Gouree, Govar, Gowar(a), Guaru, Guar khurti, Guwar, Gwar (Ind.); Kotaranga (Sri La.); Kothaverai, Kothaveray (Tam.); Kottavarai (Malays.); Kulti, Kuwara (Ind.); Mgwaru (Swah.); Pè-walee (Burm.); Siam bean; Vah, Vahkiphali (Ind.); Walee-pè (Burm.).

Botany

A robust, annual, usually ranging from 1.6 to 10 ft (0.5–3 m) in height; with a long tap-root and a well-developed lateral root-system, covered with large, light coloured nodules. There is considerable variability, some types are erect, others have numerous strong branches arising from basal nodules, some have small hairs on all parts of the plant, others are completely glabrous. The stems are angular, ribbed and hollow. The leaves are alternate, trifoliolate, and borne on long petioles at the base of which is a marked pulvinus. The leaflets are ovate, 2–4 in. (5–10 cm) long; usually branched forms have smaller leaflets. The inflorescence is an axillary raceme borne in clusters. The flowers are small, typically papilionaceous, with the standard petal and keel white and the wings pinkish-purple. The pods are generally oblong, 2–4.5 in. (5–11.25 cm) in length, although there is a sickle-shaped form. In the single-stemmed forms the pods are larger and more fleshy. They normally contain 5–12 oval, or cube-shaped seeds of variable size and colour. Some forms bear clusters of pods at the nodes, whereas in others they may be borne at alternate nodes, or even irregularly.

There are numerous cultivars and in general branched types are considered more suitable for seed production and the erect single-stemmed types for pod production. In the USA improved, high-yielding, disease-resistant cultivars such as Brooks, Hall, Mills, Kinman and Esser, have been developed for seed production. In some parts of India three main types of cluster beans are recognized: (i) *Deshi*—4–5 ft (1.2–1.5 m) tall, grown mostly as a rain-fed crop for seed; (ii) *Pardeshi*—5–6 ft (1.5–1.8 m) tall, grown mostly as a vegetable crop, for its pods; (iii) *Sotiaguvar*—8–10 ft (2.4–3 m) tall, grown mainly for fodder, or as a green manure crop, but sometimes also as a vegetable. Some crop improvement work has been undertaken and a high-yielding dual purpose cultivar FS–277 was released in 1971. Other improved cultivars include Duragapura Safed, an early-maturing type.

Origin and distribution

Many authorities consider that the cluster bean originated in the SE. Asian sub-continent, where it has been widely cultivated as a vegetable since ancient times. However, it has recently been suggested that it developed from the drought-tolerant, wild African species *C. senegalensis*, which was originally taken from Africa by Arab traders as fodder for horses. The cluster bean can tolerate a wide range of climatic and soil conditions and its cultivation has spread from Asia to many parts of the tropics, including S. and C. America, Africa, the southern USA and Australia.

Cultivation conditions

Temperature—the cluster bean is essentially a sun-loving plant, tolerant of high temperatures and for maximum growth requires a soil temperature of 77° to 86°F (25–30°C). It is intolerant of shade and very susceptible to frost, requiring a mimimum frost-free period of 110 to 130 days.

Rainfall—a drought tolerant legume, especially suited to areas of 20 to 30 in. (500–750 mm) annual rainfall. In India it is grown successfully in areas with a rainfall of 12 in./a (300 mm/a), when supplementary irrigation is available. Although widely grown as a rain-fed crop, optimum yields are usually obtained when the cluster bean is grown under irrigation. When grown under irrigation the following procedure has been recommended: pre-plant irrigation to provide 3.2–4.8 ft (1–1.5 m) of moisture penetration; providing little or no rain has fallen, irrigate again, approximately 4 in. (100 mm) about 30 days after planting and again 4 in. (100 mm) 50 days after planting; if the plants show moisture stress about 80 days after planting a last irrigation of 4 in. (100 mm) should be given. Dry weather once the pods have set is essential, since rain or high humidity can cause blackening of the pods and seeds.

Soil—the cluster bean can be grown on a wide range of soils provided that they are well-drained and not acidic; it cannot stand waterlogging. Sandy or sandy loams, pH 7.5 to 8.0, are generally preferred, but it will tolerate soils of very low fertility, high alkalinity, or salinity, and is sometimes used in the reclamation of saline or alkaline soils. Provided that the soils are reasonably fertile and there has been effective inoculation, the cluster bean does not usually respond to nitrogen, but like most legumes has a relatively high phosphorus requirement. The application of 180–220 lb/ac (200–250 kg/ha) of superphosphate has been recommended, or on soils of very low fertility, 10–20 lb/ac (11–22 kg/ha) nitrogen, 50–70 lb/ac (56–78 kg/ha) phosphorus and 50–70 lb/ac (56–78 kg/ha) potassium. Experimentally, sulphur alone, or in combination with phosphorus, has been found to increase nodulation significantly. Trace element nutrition, particularly molybdenum and zinc, is sometimes required. Spraying with molybdenum and tryptophan has been found to increase nodulation.

Altitude—in India the cluster bean is usually grown from sea level up to elevations of about 3 000 ft (900 m).

Day-length—in northern India most types of cluster bean are photo-sensitive, but certain improved Indian cultivars such as Pusa Sadabahar, and the US cultivars Brooks and Mills, are day-neutral.

Planting procedure

Material—seed, which should be plump, disease-free, free of weed seed, and preferably true to cultivar, is used. The germination rate should be at least 85 per cent. Inoculation with suitable strains of *Rhizobium*, ie a special guar inoculum, or the cowpea (Group E), is usually essential. In the USA the standard practice is to inoculate the seed with the inoculum mixed with a small quantity of molasses, diluted 50:50 with water. Scarification of the seed-coat, either by scratching, or treatment with sulphuric acid, has been recommended in order to obtain rapid and uniform germination.

Method—in SE. Asia the cluster bean is often grown in mixed cultivation; elsewhere it is more usually grown as a pure stand. The land is ploughed to a depth of about 8 in. (20 cm) to give a reasonable tilth, and cleared of weeds. When grown as a pure crop for seed it is recommended that planting is done in raised seed-beds to facilitate harvesting.

In India, when grown for fodder, or as a green manure crop, sowing is done broadcast, when grown as a vegetable or for seed, it is drilled 1–2 in. (2.5–5 cm) deep in rows. In the USA conventional two-way tractor equipment is used, in S. Africa wheat, cotton, or grain sorghum planters are used. The seed should not be planted until the soil temperature is 70°F (21°C), or more.

Although the seeds can germinate with a minimum amount of water in the soil, young cluster bean seedlings are very susceptible to moisture stress and for optimum yields the soil must be kept in a thoroughly moist condition until the plants are well established. The cluster bean is very susceptible to competition from weeds and light harrowing to firm the seed-bed and kill weeds immediately prior to planting is recommended. After germination the crop usually receives at least two weedings. In the USA, the pre-planting application of trifluralin is often recommended, pre-emergence applications of EPTA, chlorthal, naptalam or linuron have shown promise as effective herbicides, while prometryne, noruron and 2, 4–DES are reported to 'injure' the crop. Effective weed control is also reported to be obtained with the post-emergence application of 4–(2, 4–dichlorophenoxy) butyric acid. In India the use of vernolate has given promising results.

Field-spacing—in India when grown as a vegetable the cluster bean is often sown in rows 18 to 24 in. (45–60 cm) apart, with 9–12 in. (22.5–30 cm) between the plants; for seed purposes a slightly closer spacing 6–8 x 18–24 in. (15–20 x 45–60 cm) is normally used. In the USA, when grown for seed, the rows are usually 36–40 in. (90–100 cm) apart, with 4 to 12 in. (10–30 cm) between the plants. In S. Africa the recommended spacing for seed crops is 30 to 40 in. (75–100 cm) between the rows and 3 to 4 in. (7.5–10 cm) between the plants. Experiments indicate that closer spacing increases the yield of seed.

Seed-rate—in India, the seed-rate varies from 10 to 25 lb/ac (11–28 kg/ha) when grown as a vegetable, or for seed, and from 30 to 40 lb/ac (34–45 kg/ha) when grown for fodder, or green manure. In the USA the seed-rate ranges from 5 to 6 lb/ac (5.6–6.7 kg/ha) when grown for seed and drilled, to about 25 lb/ac (28 kg/ha) when sown broadcast and grown as a green manure crop. In S. Africa the recommended seed-rate is 3.6–4.4 lb/ac (4–5 kg/ha).

Pests and diseases

The cluster bean is not subject to serious attack from insects, although in the USA stink bugs, *Nezara* spp., and larvae of *Contarinia* spp. can sometimes cause considerable damage. Recently, the cluster bean pod gall midge, *Asphondylia cyamopsii*, has been reported to cause severe reduction of pod yields in India.

Cluster bean

The cluster bean is susceptible to a number of diseases, of which bacterial blight, *Xanthomonas cyamopsidis*, is the most serious and can cause considerable crop losses from the seedling stage until maturity. Symptoms include large, angular lesions at the tops of the leaves which cause defoliation and black streaking of the stems, often resulting in the death of the plant. Investigations have shown that the pathogen is seed-borne and that infection can be spread by irrigation water. Control is most effective by the growing of resistant cultivars, such as Brooks, but immersing the seeds in water at 133°F (50°C) for 10 minutes is reported to produce disease-free seedlings. When grown under humid conditions the cluster bean is susceptible to attack from leaf spot, *Alternaria* spp., which causes the pods and seeds to blacken. Other diseases reported to affect the crop in SE. Asia include: a dry root rot, *Macrophomina phaseoli*, a root rot, *Rhizoctonia solani*, anthracnose, *Colletotrichum capsici*, a leaf spot disease, *Myrothecium roridum*, wilt, *Fusarium solani* var. *coeruleum*, a fusarium blight, *Fusarium moniliforme*, and powdery mildew, *Oidium* sp. The application of organic manure is reported to reduce the incidence of disease and the use of Agrosan 'GN' as a seed-dressing has also been recommended. In the lower Rio Grande Valley of the USA serious commercial losses have been reported due to top necrosis virus, which causes the plants to lose leaves and the terminal ends of the stalk to turn brown and die. When grown under humid conditions in the USA the stems and seeds sometimes develop a purple discolouration due to *Cercospora kikuchii*.

Growth period

Harvesting of the green pods usually starts about 45 to 55 days after sowing, with a peak about 75 to 80 days. In India most cultivars produce mature seed in 110 to 165 days, in the USA in 125 to 130 days and in S. Africa in about 140 to 160 days. In the USA when the crop is grown for forage the optimum green vegetative tonnage is obtained approximately 90 days after sowing.

Harvesting and handling

The green immature pods are picked by hand; for seed production the plants are usually uprooted, or cut a few inches above ground level. The cluster bean does not shatter and can be left to stand, but for good quality seed it should be harvested as soon as the seed pods have become brown and dry. In SE. Asia the plants are usually stacked to dry and threshed by hand. In developed countries, such as the USA, the seed is successfully harvested using modified combine harvesters. The speed should be adjusted to prevent the seed from being damaged. For successful storage the seed should be dried as quickly as possible to a moisture content of 14 per cent or less. Reports suggest that it is relatively resistant to most of the legume storage pests.

Primary product

Pod—the thick, fleshy, dehiscent seed pod is usually 2–4.5 in. (5–11.25 cm) long and contains 5–12 small, shiny, oval or cube-shaped seeds, although some improved Indian cultivars such as Duragapura Safed contain 20–30 seeds per pod. The seeds vary in colour—white, pink, grey or black. On average 100 seeds weigh approximately 0.09–0.16 oz (2.6–4.7 g).

Yield

In India yields of the green pods have been reported to vary from 5 500 lb/ac to 7 400 lb/ac (6 200–8 300 kg/ha), although two improved cultivars, Pusa Sadabahar and Pusa Mausmi, are reported to be capable of yielding about 8 200 lb/ac (9 180 kg/ha) with efficient cultivation. Yields of mature dry seed average between 600 and 800 lb/ac (670–900 kg/ha) under dryland farming in India, but, with irrigation and efficient cultivation, double these yields can be achieved. In the USA on dryland farming, with improved cultivars such as Brooks, top commercial yields are reported to range from 1 070 to 1 500 lb/ac (1 200–1 700 kg/ha), and under normal conditions to average from 710 to 890 lb/ac (800–1 000 kg/ha). Under normal conditions yields from irrigated crops in the USA average between 1 500 and 2 700 lb/ac (1 700–3 000 kg/ha), although yields in excess of 2 700 lb/ac (3 000 kg/ha) can be achieved.

Main use

The green immature pods of the cluster bean have been used for centuries in Asia as a nutritious vegetable and the crop is widely grown in market gardens. The young pods are sweet and are often cooked like French beans, *Phaseolus vulgaris*, but as they mature they become bitter. In addition to being boiled as a vegetable, the pods are sometimes dried and fried like potato chips, or dried and salted for future use as a vegetable. In some areas, especially in times of food shortage, the seeds are converted into dhal and used as a substitute for lentils. In certain countries, eg the USA, the cluster bean is grown as a source of a vegetable gum, which is widely used in the food, paper and textile industries, as a fluocculant and filterant in the refinement of mineral ores, particularly those of uranium, and in the manufacture of explosives.

Subsidiary uses
The mature seeds are used in SE. Asia as a nutritious cattle food, when they are usually mixed with mustard oil and cooked. The possibility of utilizing the seeds in the preparation of protein-enriched foods has been investigated in Pakistan and an experimental product has been produced by boiling the seeds under pressure, drying, then grinding them into a powder which can be mixed with pigeon pea flour, or fish flour, to produce a low-cost protein food.

The cluster bean is sometimes grown for fodder, or as a green manure. When used for fodder, only the tender succulent parts of the plant should be fed to the cattle as the woody stalks are reported to cause tympanitis. The approximate composition of the green fodder has been given as: moisture 80.8 per cent; crude protein 3.1 per cent; ether extract 0.4 per cent; crude fibre 4.4 per cent; N-free extract 8.0 per cent; ash 3.3 per cent; calcium 0.61 per cent; phosphorus 0.07 per cent; digestible crude protein 2.4 per cent; total digestible nutrient 12.0 per cent. The composition of cluster bean hay has been reported as: moisture 9.3 per cent; crude protein 16.5 per cent; ether extract 1.3 per cent; crude fibre 19.3 per cent; ash 12.4 per cent; N-free extract 41.2 per cent; digestible crude protein 12.4 per cent; total digestible nutrients 51.6 per cent. When grown as a green manure crop the cluster bean is reported to add on average 45–135 lb/ac (50–150 kg/ha) of nitrogen to the soil.

Secondary and waste products
The cluster bean has been used in traditional medicine in Asia for centuries. The young leaves are reported to be used occasionally in Africa as a spinach-type vegetable.

The residue, or meal, left after the extraction of the gum from the endosperm is high in protein and is sometimes utilized as a livestock food. It is reported to contain an antitrypsin factor, and to be bitter and poisonous if it constitutes too high a proportion of the ration. It is often toasted to improve its palatability, and experimentally acid treatment has been shown to increase its nutritive value significantly. The Indian Standards Institute has laid down the following specifications for the meal: moisture not more than 10 per cent; crude protein minimum 40 per cent; fat minimum 3 per cent; crude fibre not more than 12 per cent.

Special features

An approximate analysis of the green pods of the cluster bean has been given as: moisture 82.5 per cent; protein 3.7 per cent; fat 0.2 per cent; fibre 2.3 per cent; carbohydrate 9.9 per cent; ash 1.4 per cent; calcium 0.13 per cent; phosphorus 0.25 per cent; vitamin A 330 iu/100g; vitamin C 49 mg/100 g; iron 5.8 mg/100 g. The young bean pods are reported to contain from 40–70 mg of hydrocyanic acid per 100 g, but mature dry pods contain only traces.

The seeds consist of 14–16 per cent testa, or seed-coat, 38–45 per cent endosperm and 40–46 per cent cotyledons. Field-run seeds in the USA contain on average 28 to 31 per cent protein; in the cotyledons it is approximately 47 per cent and in the endosperm 4–5 per cent. The approximate composition of Pakistani seeds has been given as: moisture 9.9 per cent; protein 30.5 per cent; fat 2.5 per cent; ash 3.0 per cent; crude fibre 13.0 per cent; carbohydrate 41.1 per cent. The following protein fractions have been isolated: globulins 31.8 per cent; albumins 13.9 per cent; prolamins 16.5 per cent; glutelins 21.8 per cent. The average amino acid content (mg/gN) is: isoleucine 200; leucine 369; lysine 250; methionine 88; cystine 38: phenylalanine 231; tyrosine 206; threonine 175; tryptophan 119; valine 263; arginine 781; histidine 156; alanine 263; aspartic acid 638; glutamic acid 1 256; glycine 319; proline 194; serine 306. The oil present in the seeds is edible and has the following characteristics: N_D 1.480; iod. val. 123; sap. val. 243.

The endosperm contains about 68 to 70 per cent of a polysaccharide, galactomannan gum, usually known as guar flour. It is this product, not the whole seed, which normally enters international trade. Guar flour has the following approximate composition: moisture 12.0 per cent; protein 5.0 per cent; gum 80.0 per cent; fat 0.7 per cent; fibre 1.4 per cent; ash 0.9 per cent. The gum is composed of D-galactopyranose and D-mannopyranose units, and has an exceptionally high viscosity at low concentrations. Moreover, its cold water swelling can function over a wide pH range and it has excellent thickening, film-forming and stabilizing properties. In 1974 the US Food and Drug Administration affirmed the 'generally recognized as safe' (GRAS) status of guar gum, with specific limits.

Processing

The gum-containing endosperm is separated from the rest of the seed by milling; basic flour milling equipment is used for this operation in the USA.

Production and trade

Production—India, Pakistan and the USA are the major producers of cluster beans. Indian production for the years 1970–74 is estimated to have averaged 495 000 t/a, compared with 423 000 t/a for 1965–69, an increase of 17 per cent. Production in the USA has been estimated at some 25 000 t/a, but the demand exceeds the domestic supply. No information has been traced regarding production in Pakistan.

Trade—there is very little information available relating to the trade in guar flour, the form in which cluster beans enter international trade. India and Pakistan are the major exporting countries, and the USA the major importer. Details are available of Indian exports for the period 1966 to 1973. These have fluctuated considerably from year to year, but averaged 12 091 t/a for the period 1970–73, compared with 5 211 t/a for 1966–69, an increase of about 130 per cent. More recent reports give Indian exports for the fiscal year 1974–75 as 37 128 t and for 1975–76 22 837 t. In the Pakistan trade returns details are given of exports of guar protein extract. These averaged some 37 000 t/a for the period 1970–74, compared with 24 685 t/a for 1967–69.

Prices—the average fob value of Indian exports of guar flour was £257.00/t during the fiscal year 1975–76 and averaged £150.00/t for the period 1970–74, compared with £81.75/t for 1966–69. The average fob value of Pakistani exports of guar protein extract was £3 076.00/t for the period 1970–74, compared with £1 432.00/t for the years 1967–69.

Major influences

The cluster bean is a valuable drought resistant legume, which can be grown successfully on soils of low fertility in the arid and semi-arid areas of the tropics and subtropics. Efforts have been made in recent years to stimulate production, because of its value as a source of a versatile vegetable gum, consumption of which is rising. Although the cluster bean has been used traditionally in Asia as a foodstuff and fodder crop, the toxic constituents of the seeds require further investigation.

Bibliography

ABODUNDE, S. O. 1965. Introduction of pulses from India to northern Nigeria. *Samuru Agric. Newsl.*, *7* (3), 40.

AHMAD, I., MALIK, M. Y. AND SHEIKH, A. A. 1965. Replacement of cottonseed cake with guar meal in sheep fattening rations. *J. Agric. Res.*, (*Punjab*), *3*, 150–153.

AKMAL KHAN, M., GILANI, A. H. AND SIAL, M. B. 1972. Influence of varying levels of toasted guar meal on the digestibility of mixed ration for dry Sahiwal cows. *J. Agric. Res.*, (*Punjab*), *10*, 154–157.

ALA-UD-DIN, SIAL, M. B. AND SCHNEIDER, B. H. 1965. Fattening of cattle: The effect of substituting guar meal for cottonseed cake as a protein supplement. *Agric. Pak.*, *16*, 67–74.

ALI, S. M., HANIF, M. AND ALVI, S. A. 1963. Elimination of toxicity from guar seed (*Cyamopsis tetragonoloba*) for the production of a protein rich edible guar seed flour. *Pak. J. Sci. Res.*, *15*, 153–156.

ALLEN, G. H. 1964. Guar shows promise. *Qd. Agric. J.*, *90*, 224–227.

ALVI, A. S., HANIF, M. AND ALI, S. M. 1964. Net protein utilization of 'guar' (*Cyamopsis tetragonoloba*) treated in different ways and enhancement of its protein value by supplementation. *Pak. J. Sci.*, *Res.*, *16*, 1–4.

ANON. 1974. Guar gum: A useful industrial raw material. *J. Tr. Ind.*, (*India*), *24* (12), 32–33.

BHAN, S. AND PRASAD, R. 1967. Guar has many uses. *Indian Farming*, *17* (6), 17–19.

BHATTI, M. B. AND SIAL, M. B. 1971. Guar: Its utility in food and non-food industries. *Pak. J. Sci.*, *23*, 1–5.

BROOKS, L. E. AND HARVEY, C. 1950. Experiments with guar in Texas. *Texas Agric. Exp. Stn. Circ.* 126, 10 pp.

CHARI, M. S. 1971. Nature and extent of damage caused by the larvae of cluster bean pod gall midge *Asphondylia cyamopsii* n. sp. on different varieties of cluster bean, *Cyamopsis tetragonoloba* (L.) Taub. *Madras Agric. J.*, *58*, 892–893.

CHEVALIER, A. 1939. Recherches sur les espèces du genre *Cyamopsis* plantes fourragères pour les pays tropicaux et semi-arides. [Research on species of the genus *Cyamopsis*, forage plants for tropical and semi-arid countries]. *Rev. Bot. Appl. Agric. Trop.*, *19*, 242–249.

Cluster bean

COMMONWEALTH BUREAU OF PASTURES AND FIELD CROPS. nd. Guar (*Cyamopsis tetragonoloba*). Annotated bibliography 1298, 1952–72. *Commonw. Agric. Bur., Bur. Pastures & Field Crops, Hurley, Maidenhead, Berks. England*, 12 pp.

ESSER, J. A. 1956. Guar gains after long effort for farm and market progress. *Chemurg. Dig., 15* (12), 4–7.

ESSER, J. A. 1958. Guar: Its development and uses. *Chemurg. Dig., 17* (4), 9–12.

FOOD AND AGRICULTURE ORGANIZATION OF THE UNITED NATIONS. 1970. Amino acid content of foods and biological data on proteins. *FAO Nutr. Stud.* 24, pp. 52–53. Rome: FAO, 286 pp.

GHONSIKAR, C. P. AND SAXENA, S. N. 1973. Influence of molybdenum and tryptophane on nodulation in cluster beans (*Cyamopsis tetragonoloba* (L.) Taub.). *Indian J. Agric. Sci., 43*, 938–941.

GOLDFRANK, H. 1957. Guar: Its commercial uses and agricultural characteristics. *Chemurg. Dig., 16* (2/3), 10–11.

HODGES, R. J., KINMAN, M. L., BRINTS, N. W., BORING, E. P. AND MULKEY, J. R. 1970. Keys to profitable guar production. *Texas A & M Univ. Agric. Ext. Serv. Coll. Stn., Fact Sheet* L–907, 4 pp.

HUSSAIN, A. AND ULLAH, M. 1964. Yield, seed damage and gum content of guar seed (*Cyamopsis psoralioides*) as influenced by different harvesting procedures. *West Pak. J. Agric. Res., 2*, 1–7.

HYMOWITZ, T. 1964. Guar gum in India. *Chemurg. Dig., 22* (3), 6.

HYMOWITZ, T. 1972. The trans-domestication concept as applied to guar. *Econ. Bot., 26*, 49–60.

HYMOWITZ, T. AND MATLOCK, R. S. 1963. Guar in the United States. *Okla. State Univ. Exp. Stn. Bull.* B–611, 34 pp.

ISLAM SHAH, S. S., SIAL, M. B. AND SCHNEIDER, B. H. 1966. The digestibility of guar meal. *Agric. Pak., 17*, 35–39.

JOHNSON, H. W. AND JONES, J. P. 1962. Purple stain of guar. *Phytopathology, 52*, 269–272.

KAWATRA, B. L., GARCHA, J. S., CHARANJEET, K. AND SINGH, R. 1974. Comparative study on different treatments of detoxification of guar (*Cyamopsis tetragonoloba*) meal and their subsequent effect on its nutritive value. *J. Food Sci. Technol., 11*, 263–265.

LUTTRELL, E. S. 1951. Diseases of guar in Georgia. *Plant Dis. Rep.*, *35*, 166.

MAL, B. 1969. Guar a good soil improver and food, forage and gum producer. *Indian Farming*, *18* (11), 37–38.

MATHUR, R., MENON, U. AND JOHARI, J. N. 1976. Pod set in cluster bean—*Cyamopsis tetragonoloba* (L.) Taub. *Sci. Cult.*, *42*, 376–378.

MATLOCK, R. S. AND OSWALT, R. M. 1964. Brooks-guar. *Okla. State Univ. Agric. Exp. Stn. Bull.* B–624, 7 pp.

MEHOTRA, O. N., TRIPATHI, R. D., SRIVASTAVA, G. P. AND MISRA, M. C. 1975. Food value of some guar (*Cyamopsis psoralioides*) varieties. *Indian J. Agric. Res.*, *9*, 37–42.

NAGPAL, M. L., AGRAWAL, O. P. AND BHATIA, I. S. 1971. Chemical and biological examination of guar meal (*Cyamopsis tetragonoloba* L.). *Indian J. Anim. Sci.*, *41*, 283–293.

NAGPAL, M. L. AND BHATIA, I. S. 1972. Classification of guar (*Cyamopsis tetragonoloba*) seed proteins. *Indian J. Nutr. Diet.*, *9*, 5–7.

NARANG, M. P., ARNEJA, D. V. AND RAO, A. R. 1973. Guar meal: A feed of promise in livestock rations. *Indian Farming*, *23* (9), 37.

NARAYANAN, T. R. AND DABADGHAO, P. M. 1972. Cluster beans. *Forage crops of India*. pp. 66–69. New Delhi: Indian Coun. Agric. Res., 373 pp.

NATIONAL ACADEMY OF SCIENCES. 1975. Guar. Underexploited tropical plants with promising economic value. pp. 145–149. *Washington, Comm. Int. Relat., Natl. Acad. Sci., Advis. Stud. & Spec. Rep.* 17, 188 pp.

OKE, O. L. 1964. Hydrocyanic acid content and nitrogen fixing capacity of guar. *Nature*, *204*, 405–406.

OKE, O. L. 1967. Nitrogen fixing capacity of guar bean. *Trop. Sci.*, *9*, 144–147.

OPEKE, L. K. 1964. A new crop in Western Nigeria. *Niger. Agric. J.*, *1*, 39–40.

ORELLANA, R. G., THOMAS, C. A. AND KINMAN, M. L. 1965. A bacterial blight of guar in the United States. *FAO Plant Prot. Bull.*, *13*, 9–13.

PAL, B. P., SIKKA, S. M. AND SINGH, H. B. 1956. Improved Pusa vegetables. *Indian J. Hortic.*, *13* (2), 68–69.

PHILLIPS, M. A. 1962. The detoxification of guar and soya meals. *World Crops*, *14*, 393–396.

Cluster bean

POATS, F. J. 1960. Guar, a summer row crop for the southwest. *Econ. Bot.*, *14*, 241–246.

REPUBLIC OF SOUTH AFRICA, DEPARTMENT OF AGRICULTURE. 1975. Guar a crop with a future in S. Africa. *Pretoria, Dep. Agric. Tech. Serv., Inst. Crops & Pastures, Leafl.* 137, 4 pp.

REWARI, R. B., SEN, A. N. AND SEN, A. 1965. Nitrogen fixation by cluster bean (*Cyamopsis tetragonoloba* (L.) Taub.) in relation to phosphate uptake from soil. *Indian J. Agric. Sci.*, *35*, 162–167.

SANDERSON, K. W. 1974. Guar (*Cyamopsis tetragonoloba*) in the Rhodesian lowveld. *Rhod. Agric. J.*, *71*, 17–18; 21.

SANGHI, A. K. AND SHARMA, S. K. 1965. Duragapura Safed: a new guar for Rajasthan. *Indian Farming*, *14* (10), 23.

SASTRI, B. N. (ed.). 1950. *Cyamopsis* D. C. *The wealth of India: Raw materials.* Vol. 2 (C). pp. 407–408. Delhi: Indian Counc. Sci. Ind. Res., 427 pp.

SELLSCHOP, J. P. 1967. Guar beans—a bean with a future. *Farming South Afr.*, *42* (12), 35–39.

SINGH, H. B. AND MITAL, S. P. 1970. Cluster bean *Cyamopsis tetragonoloba* (L.). Taub. *Pulse crops of India.* (Kachroo, P. and Arif, M. eds.). pp. 188–200. New Delhi: Indian Counc. Agric. Res., 334 pp.

SINGH, H. B., MITAL, S. P. AND THOMAS, T. A. 1962. Guar has a great future. *Indian Farming*, *12* (9), 23–27.

SMITH, D.T., WIESE, A. F. AND SANTELMANN, P. W. 1973. Weed control research in guar in Texas and Oklahoma 1961–72. *Texas A & M Univ. Agric. Exp. Stn.*, B–1138, 11 pp.

SOLANKI, J. S. AND SINGH, R. H. 1976. Note on the fungicidal control of powdery mildew of cluster bean. *Indian J. Agric. Sci.*, *46*, 241–243.

STAFFORD, R. E., KINMAN, M. L., BROOKS, L. E. AND LEWIS, C. R. 1976. Registration of Brooks, Hall and Mills guar. *Crop Sci.*, *16*, 309.

STAFFORD, R. E., KIRBY, J. S., KINMAN, M. L. AND LEWIS, C. R. 1976. Registration of Kinman and Esser guar. *Crop Sci.*, *16*, 310.

UNITED STATES, FOOD AND DRUG ADMINISTRATION. 1974. Guar gum: Proposed affirmation of GRAS status with specific limitations as direct human food ingredient and affirmation of GRAS status as indirect human food ingredient. *Fed. Regist.*, *39*, 34201–34203.

WILLIAMS, L. O. 1961. Guar, un cultivo para America Central. [Guar, a crop for Central America]. *Ext. Am.*, *6* (1), 17–19.

Common names	**COWPEA, Black-eye bean or pea, Catjang, China pea, Cowgram, Southern pea.**
Botanical name	*Vigna unguiculata* (L.) Walp., syn. *V. sesquipedalis* Fruhw., *V. sinensis* (L.) Savi ex Hassk.
Family	Leguminosae.
Other names	Adagura (kwolla), Adonguari (Eth.); Afunat habakar (Is.); Agwa, Akide enu, Akidiani, (Nig.); Ambelophassula (Gr.); Amuli (Ug.); Atera argobba (Eth.); Bannette (Fr.); Barbata/i (Ind.); Batong (Philipp.); Bean[1] (Nig.); Boo-ngor (Ug.); Börülce (Turk.); Calavance (W.I.); Callivance (USA); Catjangbohne (Ger.); Caupí (Ang.); Chaula, Chavli (Ind.); Chicharo de vaca (Mex.); Chowlee, Chowli (Ind.); Cornfield pea (USA); Coupé, Dagarti bean (W.Afr.); Dâu den, D. trang, D. tua, D. xa (Viet.); Digir (Eth.); Dinawa[2] (Bots.); Dolico (It.); Dolique de Chine (Fr.); D. indigène (Tog.); D. mongette (Zar.); D. mougette (Alg.); Eka-wohe (Eth.); Enkoole, Enkoore (Ug.); Ere(e) (Nig.); Ervilha de vaca (Braz.); Fagiolino dall'occhio (It.); Fasolea-dima (Eth.); Feijão brabham (Braz.); F. de China (Port.); F. de corda (Braz.); F. fradinho, F. makunda (Ang.); Frijol(e)[3] (Lat. Am.); F. carita (Cuba); F. de ojo negro (Sp.); F. precioso (Cuba); Gaisa (Eth.); Halifax pea (P.R.); Haricot à oeil noir, H. dolique (Fr.); H. indigène (Benin); H. kunde (Zar.); Hindu pea (Ind.); Ilanda (Zam.); Imare (Ug.); Indian pea[4] (USA); Kachang

1 Used in many countries for *Phaseolus* spp., especially for the haricot bean, *P. vulgaris*, or the lima bean, *P. lunatus*.
2 Also used for the tepary bean, *Phaseolus acutifolius*.
3 Also used for the haricot bean, *Phaseolus vulgaris*.
4 More commonly used for the grass pea, *Lathyrus sativus*.

bol, K. panjang[5], K. toonggak (Malays.); Kaffer boon, Kaffir bean/pea[6] (S. Afr.); Karakala (Philipp.); Katjang merah[3], K. panjang[5] (Malays.); Kibal (Philipp.); Kunde[3] (Zan.); Laputu, Liboshi, Likote, Likotini (Ug.); Lobia[7] (Egy., Ind.); Loputa (Ug.); Lubia (Ar.); L. beida, L. helu, L. kordofani, L. tayiba (Sud.); Lubya baladi, L. msallat (Ar.); L. tarbutit (Is.); Marble pea, Me-karal (Sri La.); Ngeri (Ug.); Nguno (Eth.); Niébé (Fr.); Nori (Eth.); Nyemba bean (Rhod.); Nyoari (Eth.); Omugobe, Osu (Ug.); Otong, Paayap (Philipp.); Paythenkai (Tam.); Pois de Brazil, P. de canne, P. poona, P. vache (Fr.); Poncho (Col.); Poona pea (Aust.); Porotito del ojo (Arg.); Sai dau-kok (Haw.); Sasage (Jpn.); Tau-kok[5] (China); Thattapayru (Tam.); Tonkin pea; Tua dam (Thai.); Vigna einese (It.); Voamba (Malag.); Voehm (Maur.); Voeme (Re.); Wuch (Eth.).

Botany

An annual herb showing great variation, according to cultivar, climatic and soil conditions. Trailing, climbing, bushy and erect forms exist. Most cultivars are indeterminate, ie produce flowers and seeds over a long period, but some are determinate, producing flowers and seeds within a very short period. The tap-root is well developed, with numerous spreading laterals near the soil surface and large nodules often collected in groups. The thin, rounded, slightly ribbed stems may be rough or smooth and are sometimes tinged with purple. The leaves are alternate, trifoliolate, with a petiole 2–10 in. (5–25 cm) long; the leaflets are large, 2.5– 6.3 x 1.6–4.3 in. (6.25–16 x 4–11 cm), usually dark-green and ovate acute in shape. The inflorescence is an unbranched axillary raceme, with a long peduncle, 2–24 in. (5–60 cm) long. The flowers are borne in alternate pairs and although numerous pairs may occur per inflorescence, frequently only the first two

---

[5] Also used for the asparagus bean, *Vigna unguiculata* ssp. *sesquipedalis*.
[6] Sometimes used for the bambara groundnut, *Voandzeia subterranea*.
[7] Also used for the lima bean, *Phaseolus lunatus*.

develop. The flowers, which are conspicuous and self-pollinated, are borne on short pedicels and may be white, dirty-yellow, pale-blue, pink or violet. They commonly open early in the day and close around mid-day; after opening once, the flowers wilt and collapse. The pods can vary greatly in size, shape, colour and texture. They may be linear, crescent-shaped or coiled, and normally vary from 3 to 18 in. (7.5–45 cm) in length, but can reach 40 in. (100 cm). They are indehiscent, usually yellow when ripe, although brown or purple-coloured ones can occur, and normally contain 8–20 seeds. These also vary considerably in size, shape and colour, but the commonest forms are white, creamy-white or black.

There is a great deal of confusion and disagreement on the proper classification of the cowpea, because of the large number of distinct forms which exist and the fact that hybridization is readily achieved, so that it is probable that some, if not all, of the cultivated forms are in fact hybrids. Numerous specific names have been given to the cowpea and its various forms. Some authorities consider that there should be three separate species: (i) the catjang pea, *Vigna unguiculata*, a primitive form, which although found in Africa, is more common in Asia; (ii) the common cultivated cowpea, *Vigna sinensis*, commonly found in Africa; (iii) the asparagus bean, *Vigna sesquipedalis*, grown mostly for its immature pods, and widely cultivated in Asia. However, it is now widely accepted that there should be no distinction between these so-called species and that there should be one single species *Vigna unguiculata*, with *V. sesquipedalis* and *V. sinensis* as synonyms. Recently Verdcourt (1971) has suggested that five subspecies of *Vigna unguiculata* should be recognized:

(i) ssp. *mensensis*, the wild cowpea of the African forest;

(ii) ssp. *dekindtiana*, the wild cowpea of the African savanna zone;

(iii) ssp. *unguiculata*, the common and very variable cultivated cowpea of Africa, America and Asia;

(iv) ssp. *sesquipedalis*, the asparagus bean, widely cultivated in Asia;

(v) ssp. *cylindrica*, the catjang or Hindu cowpea, common in Asia, especially India, but also occurring in tropical Africa.

In this digest the common cowpea and the catjang pea have been reviewed together and the asparagus bean has been treated as a separate legume crop.

In recent years considerable efforts have been made in many African countries to improve the cowpea and since 1971 the International Institute of Tropical Agriculture, (IITA), in Nigeria, has been attempting to develop high-yielding strains with resistance to many of the major pests and diseases which affect the crop. Several elite strains, notably VITA–1, VITA–2, VITA–3, VITA–4, and VITA–5, with moderately high-yielding ability, a high disease-resistance and a degree of resistance to attack from certain insect pests, have been released.

Origin and distribution

The cowpea is an ancient food crop, whose centre of origin is uncertain, having been reported as possibly Asia—Hindustan and Iran, Africa—Nigeria and Ethiopia, and even S. America. It is now widely distributed throughout the tropics and subtropics and is an important food legume crop in Africa, south of the Sahara, particularly in the W. African savanna zone. Africa in fact produces about 95 per cent of the world crop with Nigeria, Niger, Upper Volta and Uganda being the more important producing countries. Outside Africa the cowpea is also grown in Asia, especially India, Australia, the Caribbean, the southern USA, and the lowland and coastal areas of S. and C. America.

Cultivation conditions

Temperature—the cowpea is predominantly a hot weather crop well adapted to semi-arid regions and preferring temperatures of between 68° and 95°F (20–35°C). Although it can tolerate temperatures as low as 59°F (15°C), for good germination a minimum soil temperature of 68°F (20°C) is required, but temperatures above 90°F (32°C) are reported to reduce root growth significantly. It cannot tolerate frost and young plants are susceptible to various injuries when exposed to temperatures of between 41° and 50°F (5–10°C) for periods as short as 24 hours. At temperatures above 95°F (35°C) yields are liable to be reduced because of flower and pod shedding. Recently it has been demonstrated that the amount of solar radiation the cowpea receives has a direct effect upon growth and dry matter yields, but air temperatures would appear to exert a far greater influence. Experiments with 30 cultivars showed that maximum dry matter production occurred at 80°F (27°C) day temperature and 72°F (22°C) night temperature.

Rainfall—most cowpea crops are grown under rain-fed conditions; short duration determinate types can be grown in semi-arid regions with a rainfall of less than 24 in./a (600 mm/a); medium and long duration types are grown in regions with a rainfall of between 24 and 60 in. (600–1 500 mm). However, excessive rain or atmospheric humidity is liable to reduce yields because of the high incidence of fungal diseases. The cowpea is sometimes grown under irrigation when 8–10 ac/in. (822–1 028 m^3) in three or four irrigations is generally recommended. Short duration types are also grown on residual moisture on soils with a high water-holding capacity, eg after rice. Although generally regarded as being a drought resistant crop, moisture stress during the period from emergence to first flower can reduce productivity significantly, but thereafter does not have a significant effect upon determinate types.

Soil—the cowpea can be grown over a wide range of soil types, provided that they are well drained as it cannot tolerate waterlogging. For optimum yield of seed medium loams are preferred; heavy clays and soils of high fertility usually result in high yields of hay, but poor seed yields, while on light sandy soils heavy infestation with nematodes is liable to occur. It cannot tolerate salinity and although reasonably tolerant of acidity, a pH of 5.5 to 6.5 is preferred.

Many fertilizer experiments that have been reported on the cowpea in the tropics are invalid because losses due to insect pests and diseases are so high in relation to the response to the fertilizer. However, experiments in W. Africa have shown low, but significant, responses to nitrogen, phosphorus and potassium. The total nutrient requirement per 220 lb of seed has been given as approximately 11 lb nitrogen, 3.7 lb phosphorus (P_2O_5), 10.5 lb potassium (K_2O), 3.5 lb calcium (CaO), 3.3 lb magnesium (MgO) and 0.9 lb sulphur [5 kg N, 1.7 kg P_2O_5, 4.8 kg K_2O, 1.6 kg CaO, 1.5 kg MgO, 0.4 kg S/100 kg seed]. A most common fertilizer recommendation is for phosphorus at 18–53 lb/ac (20–60 kg/ha) P_2O_5, but potash may also be included at the rate of 27–53 lb/ac (30–60 kg/ha) K_2O, if known to be deficient, and on soils of low fertility the application of 13–27 lb/ac (15–30 kg/ha) of nitrogen has been suggested as being profitable. The application of 1 680–2 800 lb/ac (1 880–3 140 kg/ha) of agricultural lime has been recommended to growers in the USA and for growers in southern Georgia, nitrogen 24–49 lb/ac (27–55 kg/ha), phosphorus 11–22 lb/ac (12–24 kg/ha) and not more than 41 lb/ac (46 kg/ha) of potassium. In India the foliar application of 50 lb/ac (56 kg/ha) P_2O_5 has been found to be more effective than the same application banded beside the row. In W. Nigeria it has been reported that the application of a general fertilizer of 0–15 nitrogen, 36–48 phosphorus and 60–60 potassium drilled at planting at the rate of 250 lb/ac (280 kg/ha), results in an increase in seed yield of 40 to 60 per cent, although recent experiments have shown that seed yields of nodulated cowpeas are not increased by liberal applications of inorganic nitrogen. In areas where there is likely to be a molybdenum deficiency in the soil the application every 5 years of 200 lb/ac (225 kg/ha of molybdenized superphosphate has been recommended.

Day-length—short-day, day-neutral and long-day types of cowpea exist and it has been suggested that there is a relationship between late-maturing types and large-seeded types, and short-day types and the spreading habit. The optimal photoperiod for induction of flowering in the cowpea is from 8 to 14 hours. Field trials carried out in relation to the equator demonstrated that seed yield, dry matter production and nodulation were all reduced in photoperiods of less than 12 hours 13 minutes and that differences of as little as 12 minutes could have a significant effect upon flowering and seed yield.

Altitude—in E. Africa the cowpea is normally grown at elevations up to 5 000 ft (1 500 m).

Planting procedure

Material—seed is used, germination is epigeal and the germination rate is usually between 85 and 95 per cent. Provided precautions have been taken against insect infestation the seed will retain its viability for several years. Treatment of the seed prior to planting with an insecticide/fungicide dressing such as aldrin and Thiram, is often recommended. Inoculation with efficient strains of rhizobia, if nodulated cowpeas or related leguminous species have not been grown recently, has been found to be very beneficial.

Method—most of the world's cowpea crop is grown by traditional agricultural methods using hand tools, often in association with cereal crops such as maize, sorghum or millet, when it is frequently planted broadcast after the cereal is about 20 in. (50 cm) tall. When grown as a pure stand the cowpea is planted in rows, 0.8–2 in. (2–5 cm) deep. In francophone Africa, hill plantings (2–3 seeds per drop) are frequently used. In mechanized agriculture maize planters with 0.4 in. (9.5 mm) plates are sometimes employed.

The seed-bed should be firm and moist, free of clods and weeds. If artificial fertilizers are used they should be placed in bands a little to the side of the seed-row. Cowpeas must be kept free from weeds during the early stages of growth, either by hand weeding, or by mechanical cultivation. They are more sensitive to herbicides than many other crops. Moreover, several herbicides which have shown promise have been found to lack persistence, and herbicide treatment followed by a tillage operation has been suggested for effective weed control. Trifluralin applied presowing and immediately harrowed, or rotavated in, and followed by irrigation immediately after planting, has given good control in the USA. However, in trials at the International Institute of Tropical Agriculture (IITA) in Nigeria, it was found to be phytotoxic to the crop when applied at 2.7 lb/ac (3 kg/ha). Lower rates although safe to the crop did not give effective weed control. Applied in association with chloropropham, diphenamid or linuron, trifluralin at 0.4–0.9 lb/ac (0.5–1.0 kg/ha) has given promising results at IITA. Chloramben is also reported to give satisfactory results. The use of ester formulations of the phenoxyacetic acid herbicides (2, 4–D and MCPA) is not recommended as they can cause retarded growth and leaf abnormalities.

Field-spacing—varies according to the growth habit of the plants. In mechanized agriculture when grown for seed production erect types are usually planted 2–5 in. (5–12.5 cm) apart, with 24–40 in. (60–100 cm) between the rows; the difference between spreading types may reach 12 in. (30 cm). In francophone Africa, the recommended hill spacings are 20 x 20 in. (50 x 50 cm) or 20 x 24 in. (50 x 60 cm) for early erect cultivars, and somewhat wider spacings for late or spreading ones. Experiments in Ghana indicate that plant spacing and arrangement could have a significant effect upon seed yield.

Seed-rate—when grown as a pure crop for seed approximately 15–25 lb/ac (17–28 kg/ha) of seed is required. When grown for forage, or mixed with cereal crops, seed-rates of up to 89 lb/ac (100 kg/ha) may be used, although in African mixed cropping systems the seeds are frequently broadcast at a seed-rate of 20–29 lb/ac (22–33 kg/ha).

Pests and diseases

The cowpea is susceptible to a wide range of pests and diseases, but there is a great deal of variation in the susceptibility and resistance of the main cultivars from country to country.

Pests—insects are probably the major limiting factor in cowpea production, particularly in the low humid tropics and their effective control can result in an increase in productivity of 10 to 30 times, compared with the unprotected crop. In Africa and Asia there are about 15 major and more than 100 minor insect pests of the cowpea. In Nigeria the more important ones attacking the plant during its growth cycle are: *Melanagromyza vignalis*, a small fly which attacks the green pods, the weevil, *Piezotrachelus varius*, the bruchids, *Bruchidius atrolineatus* and *Callosobruchus maculatus*, the foliage beetle, *Ootheca mutalbis*, the borers *Maruca testulalis* and *Cydia ptychora*, which attack the flower buds and pods, and the bugs, *Anoplocnemis curvipes*, *Riptortus dentipes* and *Acanthomia horrida*, which also attack the pods. Spraying at weekly intervals with DDT or BHC has been suggested for effective control, or two post-harvest applications of dieldrin. In Senegal crop losses due to insect pests are reported to range from 10 to 60 per cent. The major pests attacking the cowpea crop in Senegal include bean fly, *Ophiomyia* (*Melanagromyza*) *phaseoli*, which is also of considerable economic importance in other areas of W. Africa. In India the major pests are the cotton tree beetle, *Sphenoptera khartoumensis*, the hairy groundnut caterpillar, *Amsacta moloneyi*, and the pod borers, *Maruca testulalis*, *Spodoptera exempta*, *Deudorix antalus*, and *Cosmolyce boetica*. Reasonable control over most of these pests is reported to be obtained by spraying with DDT. In the USA the most serious pest is the cowpea curculio, *Chalcodermus aeneus*. In 1973 it was estimated that losses due to this one pest exceeded £500 000. Three applications of 5 per cent toxaphene dust is reported to give satisfactory control, without the development of objectionable off-flavours in the seeds. Other pests that can be troublesome occasionally in the USA are the green stink bug, *Nezara viridula*, the Mexican bean beetle, *Epilachna varivestis*, the beet armyworm, *Spodoptera exigua*, the harlequin bug, *Murgantia histrionica*, the serpentine leaf miner, *Liriomyza pusilla*, and thrips, *Frankliniella* spp. Cutworms, particularly the black cutworm, *Agrotis ipsilon*, can also be troublesome on occasions. The bean leaf beetle, *Cerotoma ruficornis*, is of considerable economic importance in the Caribbean area as the vector for cowpea mosaic. Aphids, especially *Aphis craccivora*, are also of economic importance in many areas as the vectors of virus diseases.

During storage the most important insects attacking cowpea seed are the bruchids, *Callosobruchus maculatus* and *C. chinensis*. Although in W. Africa, *Bruchidius atrolineatus* is also of considerable importance and in southern Africa, *C. rhodesianus*. In the USA stored cowpeas are also liable to attack from the Chinese weevil, *Mylabris chinensis*, and the bean weevil, *Acanthoscelides obtectus*.

Cowpea

The cowpea is very susceptible to attack by nematodes, although efforts are being made to develop resistant strains. The root-knot nematode, *Meloidogyne javanica*, is troublesome, especially on sandy soils in Africa; in the USA the root-knot nematodes, *M. incognita* var. *incognita* and *M. incognita* var. *acrita*, and the sting nematode *Belonolaimus gracilis*, are of considerable economic importance. Control measures include efficient crop rotation, flooding, and the use of soil fumigants, such as nemagon, but the last measure is expensive and is not very effective unless it is carried out very carefully. In addition, *Striga gesnerioides* is a root parasite causing considerable crop losses in tropical Africa.

Diseases—the more important and widespread diseases affecting the cowpea are rust, *Uromyces phaseoli*, bacterial canker, *Xanthomonas vignicola*, fusarium wilt, *Fusarium oxysporum*, mildew, *Erysiphe polygoni*, charcoal rot, *Sclerotium bataticola*, and several virus diseases particularly, yellow mosaic, green mottle virus and tobacco mosaic. In addition, anthracnose, *Colletotrichum lindemuthianum*, can cause serious losses during the rainy season in many areas.

In S. Nigeria, the seedling diseases, *Rhizoctonia solani*, *Phythium aphanidermatum* and *Botryodiplodia theobromae*, and the leaf spot diseases, *Cercospora cruenta* and *C. canescens*, are also of economic importance. In certain areas of Africa the seed-borne disease, leaf and pod spot, *Ascochyta phaseolorum*, is reported to cause considerable crop loss. In the drier parts of francophone Africa a wilting disease, *Neoscosmopora vasinfecta*, and two diseases affecting the foliage and pods, *Leptosphaerulina* sp. and *Choenephora* sp., have been reported. In Australia an ashy stem blight caused by *Macrophomina* spp. is sometimes troublesome, especially if the plant has been weakened by drought or mechanical injury. In India the more important diseases affecting the crop include rust, *Uromyces appendiculatus*, a fungal disease, *Glomerella cingulata*, and root rots, *Rhizoctonia* spp. The most common of these is reported to be a violet root rot, *R. violaceae*, another, a root canker, *R. solani*, mainly affects the crop when it is grown in low-lying areas subject to flooding.

The use of potassium azide as a seed-dressing has proved effective against *Rhizoctonia* spp., *Sclerotium* spp. and *Phythium* spp. and other fungicides such as benomyl, carboxin, perenox or kocide give reasonable disease control when used as foliar sprays. However, the most effective solution to the disease problem of the cowpea lies in the development of resistant cultivars, the destruction of all diseased plant material in the field, and crop rotation.

Growth period

Depending upon the cultivar and environment cowpeas may take from about 60 to between 210 and 240 days to produce mature seeds.

Harvesting and handling

The green pods are usually harvested by hand when they are still immature and tender, before the seeds are fully developed. When grown as a grain legume harvesting is complicated by the prolonged and uneven ripening of many cultivars. Bushy and tall types of cowpea can usually be harvested successfully by combine, but the trailing types normally have to be harvested by hand. Time of harvesting is critical as the mature pods easily shatter and the seeds readily germinate if there is adequate moisture, as can occur if there is rain. In India, cowpeas are frequently hand-picked at intervals of 3–4 days at the beginning and end of the season, and every alternate day during the middle. When harvested mechanically, the whole plant is cut or uprooted when the leaves have begun to fall and at least two-thirds of the pods are dry and the yellow colour has disappeared. Wheat-threshers are sometimes used, but the drum speed must be reduced to between 200 and 400 r/m to avoid damaging the seeds.

After threshing the seed should be thoroughly dried to a moisture content of 14 per cent, or less, before being stored. Attention has recently been drawn to the possibility of the seed being contaminated with aflatoxin. Cowpeas are extremely susceptible to insect infestation during storage. In Nigeria it has been estimated that currently some 30 000 t/a are lost, mainly due to damage by bruchids during storage. Traditional protection methods include lightly roasting the skins soon after threshing and storing in closed tins and mixing with ashes. In Nigeria it is recognized that harvesting promptly reduces the initial damage and that storage in the pod affords a degree of protection. Fumigation is effective, but difficult for small farmers to perform efficiently. The hermetic storage of cowpeas in small granaries, silos and pits is being developed in Nigeria, where a very encouraging development has been the use of plastic liners in traditional dried-earth granaries. Investigations in Senegal have shown that cowpeas may be stored satisfactorily for up to one year in plastic sacks using soft capsules of carbon tetrachloride as a

fumigant. Cowpeas treated with palm, groundnut or coconut oils are reported to be protected against insect infestation during storage for periods of up to 6 months, but to lose their viability.
In the USA considerable quantities of immature Blackeye cowpeas are eaten fresh and these have a very limited shelf-life if stored at ambient temperatures, but may be stored satisfactorily for 8 days at 46°F (8°C) if in the pod; the shelled peas have a life of less than 5 days.

Primary product

Seed—cowpeas can vary very considerably in size, shape, colour and eye patterns, ie the pigmented area around the hilum. Normally they range from 0.08 in. to 0.5 in. (2–12 mm) in length and can be globular or reniform in shape. The skin-coat may be smooth, rough or wrinkled, and can vary in colour from white, through various shades of buff, green, brown, red and purple, to black, sometimes with mottled, blotched or speckled patterns. The hilum is white, approximately 0.12 in. (3 mm) in length, and in the black-eyed types surrounded by a dark ring. The seed weight averages between 0.18 oz and 1.05 oz (5–30 g)/100 seeds.

Yield

Under subsistence agriculture in Africa the average yield of dry seed normally ranges between 89 lb/ac and 270 lb/ac (100–300 kg/ha). In E. Africa however, with reasonably efficient crop management, yields can reach 600–800 lb/ac (670–900 kg/ha) and in Uganda yields of 2 000 lb/ac (2 240 kg/ha) have been obtained, using insecticides and improved cultivars. In Nigeria the average yield is about 200 lb/ac (225 kg/ha), but with an efficient insect spraying programme it is estimated that average yields of 1 600 lb/ac (1 790 kg/ha) could be achieved, and at IITA yields in excess of 2 800 lb/ac (3 140 kg/ha) have been obtained in experimental trials. In India, when cowpeas are grown as a mixed crop, yields of 150 lb/ac (170 kg/ha), or less, are by no means unusual, but when grown as a pure crop certain cultivars, eg Bombay, are reported to yield approximately 1 000 lb/ac (1 120 kg/ha). In the USA yields fluctuate very considerably, but have averaged about 530 lb/ac (600 kg/ha) in recent years.

Main use

The dried seeds are an important food legume crop in the tropics and subtropics providing an inexpensive source of protein in many diets. In Africa, where they are the preferred food legume, they are consumed in three basic forms, although there are many local variations. Most frequently they are cooked together with vegetables, spices and often palm oil, to produce a thick bean soup, which accompanies the basic staple food, such as cassava, yams, plantains, etc. They are also decorticated, ground into a flour and mixed with chopped onion and spices and made into cakes, which are either deep-fried (akara balls), or steamed (moin-moin). For the preparation of the bean cakes white or pale coloured seeds with a rough skin-coat are preferred as they require a shorter period of soaking before decortication. In India the cowpea is used mostly as a pulse, either whole, or as dhal.

Subsidiary uses
The fresh immature seeds and the immature seed pods are sometimes eaten boiled as a vegetable. In some countries, eg the USA, certain cultivars, such as Black-eye, Crowder and Purple-eye, are grown for canning and quick-freezing. The possibility of utilizing cowpeas as a source of protein concentrates is being investigated. Recently an experimental method has been described for the production of an acceptable methionine-supplemented cowpea powder suitable for use in the preparation of local W. African foods such as moin-moin. In certain countries, eg India, the cowpea is grown as a dual purpose crop, the green pods are harvested for use as a vegetable and the residual plant material, which contains 11--12 per cent protein, on a dry matter basis, is used for feeding livestock.

Secondary and waste products
In many parts of the tropics the young shoots are frequently boiled and eaten as spinach. In some areas the mature leaves are boiled for approximately 15 minutes, drained, and then dried in the sun, before being stored for use as a relish, when fresh vegetables are scarce.

Cowpeas are sometimes grown for green manure, as a cover or anti-erosion crop, or for the production of hay. When grown for hay yields of about 2 T/ac (5t/ha) may be obtained. Cowpea hay has the following approximate composition: moisture 9.6 per cent; crude protein 18.6 per cent; crude fibre 23.3 per cent; fat 2.6 per cent; N-free extract 34.6 per cent; ash 11.3 per cent. Culled, broken and surplus seeds are used for livestock feeding. The seeds are also reported to be roasted occasionally and used as a coffee substitute. In the USA, immature green cowpeas are sometimes roasted and eaten as a snack food similarly to groundnuts. The seeds have a relatively high starch content and it has been suggested that this could be extracted as a by-product in the preparation of protein concentrates, and used by the food manufacturing, textile and paper industries.

Special features
The cowpea is highly palatable, very nutritious, and relatively free of metabolites and other toxic principles. The composition of the seed, particularly the protein, starch and vitamin B content, varies considerably according to the cultivar and the origin of the seed. An approximate analysis is: moisture 11.0 per cent; protein 23.4 per cent; fat 1.3 per cent; carbohydrate 56.8 per cent; fibre 3.9 per cent; ash 3.6 per cent; calcium 76 mg/100 g; iron 5.7 mg/100 g; vitamin A 40 iu/100 g; thiamine 0.92 mg/100 g; riboflavin 0.18 mg/100 g; niacin 1.9 mg/100 g; folic acid 0.15–0.16 mg/100 g; ascorbic acid 2 mg/100 g. Variations of from 18 to 29 per cent in the crude protein content have been reported and recent research indicates that in some cultivars the protein content may reach 35 per cent.

The proteins consist of 90 per cent water-insoluble globulins and 10 per cent water-soluble albumins. The amino acid content (mg/g N) has been reported as: isoleucine 239; leucine 440; lysine 427; methionine 73; cystine 68; phenylalanine 323; tyrosine 163; threonine 225; tryptophan 68; valine 283; arginine 400; histidine 204; alanine 257; aspartic acid 689; glutamic acid 1 027; glycine 234; proline 244; serine 268. There is, however, considerable variation in the amino acid content according to the cultivar and origin of the seed, but methionine, cystine and tryptophan are deficient. Infestation with *Callosobruchus chinensis* has been found to reduce the lysine and threonine content significantly.

The total sugars present in cowpeas can range from 13.7 to 19.7 per cent and include sucrose 1.5 per cent; raffinose 0.4 per cent; stachyose 2.0 per cent and verbascose 3.1 per cent. Starch contents ranging from 50.6 to 67.0 per cent have been reported; amylose 20.9 to 48.7 per cent, amylopectin 11.4 to 36.6 per cent. The starch granules present in the seed have an average diameter of 30 to 32 microns, with a high viscosity over a wide range of temperature.

The fatty acid composition of the oil from Pakistani seeds has been given as: linolenic 12.3 per cent; linoleic 27.4 per cent; oleic 12.2 per cent; lignoceric 1.1 per cent; behenic 4.0 per cent; arachidic 0.9 per cent; stearic 7.1 per cent; palmitic 33.4 per cent. The seeds are also reported to contain 0.025 per cent of a sterol, identified as stigmasterol, $C_{29}H_{48}O$. Phytin and pectin contents vary and appear to affect the culinary quality; easier cooking and softening appears to be associated with high phytin and pectin content (0.017 and 0.79 mg/100 g seed, respectively). Cowpeas with low phytin and pectin contents (0.006–0.009 mg and 0.35–0.41 mg/100 g seed, respectively), have been found to have very inferior culinary qualities. In addition, the raw seeds have a trypsin inhibitor activity of 20.9 per cent, which is destroyed by heating. The seed-coat which constitutes approximately 11 per cent ($\pm$2 per cent) of the seed, contains anthocyanins which cause cowpeas to discolour when they are processed, a factor which has handicapped their acceptance as a foodstuff.

Processing

Considerable quantities of cowpeas (Southern, Black-eye, Crowder peas) are processed in the USA. The dried cowpeas are usually soaked in cold, soft water for 10–12 hours, after which they are inspected and graded, all shrivelled, blemished, discoloured or broken peas, and any extraneous matter, being removed at this stage. The soaked cowpeas are then blanched, usually for 2 to 6 minutes at a temperature of 190–200°F (88–93°C), although it has been suggested that 10 minutes at 212°F (100°C) results in a better coloured product. Immediately after blanching the cowpeas are washed with clean cold water and packed into lacquered cans. Control of the fill weight is a critical factor for the quality and overall consumer acceptability of the final product. Overfilling results in severe clumping and starch gelation, underfilling can produce a discoloured product. Boiling 2 per cent brine is added to a constant headspace, the cans are closed immediately, while the centre temperature is about 180°F (82°C). The addition of sodium EDTA (ethylene diamine tetra-acetate) 145 ppm, or calcium EDTA (300 ppm), to the brine helps to prevent the discolouration of the final product. Small cans (401 x 411 & smaller) are processed for 45 minutes at 240°F (116°C), 30 minutes at 245°F (118°C) and 20 minutes at 250°F (121°C); No 10 size cans (603 x 700) for 70 minutes at 240°F (116°C), 55 minutes at 245°F (118°C) and 40 minutes at 250°F (121°C). Immediately after processing the cans are water-cooled until the average temperature of the contents reaches 95–100°F (35–38°C). Green cowpeas may also be processed in a similar manner with the omission of the soaking process; for a palatable product the alcohol insoluble solids (AIS) content of the seeds should be less than 26 per cent and to facilitate shelling storage at 70–100°F (21–38°C) in bins for 24 hours is frequently practised.

Cowpea

Production and trade

Production—calculated on available statistical data, cowpeas constitute about 2 per cent of the total world output of grain legumes. World production for the period 1970–74 averaged 1 100 000 t/a, compared with 898 000 t/a for the period 1965–69, an increase of almost 23 per cent. Production in 1975 was estimated to be about 1 097 000 t. Since many countries do not classify cowpeas as a separate item but include them under a general item 'dry beans', and a high proportion of the crop is produced by smallholders in kitchen gardens, or in village compounds, it is considered that reported production figures are under-estimated by at least 10 to 15 per cent.

Cowpeas: Major producing countries

Quantity tonnes

	Annual average 1965–69	Annual average 1970–74	1975
Nigeria	653 000	871 000	850 000
Upper Volta	63 000	58 000	75 000
USA	22 000	20 000	20 000
Uganda	47 000	57 000	9 000

Trade—cowpeas are produced almost entirely for local consumption and are of negligible importance in international trade. There is, however, some inter-regional trade in Africa of production in excess of the requirements of individual localities.

Prices—in Africa prices fluctuate considerably from year to year, and from month to month, according to the availability of local supplies. For example, the average annual price in Kano market, Nigeria, varied from 4.8 to 11.1 d/kg between 1957 and 1965, but during the same period the mean price per month ranged from 6.5 to 9.8 d/kg and was highest from June to October (8.7 to 9.8 d/kg). A high proportion of the cowpeas marketed in W. Africa are a mixture of seed colours and sizes and are heavily infested with bruchid beetles. There is, however, a marked preference for white cowpeas in N. Nigeria and these often fetch a premium of 10 to 20 per cent compared with the multicoloured lots. In S. Nigeria, however, the preference is for red coloured cowpeas and these can command a premium of 10 to 15 per cent over the white seeds in Kano markets.

Major influences

The cowpea is of considerable importance, particularly in W. Africa, as a nutritious legume crop, low in antinutritional factors. It is a crop with a wide range of ecological adaption and could be more widely grown. In fact it probably has the greatest potential of all food legumes in the semi-arid to subhumid tropical areas.

Although average yields are currently very low, approximately 180 lb/ac (200 kg/ha), trials at IITA with improved day-neutral cultivars, which have a high degree of resistance to many of the major pests and diseases affecting the cowpea, indicate that this food legume has a high potential level of productivity. However, any significant increase in production at the farmer level could probably only be achieved by the widespread cultivation of the cowpea as a sole crop instead of as an intercrop, as at present, the use of improved, pest and disease resistant cultivars, and the implementation of integrated programmes for the control of insect pests, both in the field and in storage.

There is considerable variation in the protein content of various cultivars and it should be possible to develop cultivars with a high protein content and to improve the methionine content by breeding.

Bibliography

ABEYGUNAWARDENA, D. V. W. AND PERERA, S. M. D. 1964. Virus disease affecting cowpea in Ceylon. *Trop. Agric.* (*Ceylon*), *120*, 181–204.

ABO EL-SAOUD, E. I., OSMAN, H. O., ZOUEIL, M. E. AND HAMDY, S. A. 1974. Physical, chemical and technological studies on the adaptability of green cowpeas (*Vigna sinensis*) varieties to freezing and canning. *Proc. 4th Veg. Conf.*, 1973. pp. 293–306. Egy., Alexandria Univ., 306 pp.

ADAMS, J. M. 1977. A bibliography on post-harvest losses in cereals and pulses with particular reference to tropical and subtropical countries. *Rep. Trop. Prod. Inst.*, G110, iv + 23 pp.

AKINGBOHUNGBE, A. E. 1976. A note on the relative susceptibility of unshelled cowpeas to the cowpea weevil (*Callosobruchus maculatus* Fabricius) (Coleoptera Bruchidae). *Trop. Grain Legume Bull.*, (5), pp. 11–13.

AKINPELU, M. A. 1974. Consumer preference in grain quality of the cowpea. *Samuru Agric. Newsl.*, *16*, 7–8.

Cowpea

AMMERMAN, G. R. AND SEALE, A. D. 1970. Canned southern pea quality as affected by fill weight and time and temperature of blanch. *Food Technol.*, *24*, 478–481.

AMOSU, J. O. 1974. The reaction of cowpea (*Vigna unguiculata* (L.) Walp.) to the root-knot nematode (*Meloidogyne incognita*) in western Nigeria. *Niger. Agric. J.*, *11*, 165–169.

ANON. 1966. Intérêt de la culture du niébé en Afrique tropicale et modalités de culture. [Interest in the cultivation of cowpea in tropical Africa and methods of cultivation]. *Cameroun Agric. Past. For.*, (98), pp. 56–60.

ANON. 1976. Control of cowpea weevil in storage. *Trop. Grain Legume Bull.*, (6), p. 36.

APPERT, J. 1964. Faune parasitaire du niébé (*Vigna unguiculata* (L.) Walp. = *Vigna catjang* (Burm.) Walp.) en République du Sénégal. [Parasitic fauna of cowpea (*Vigna unguiculata* (L.) Walp. = *Vigna catjang* (Burm.) Walp.) in the Republic of Senegal]. *Agron. Trop.*, *19*, 788–799.

ARAULLO, E. V. 1974. Processing and utilization of cowpea, chickpea, pigeon pea and mung bean. *Proc. symp. interaction of agriculture with food science, Singapore*, 1974. (MacIntyre, R. ed.). pp. 131–142. Ottawa, Int. Dev. Res. Cent., IDRC-033e, 166 pp.

ARNON, I. 1972. Pulses or grain legumes. *Crop production in dry regions.* Vol. 2. pp. 217–260. London: Leonard Hill Books, 683 pp.

ARORA, S. K. AND DAS, B. 1976. Cowpea as potential crop for starch. *Die Stärke*, 1976, *28*, 158–160.

ARYEETEY, A. N. 1971. Increasing cowpea production in Ghana. *Ghana Farmer*, *15*, 51–55; 83.

BHAID, M. U. AND TALAPATRA, S. K. 1965. Cowpea as a dual purpose crop. *Indian J. Dairy Sci.*, *18*, 153–155.

BINDRA, O. S. AND SAGAR, P. 1976. Comparison of Vita 1: A pest-resistant cowpea with local cultivars under different sowing dates and distances with minimal pesticide application. *Trop. Grain Legume Bull.*, (6), pp. 8–9.

BLISS, F. A. 1975. Cowpeas in Nigeria. *Nutritional improvement of food legumes by breeding.* (Milner, M. ed.). pp. 151–158. New York/London: John Wiley and Sons, 399 pp.

BOOKER, R. H. 1967. Observations on three bruchids associated with cowpea in northern Nigeria. *J. Stored Prod. Res.*, *3*, 1–15.

BOTT, W. 1970. Cowpea seed growing on Darling downs. *Qd. Agric. J.*, *96*, 722–726.

BOULTER, D., EVANS, I. M., THOMPSON, A. AND YARWOOD, A. 1975. The amino acid composition of *Vigna unguiculata* (cowpea) meal in relation to nutrition. *Nutritional improvement of food legumes by breeding.* (Milner, M. ed.). pp. 205–215. New York/London: John Wiley and Sons, 399 pp.

BRANTLEY, B. B. 1976. Coronet a new southern pea variety. *Georgia, Agric. Exp. Stn., Dep. Hortic. Res. Rep.* 220, 3 pp.

BRENNAN, J. G., JOWITT, R. AND OSSAI, G. E. A. 1974. A study of the application of modern industrial techniques to a traditional process based on the cowpea (*Vigna unguiculata*). *Madrid, 4th Int. Congr. Food Sci. & Technol., Work Doc. 8b, Exploiting local food raw materials, Pap.* 19, pp. 49–51.

BRESSANI, R., ELIAS, L. G. AND NAVARRETE, D. A. 1961. Nutritive value of Central American beans: (iv) The essential amino acid content of samples of black beans, red beans, rice beans and cowpeas of Guatemala. *J. Food Sci.*, *26*, 525–528.

CASWELL, G. H. 1961. The infestation of cowpeas in the western region of Nigeria. *Trop. Sci.*, *3*, 154–158.

CASWELL, G. H. 1968. The storage of cowpea in the Northern States of Nigeria. *Proc. Agric. Soc. Nigeria.*, *5*, 4–6.

CASWELL, G. H. 1973. The storage of cowpea. *Samaru Agric. Newsl.*, *15*, 73–75.

CASWELL, G. H. 1974. The development and extension of nonchemical control techniques for stored cowpeas in Nigeria. *Proc. 1st Int. Working Conf. Stored Prod. Entomol.*, pp. 63–67. Savannah, Georgia, USA, 706 pp.

CHADA, Y. R. (ed.). 1976. *V. unguiculata* (Linn.) Walp. *The wealth of India: Raw materials.* Vol. 10 (Sp–W). pp. 497–510. New Delhi: Counc. Sci. Ind. Res., Publ. Inf. Dir., 591 pp.

CHALFANT, R. B. 1976. Chemical control of insect pests of the southern pea in Georgia. *Georgia, Agric. Exp. Stn. Res. Bull.* 179, 31 pp.

CHESNEY, H. A. D. 1974. Performance of cowpeas cv. Black-eye in Guyana as affected by phosphorus and potassium. *Turrialba*, *24*, 193–199.

Cowpea

COMMONWEALTH BUREAU OF PASTURES AND FIELD CROPS. nd. Cowpeas (*Vigna unguiculata*). Annotated bibliography 1255, 1962–70. *Commonw. Agric. Bur., Bur. Pastures & Field Crops, Hurley, Maidenhead, Berks., England,* 15 pp.

CORREA, R. T. AND STEPHENS, T. S. 1956. Variety and strain evaluation of southern peas. *J. Rio Grande Valley Hortic. Soc., 10,* 90–95.

CULVER, W. H. AND CAIN, R. F. 1952. Nature, causes and correction of discolouration of canned black-eye and purple hull peas. *Texas Agric. Exp. Stn. Bull.* 748, 23 pp.

CUTHBERT, F. P. (JR.), FERY, R. L. AND CHAMBLISS, O. L. 1974. Breeding for resistance to the cowpea curculio in southern peas. *Hortscience, 9,* 69–70.

DART, P. J., HUXLEY, P. A., EAGLESHAM, A. R. J., MINCHIN, F. R., SUMMERFIELD, R. J. AND DAY, J. M. 1977. Nitrogen nutrition of cowpea (*Vigna unguiculata*): II Effects of short-term applications of inorganic nitrogen on growth and yield of nodulated and non-nodulated plants. *Exp. Agric., 13,* 241–252.

DINA, S. O. 1973. Insecticidal control of cowpea pests. *Proc. 1st Int. Inst. Trop. Agric., Grain legume improvement workshop.* pp. 282–294. Ibadan, Nigeria, Int. Inst. Trop. Agric., 325 pp.

DINA, S. O. 1977. Effects of monocrotophos on insect damage and yield of cowpea (*Vigna unguiculata*) in southern Nigeria. *Exp. Agric., 13,* 155–159.

DOKU, E. V. 1970. Effect of day-length and water on nodulation of cowpea (*Vigna unguiculata* (L.) Walp.) in Ghana. *Exp. Agric., 6,* 13–18.

DOVLO, F. E., WILLIAMS, C. E. AND ZOAKA, L. 1976. *Cowpeas: Home preparation and use in West Africa.* Ottawa, Int. Dev. Res. Cent., IDRC–055e, 96 pp.

EBONG, U. U. 1968. Cowpea production in Nigeria. *Niger. J. Sci., 2,* 67–72.

EBONG, U. U. 1970. A classification of cowpea varieties (*Vigna sinesis* Endl.) in Nigeria into subspecies and groups. *Niger. Agric. J., 7,* 5–18.

EBONG, U. U. 1971. Strategies for cowpea improvement in Nigeria. *Samaru Agric. Newsl., 13,* 25–27.

EL-BARADI, T. A. 1975. Cowpeas. *Abstr. Trop. Agric., 1* (12), 9–19.

ELIÁS, L. G., HERNÁNDEZ, M. AND BRESSANI, R. 1976. The nutritive value of precooked legume flours processed by different methods. *Nutr. Rep. Int., 14,* 385–403.

ENE, L. S. O. 1973. Observation on some characteristics that affect the acceptability and yield of cowpea (*Vigna unguiculata* L. Walp.) in Nigeria (East Central State). *Proc. 1st Int. Inst. Trop. Agric., Grain improvement workshop*. pp. 62–71. Ibadan, Nigeria, Int. Inst. Trop. Agric., 325 pp.

EZEDINMA, F. O. C. 1964. Effects of inoculation with local isolates of cowpea *Rhizobium* and application of nitrate-nitrogen on the development of cowpeas. *Trop. Agric., (Trinidad), 41*, 243–249.

EZEDINMA, F. O. C. 1966. Some observations on the effect of time of planting on the cowpea (*Vigna unguiculata* (L.) Walp.) in southern Nigeria. *Trop. Agric., (Trinidad), 43*, 83–87.

EZEDINMA, F. O. C. 1973. Seasonal variations in vegetative growth of cowpeas (*Vigna unguiculata* (L.) Walp.) in relation to insolation and ambient temperatures in southern Nigeria. *Proc. 1st Int. Inst. Trop. Agric., Grain legume improvement workshop*. pp. 138–154. Ibadan, Nigeria, Int. Inst. Trop. Agric., 325 pp.

EZEDINMA, F. O. C. 1973. Non-parametric evaluation of optimum planting date for cowpeas (*Vigna unguiculata* (L.) Walp.) in southern Nigeria. *Niger. Agric. J., 10*, 270–275.

EZEDINMA, F. O. C. 1974. Effects of close spacing on cowpeas (*Vigna unguiculata*) in southern Nigeria. *Exp. Agric., 10*, 289–298.

EZUEH, M. I. 1976. An evaluation of ULV sprays for the control of cowpea insects in southern Nigeria. *Trop. Grain Legume Bull.*, (4), pp. 15–18.

FARIS, D. G. 1967. A bibliography of cowpeas. *Nigeria, Inst. Agric. Res., Ahmadu Bello Univ., Samaru Misc. Pap.* 22, 105 pp.

FERY, R. L., DUKES, P. D. AND CUTHBERT, F. P. (JR.). 1976. The inheritance of cercospora leaf spot resistance in the southern pea (*Vigna unguiculata* (L.) Walp.). *J. Am. Soc. Hortic. Sci., 101*, 148–149.

FOOD AND AGRICULTURE ORGANIZATION OF THE UNITED NATIONS. 1970. Amino acid content of foods and biological data on proteins. *FAO Nutr. Stud.* 24, pp. 52–53. Rome: FAO, 285 pp.

GATES, J. E., COOLER, J. C., KRAMER, A. AND YEATMAN, J. N. 1964. Development of objective methods for measuring the character factor of quality in canned southern peas *Vigna sinensis*. *Proc. Am. Soc. Hortic. Sci., 84*, 399–408.

GODFREY-SAM-AGGREY, W. 1973. Effects of fertilizer on harvest time and yield of cowpeas (*Vigna unguiculata*) in Sierra Leone. *Exp. Agric.*, *9*, 315–320.

GODFREY-SAM-AGGREY, W., FRANCIS, B. J. AND KAMARA, C. S. 1976. The protein evaluation of cowpea (*Vigna unguiculata*) and benniseed (*Sesamum indicum*) from Sierra Leone. *Trop. Sci.*, *18*, 147–154.

HABISH, H. A. 1972. Aflatoxin in haricot bean and other pulses. *Exp. Agric.*, *8*, 135–137.

HABISH, H. A. AND MAHDI, A. A. 1976. Effect of soil moisture on nodulation of cowpea and hyacinth bean. *J. Agric. Sci.*, (*Camb.*), *86*, 553–560.

HAIZEL, K. A. 1972. The effects of plant density on the growth, development and grain yield of two varieties of cowpea, *Vigna unguiculata* (L.) Walp. *Ghana J. Agric. Sci.*, *5*, 163–171.

HALSEY, L. H. 1961. Southern pea varieties, culture and harvesting as related to production for handling and processing. *Proc. Florida State Hortic. Soc.*, *74*, 233–237.

HANNAH, L. C., FERRERO, J. AND DESSAUER, D. W. 1976. High methionine lines of cowpea. *Trop. Grain Legume Bull.*, (4), p. 9.

HAYS, H. M. AND RAHETA, A. K. 1977. Economics of sole crop cowpea production in Nigeria at the farmers' level using improved practices. *Exp. Agric.*, *13*, 149–154.

INTERNATIONAL INSTITUTE OF TROPICAL AGRICULTURE. 1973. Grain legume improvement program. *Ibadan*, *Nigeria*, *Int. Inst. Trop. Agric.*, 78 pp.

INTERNATIONAL INSTITUTE OF TROPICAL AGRICULTURE. 1975. Grain legume improvement program. *Int. Inst. Trop. Agric.*, *Annu. Rep.* 1974, pp. 72–77; 88–123. *Ibadan*, *Nigeria*, 199 pp.

INTERNATIONAL INSTITUTE OF TROPICAL AGRICULTURE. 1976. Grain legume improvement program. *Int. Inst. Trop. Agric.*, *Annu. Rep.* 1975, pp. 75–125. *Ibadan*, *Nigeria*, 219 pp.

ISBELL, C. L. 1959. Southern table peas. *Ala. Polytech. Inst. Agric.*, *Exp. Stn. Bull.* 317, 38 pp.

IWAKI, M., ROECHAN, M. AND TANTERA, D. M. 1975. Virus diseases of legume plants in Indonesia. *Bogor. Contrib. Cent. Res. Inst. Agric.*, (13), pp. 1–14.

JOHNSON, D. T. 1970. The cowpea in the African areas of Rhodesia. *Rhod. Agric. J.*, *67*, 61–69.

JOHNSON, R. M. AND RAYMOND, W. D. 1964. The chemical composition of some tropical food plants: pigeon peas and cowpeas. *Trop. Sci.*, *6*, 68 73.

KRISHNASWAMY, N. 1970. Cowpea *Vigna unguiculata* (L.) Walp. *Pulse crops of India* (Kachroo, P. and Arif, A. eds.). pp. 201–232. New Delhi: Indian Counc. Agric. Res., 334 pp.

KUHN, C. W., BRANTLEY, B. B. AND SOWELL, G. (JR.). 1966. Southern pea viruses: identification, symptomatology and sources of resistance. *Univ. Georgia Col. Agric., Agric. Exp. Stn. Bull. N.S.* 157, 22 pp.

KUHN, G. D. 1961. A study of the microbiological activity and other deterioration in cold storage of fresh southern peas. *Proc. Florida State Hortic. Soc.*, *74*, 259–262.

LE DIVIDICH, J. 1974. Protein value of *Vigna sinensis* seed: comparison with soybean-oil meal. *J. Agric. Univ. P.R.*, *58*, 230–236.

LELEJI, O. I. 1973. Cowpea breeding and testing at the Institute for Agricultural Research, Ahmadu Bello University, Zaria. *Proc. 1st Int. Inst. Trop. Agric., Grain legume improvement workshop*. pp. 59–61. Ibadan, Nigeria, Int. Inst. Trop. Agric., 325 pp.

LELEJI, O. I. 1976. Inheritance of three agronomic characters in cowpea (*Vigna unguiculata* L.). *Nigeria, Inst. Agric. Res., Ahmadu Bello Univ., Samaru, Res. Bull.* 252, 12 pp.

LEPIGRE, M. 1965. Étude sur les possibilités d'amélioration de la conservation des haricots au Togo en milieu rural. [A study of the possibilites of improving the storage of beans in rural areas of Togo]. *Agron. Trop.*, *20*, 388–430.

LOPEZ, A. 1975. Dried field peas. *A complete course in canning*. 10th ed. pp. 592–595. Baltimore, Maryland: The Canning Trade Inc, 755 pp.

LORZ, A. P. 1961. Breeding southern peas for processing. *Proc. Florida State Hortic. Soc.*, *74*, 282–284.

MAHDI, A. A. AND HABISH, H. A. 1975. Effects of light and temperature on nodulation of cowpea and hyacinth bean. *J. Agric. Sci.*, (*Camb.*), *85*, 417–425.

MAMPICPIC, N. G. AND AYCARDO, H. B. 1976. Cowpea breeding for fresh market and processing. *NSDB Technol. J.*, *1* (4), 72–79.

Cowpea

MEHTA, P. N. AND NYIIRA, Z. M. 1973. An evaluation of five insecticides for use against pests of cowpeas (*Vigna unguiculata* (L.) Walp.) with special reference to green pod yield. *East Afr. Agric. For. J.*, *39*, 99–104.

MINCHIN, F. R., HUXLEY, P. A. AND SUMMERFIELD, R. J. 1976. Effect of root temperature on growth and seed yield in cowpea (*Vigna unguiculata*). *Exp. Agric.*, *12*, 279–288.

MITAL, S. P., DABAS, B. S. AND THOMAS, T. A. 1975. A case for grain cowpea (*Vigna unguiculata* (L.) Walp.). *Trop. Grain Legume Bull.*, (2), pp. 7–9.

MOHAMED ALI, A., BALAKRISHNAN, V. K., SANKARAN, S., RETHINAM, P. AND MORACHAN, Y. B. 1974. A note on chemical weed control in cowpea (var. Co–2) (*Vigna sinensis* Savi). *Madras Agric. J.*, *61*, 799–801.

MOLINA, M. R., ARGUETA, C. E. AND BRESSANI, R. 1976. Protein-starch extraction and nutritive value of the black-eye pea *Vigna sinensis* and its protein concentrates. *J. Food Sci.*, *41*, 928–932.

MOODY, K. 1973. Weed control in tropical grain legumes. *1st Int. Inst. Trop. Agric., Grain legume improvement workshop.* pp. 162–183. Ibadan, Nigeria, Int. Inst. Trop. Agric., 325 pp.

MURUGESAN, S. AND JANAKI, I. P. 1972. Studies on the relationship of the cowpea mosaic virus with its vector *Myzus persicae* Sulz. *Madras Agric. J.*, *59*, 280–286.

NANGJU, D. 1973. Progress in grain legume agronomic investigations at IITA. *Proc. 1st Int. Inst. Trop. Agric., Grain legume improvement workshop.* pp. 122–136. Ibadan, Nigeria, Int. Inst. Trop. Agric., 325 pp.

NANGJU, D. 1976. Effect of harvest frequency on yield quality and variability of indeterminate cowpea seed. *J. Agric., Sci.*, (*Camb.*), *87*, 225–235.

NANGJU, D. 1976. Effect of fertilizer management on seed sulphur content of cowpea (*Vigna unguiculata* L. Walp.). *Trop. Grain Legume Bull.*, (4), pp. 6–8.

O'DOWD, E. T. 1971. Hermatic storage of cowpea (*Vigna unguiculata* Walp.) in small granaries, silos and pits in northern Nigeria. *Nigeria, Inst. Agric. Res., Ahmadu Bello Univ., Samaru, Misc. Pap.* 31, 38 pp.

OGUNMODEDE, B. K. AND OYENUGA, V. A. 1969. Vitamin B content of cowpeas (*Vigna unguiculata* Walp.): (i) Thiamine, riboflavin and niacin. *J. Sci. Food Agric.*, *20*, 101–103.

Ogunmodede, B. K. and Oyenuga, V. A. 1970. Vitamin B content of cowpeas (*Vigna unguiculata* Walp.): (ii) Pyridoxine, pantothenic acid, biotin and folic acid. *J. Sci. Food Agric., 21*, 87–91.

Ojomo, A. 1973. Breeding and improvement of cowpeas in Western State of Nigeria. *Proc. 1st Int. Inst. Trop. Agric., Grain legume improvement workshop.* pp. 21–25. Ibadan, Nigeria, Int. Inst. Trop. Agric., 325 pp.

Ojomo, O. A. 1970. Pollination, fertilization and fruiting characteristics of cowpeas (*Vigna unguiculata* (L.) Walp.). *Ghana J. Sci., 10*, 33–37.

Ojomo, O. A. 1972. Effects of photoperiod on growth, flowering and yield of some local varieties of cowpeas (*Vigna unguiculata* (L.) Walp.) in western Nigeria. *Niger. J. Sci., 5*, 161–166.

Ojomo, O. A. 1974. The cowpea crop in Nigeria. *Proc. 1st FAO/SIDA Semin., Improvement and production of field food crops for plant scientists from Africa and the Near East.* pp. 119–123. Rome: FAO, 684 pp.

Ojomo, O. A. 1974. Yield potential of cowpeas *Vigna unguiculata* (L.) Walp.: Results of mass and bulk pedigree selection methods in western Nigeria. *Niger Agric. J., 11*, 150–156.

Ojomo, O. A. 1974. Breeding for improved nutritional quality in cowpeas (*Vigna unguiculata* (L.) Walp.). *Proc. 1st FAO/SIDA Semin., Improvement and production of field food crops for plant scientists from Africa and the Near East.* pp. 305–311. Rome: FAO, 684 pp.

Ojomo, O. A. and Chheda, H. R. 1972. Physico-chemical properties of cowpeas *Vigna unguiculata* L. (Walp.), influencing varietal differences in culinary values. *J. West Afr. Sci. Assoc., 17*, 3–10.

Oke, O. L. 1966. The fixation and transfer of nitrogen in Nigerian cowpeas. *West Afr. J. Biol. Appl. Chem., 9*, 17–19.

Onayemi, O. 1976. Dehydration, storage and nutritional properties of methionine supplemented cowpea powders. *New York State, Coll. Agric. Life Sci., Cornell Inst. Agric. Diss. Abstr. Ser.* 1, *3*, 28–29.

Onayemi, O., Pond, W. G. and Krook, L. 1976. Effects of processing on the nutritive value of cowpeas (*Vigna sinensis*) for the growing rat. *Nutr. Rep. Int., 13*, 299–305.

Owusu-Domfeh, K. 1972. Trypsin inhibitor activity of cowpeas (*Vigna unguiculata*) and bambara beans (*Voandzeia subterranea*). *Ghana J. Agric. Sci., 5*, 99–102.

PAWAR, N. B., SHIRSAT, A. M. AND GHULGULE, J. N. 1977. Effect of seed inoculation with *Rhizobium* on grain yield and other characters of cowpea (*Vigna unguiculata*). *Trop. Grain Legume Bull.*, (7), pp. 3–5.

POINTEL, J–G. 1967. Contribution à la conservation du niébé. [A contribution to the storage of cowpea]. *Agron. Trop.*, *22*, 925–932.

POWERS, J. J., PRATT, D. E. AND JOINER, J. B. 1961. Gelation of canned and pinto beans as influenced by processing conditions, starch and pectic content. *Food Technol.*, *15*, 41–47.

PURSEGLOVE, J. W. 1968. *Vigna unguiculata* (L.) Walp. *Tropical crops: Dicotyledons*. Vol. 1. pp. 321–328. London: Longmans, Green and Co Ltd, 332 pp.

RACHIE, K. O. 1973. Relative agronomic merits of various food legumes for the lowland tropics. Potentials of field beans and other food legumes in Latin America. (Wall, D. ed.). pp. 123–139. *Cali, Colombia, CIAT Ser. Semin.*, 2E, 388 pp.

RACHIE, K. O. (ed.). 1974. Cowpeas. *Guide for field crops in the tropics and the subtropics*. (Litzenberger, S. C. ed.). pp. 109–114. Washington, Off. Agric. Tech. Assist. Bur. Agency Int. Dev., 321 pp.

RACHIE, K. O. 1975. Improvement of food legumes in tropical Africa. *Nutritional improvement of food legumes by breeding*. (Milner, M. ed.). pp. 83–92. New York/London: John Wiley and Sons, 399 pp.

RACHIE, K. O. AND RAWAL, K. 1976. Integrated approaches to improving cowpeas *Vigna unguiculata* (L.) Walp. *Ibadan, Nigeria, Int. Inst. Trop. Agric., Tech. Bull.* 5, 36 pp.

RACHIE, K. O. AND ROBERTS, L. M. 1974. Grain legumes of the lowland tropics. *Adv. Agron.*, *26*, 1–132.

RACHIE, K. O. AND SILVESTRE, P. 1977. Grain legumes. *Food crops of the lowland tropics*. (Leakey, C. L. A. and Wills, J. B. eds.). pp. 41–74. Oxford: Oxford University Press, 345 pp.

RAHEJA, A. K. 1973. A report on the insect pest complex of grain legumes in northern Nigeria. *Proc. 1st Int. Inst. Trop. Agric., Grain legume improvement workshop*. pp. 295–301. Ibadan, Nigeria, Int. Inst. Trop. Agric., 325 pp.

RAHEJA, A. K. AND HAYS, H. M. 1975. Sole crop cowpea production by farmers using improved practices. *Trop. Grain Legume Bull.*, (*1*), p. 6.

RAHEJA, A. K. AND LELEJI, O. I. 1974. An aphid-borne virus disease of irrigated cowpea in northern Nigeria. *Plant Dis. Rep.*, *58*, 1080–1084.

RAJAN, P., DANIEL, V. A., PADMARANI, R. AND SWAMINATHAN, M. 1975. Effect of insect infestation on the protein efficiency ratio of the protein of maize and cowpea. *Indian J. Nutr. Diet.*, *12*, 354–357.

RANJHAN, S. K. 1970. Chemical composition and nutritive value of green cowpea pods for feeding cattle. *Indian J. Dairy Sci.*, *23*, 190–192.

RANJHAN, S. K. AND MAHESHWARI, M. L. 1969. Yield, chemical composition and outturn of nutrients of various varieties of cowpea fodder. *Indian J. Dairy Sci.*, *22*, 200–201.

RANJHAN, S. K., TALPATRA, S. K. AND KALA, A. C. 1967. Yield and nutritive value of dual purpose crop cowpea (*Vigna catjang*) cowpea hay as a growth production ration. *Indian J. Dairy Sci.*, *20*, 146–149.

RUBAIHAYO, P. R., RADLEY, R. W., KHAN, T. N., MUKIIBI, J., LEAKEY, C. L. A. AND ASHLEY, J. M. 1975. The Makerere program—The cowpea program. *Nutritional improvement of food legumes by breeding.* (Milner, M. ed.). pp. 117–130. New York/London: John Wiley and Sons, 399 pp.

SALLEE, W. R. AND SMITH, F. L. 1969. Commercial Blackeye bean production in California. *Univ. Calif.*, *Calif. Exp. Stn. Ext. Serv. Circ.* 549, 15 pp.

SCHILLER, J. M. AND DOGKEAW, P. 1976. Influence of planting date on rainfed mungbean and cowpea in northern Thailand. *Thai. J. Agric. Sci.*, *9*, 199–220.

SELLSCHOP, J. P. F. 1962. Cowpeas, *Vigna unguiculata* (L.) Walp. *Field Crop Abstr.*, *15*, 259–266.

SELLSCHOP, J. P. F. AND SALMON, S. C. 1928. The influence of chilling, above the freezing point, on certain plants. *J. Agric. Res.*, (*USA*), *37*, 315–338.

SÈNE, D. 1966. Inventaire des principales variétés de niébé (*Vigna unguiculata* Walpers) cultivées au Sénégal. [Survey of the principal cowpea varieties (*Vigna unguiculata* Walpers) grown in Senegal]. *Agron. Trop.*, *21*, 927–933.

SÈNE, D. AND N'DIAYE, S. M. 1974. L'amélioration du niébé (*Vigna unguiculata*) au CNRA de Bambey: de 1959 à 1973. Résultats obtenus entre 1970 et 1973. [Improvement of cowpea (*Vigna unguiculata*) at CNRA, Bambey, 1959 to 1973. Results obtained from 1970 to 1973.] *Agron. Trop.*, *29*, 772–802.

Cowpea

Shoyinka, S. A. 1973. Status of virus diseases of cowpeas in Nigeria. *Proc. 1st Int. Inst. Trop. Agric., Grain legume improvement workshop*. pp. 270–273. Ibadan, Nigeria, Int. Inst. Trop. Agric., 325 pp.

Siegel, A. and Fawcett, B. 1976. *Food legume processing and utilization (with special emphasis on application in developing countries)*. Ottawa, Int. Dev. Res. Cent., IDRC–TS1, 88 pp.

Simmons, E. A. and Riley, J. 1967. An assessment of various packaging materials for packaging cowpeas (*Vigna unguiculata*). *Lagos, Nigeria, Rep. Nigeria Stored Prod. Res. Inst. Tech. Rep.* 14, pp. 123–125.

Singh, S. R. 1976. Co-ordinated minimum insecticide trials: Yield performance of insect resistant cowpea cultivars from IITA compared with Nigerian cultivars. *Trop. Grain Legume Bull.*, (5), p. 4.

Singh, S. R., Williams, K. O., Rachie, K., Rawal, K., Nangju, D., Wien, H. C. and Luse, R. A. 1975. Vita–3 cowpea (GP–3). *Trop. Grain Legume Bull.*, (1), pp. 18–19.

Sistrunk, W. A. and Bailey, F. L. 1965. Relationship of processing procedure to discolouration of canned blackeye peas. *Food Technol.*, *19*, 871–873.

Sistrunk, W. A., Bailey, F. L. and Kattan, A. A. 1965. Influence of maturity on yield and quality of fresh and canned southern peas. *J. Am. Soc. Hortic. Sci.*, *86*, 491–497.

Smittle, D. A. and Kays, S. J. 1976. Quality deterioration of southern peas in commercial operations. *Hortscience*, *11*, 151–153.

Steele, W. M. 1976. Cowpeas. *Evolution of crop plants*. (Simmonds, N. W. ed.). pp. 183–185. London: Longmans Group Ltd, 339 pp.

Suard, C. and Degras, L. 1975. Études pour la conservation des semences du pois d'Angole (*Cajanus cajan*) et du niébé (*Vigna sinensis*). [Studies on the storage of seeds of pigeon pea (*Cajanus cajan*) and cowpea (*Vigna sinensis*)]. *Nouv. Agron. Ant. Guy.*, (1), pp. 92–97.

Summerfield, R. J., Dart, P. J., Huxley, P. A., Eaglesham, A. R. J., Minchin, F. R. and Day, J. M. 1977. Nitrogen nutrition of cowpea (*Vigna unguiculata*): (1) Effects of applied nitrogen and symbiotic nitrogen on growth and seed yield. *Exp. Agric.*, *13*, 129–142.

SUMMERFIELD, R. J., DART, P. J., MINCHIN, F. R. AND EAGLESHAM, A. R. J. 1975. Nitrogen nutrition of cowpea (*Vigna unguiculata*). *Trop. Grain Legume Bull.*, (1), pp. 3–5.

SUMMERFIELD, R. J., HUXLEY, P. A. AND STEELE, W. 1974. Cowpea (*Vigna unguiculata* (L.) Walp.). *Field Crop Abstr.*, *27*, 301–312.

SUMMERFIELD, R. J., WIEN, H. C. AND MINCHIN, F. R. 1976. Integrated field and glasshouse screening for environmental sensitivity in cowpea (*Vigna unguiculata*). *Exp. Agric.*, *12*, 241–248.

SUNDARAM, P., THANGAMUTHU, G. S. AND KANDASAMY, P. 1974. Performance of cowpea varieties under different levels of phosphorus and potassium manuring. *Madras Agric. J.*, *61*, 796–797.

TAMBOLIYA, T. S., PATEL, C. A. AND PATEL, A. S. 1975. Nutritional evaluation of cowpea varieties. *Indian J. Nutr. Diet.*, *12*, 366–371.

TARDIEU, M. AND SÈNE, D. 1966. Le haricot niébé (*Vigna unguiculata* Walpers) au Sénégal. [The cowpea (*Vigna unguiculata* Walpers) in Senegal]. *Agron. Trop.*, *21*, 918–926.

TAYLOR, T. A. 1973. Crop protection and legume production in West Africa. *J. Assoc. Adv. Agric. Sci. Afr.*, *1* (*Supl.*), 5–8.

TOLMASQUIM, E., CORREA, A. M. N. AND TOLMASQUIM, S. T. 1971. New starches: Properties of five varieties of cowpea starch. *Cereal Chem.*, *48*, 132–139.

VEERASWAMY, R., RATHNASWAMY, R., PALANISWAMY, G. A. AND RAJASEKARAN, V. P. A. 1972. Cowpea Co. 1.—a high yielding strain for Tamil Nadu. *Madras Agric. J.*, *59*, 252.

VERDCOURT, B. 1970. Studies in the *Leguminosae–Papilionoideae* for the 'Flora of tropical East Africa': IV. *Kew Bull.*, *24*, 507–569.

WESTPHAL, E. 1974. *Vigna unguiculata* (L.) Walp. Pulses in Ethiopia, their taxonomy and agricultural significance. pp. 213–232. *Wageningen, PUDOC, Cent. Agric. Publ. Doc.*, *Agric. Res. Rep.* 815, 278 pp.

WILLIAMS, C. E. 1974. A preliminary study of consumer preferences in the choice of cowpeas—Western and Kwara States Headquarters and areas of Nigeria. *Nigeria, Univ. Ibadan.*, *Dep. Agric. Econ. Ext.*, 103 pp.

WILLIAMS, R. J. 1975. Diseases of cowpea (*Vigna unguiculata* (L.) Walp.) in Nigeria. *PANS*, *21*, 253–267.

Cowpea

WOOLEY, J. N. 1976. Breeding cowpea for resistance to *Muruca testulalis:* Methods and preliminary results. *Trop. Grain Legume Bull.*, (4), pp. 13–14.

WORTHINGTON, J. W. AND BURNS, E. E. 1971. Post-harvest changes in southern peas. *J. Am. Soc. Hortic. Sci.*, *96*, 691–695.

YARNELL, S. H. 1965. Cytogenetics of the vegetable crops (iv) Legumes: Southern pea, *Vigna sinensis* Savi. *Bot. Rev.*, *31*, 300–311.

Common names — **GRASS PEA, Chickling pea, or vetch, Indian vetch, Lathyrus pea.**

Botanical name — *Lathyrus sativus* L.

Family — Leguminosae.

Other names — Almorta (Sp.); Alverjas (Venez.); Batura (Ind.); Charal (Pun.); Chural (Ind.); Cicerchia coltivata (It.); Dog-toothed pea; Fovetta (Cyp.); Frijol gallinazo, Garbanzo[1] (Venez.); Gesse blanche, G. chiche, G. commune, Gesette (Fr.); Gilban(eh) (Sud.); Indian pea[2]; Kansari, Karas, Karil, Kassar, Kesari, Khesari dhal, or meh, Khesra, Khessary pea, Kisari, Lakh, Lakhodi, Lakhori, Lang, Latri (Ind.); Lentille d'Espagne (Fr.); Matri (Pak.); Matur (Ind.); Mutter pea; Pè-kyin-baung, Pè-sa-li (Burm.); Pharetta (Cyp.); Pisello bretonne, P. cicerchia (It.); Pois carré[3] (Fr.); Saat platterbse (Ger.); Sabberi (Eth.); Santal, Teora, Tiuri (Ind.).

Botany

A much-branched, straggling, or climbing, herbaceous annual, with a well-developed tap-root system; the rootlets of which are covered with small, cylindrical, branched nodules, usually clustered together in dense groups. The stems are normally 15–30 ft (4.5–9 m) long, quadrangular with winged margins. The pinnate leaves consist of one or two pairs of linear-lanceolate leaflets, 2–3 x 0.4 in. (5–7.5 x 1 cm), and a simple or much branched tendril. The flowers are axillary, solitary, about 0.6 in. (1.5 cm) long, and may be bright-blue, reddish-purple, or white in colour. The pods are oblong, flat, about 1–1.6 in. (2.5–4 cm) in length, slightly curved, dorsally two-winged and containing 3–5 small seeds, which are frequently white, brownish-grey or yellow in colour, although spotted or mottled forms also exist. There are many different cultivars which differ considerably in the growth habit of the

[1] Frequently used for the chick pea, *Cicer arietinum*, in Latin America.
[2] Also used for the cowpea, *Vigna unguiculata*.
[3] Also used for the winged bean, *Psophocarpus tetragonolobus* and for marrowfat peas, *Pisum sativum*.

plant, the colour of the flowers and the size, colour, shape and composition of the seed. Indian workers have identified 56 types, the most common are the blue-flowered, *cyaneus*, with 45 types; followed by 10 pink or red-coloured types, *roseus*, and 1 white-flowered type, *albus*. Some varietal improvement work on the grass pea has been carried out in India and improved cultivars such as T2–12, LC–76, Rewa–1 and 2 have been available for general cultivation for a number of years.

Origin and distribution

The grass pea is indigenous to S. Europe and W. Asia, but is now grown principally in India and to a much lesser extent in the Middle East, S. Europe and parts of S. America.

Cultivation conditions

Temperature—the grass pea is grown as a cold season grain legume in India and thrives in areas where the average temperature ranges between 50° and 77°F (10–25°C).

Rainfall—it is very tolerant of drought conditions and is grown successfully in areas with an average annual rainfall of 15–25 in./a (380–650 mm/a). In India it is often grown under extreme drought conditions after the rains have failed in October. Despite its tolerance of drought the grass pea is not affected by excessive rain and can be grown on land subject to flooding.

Soil—in India the grass pea is usually grown in rice fields, but it can be grown on a wide range of soil types, including very poor soils and heavy clays.

Altitude—in India it is grown from sea level up to elevations of 4 000 ft (1 200 m).

Planting procedure

Material—seed is used and inoculation does not appear to be necessary. Germination is hypogeal.

Method—the grass pea is seldom grown as the sole crop; often it is sown broadcast in a standing crop of rice, sometimes it is grown mixed with barley, linseed or chick peas. For optimum yields it should be drilled in a fairly clean seed-bed and kept reasonably free from weeds, especially during the early stages of growth. Frequently, the crop receives virtually no attention after sowing.

Field-spacing—when sown in furrows the seeds are drilled about 1 in. (2.5 cm) apart.

Seed-rate—in India it can range from 12–50 lb/ac (13–56 kg/ha), according to local conditions. When sown broadcast as a pure crop the seed-rate normally averages 40–50 lb/ac (45 kg/ha–56 kg/ha), when grown as a mixed crop, the seed-rate is often about 30 lb/ac (34 kg/ha).

Pests and diseases

When grown experimentally in Australia the grass pea was found to be susceptible to attack from the red-legged mite, *Halotydeus destructor*. Aphids can sometimes be troublesome in India.

In India the crop is reported to be infected by the following diseases: the mildews, *Oidium erysiphoides* and *Peronospora lathyri-palustris*, the rusts, *Uromyces pisi* and *U. fabae*, and a wilt, *Fusarium orthoceras* var. *lathyri*. Wilt is reported to have caused serious crop losses in Gujarat, until a resistant strain T2–12 was released to the local farmers.

Growth period

The seed normally ripens in 150–180 days, with a few early-maturing types in 120 days.

Harvesting and handling

The pods are harvested as soon as they begin to turn yellow, as fully ripe they dehisce and there is considerable seed loss. The plants are either cut with a sickle, or uprooted, and then left to dry for approximately 7 days, before being threshed, winnowed, and stored similarly to other grain legumes.

Primary product

Seed—this normally ranges from 0.14–0.56 in. (3.5–15 mm) in length, and is wedge-shaped, although some cultivars have small, globular seed. On the broad edge of the seed is a small, oval, pit-like, hilum. The seed may be white, yellow, greenish-yellow, brown, grey or black, according to the cultivar and sometimes it is spotted or mottled. White seed is the most popular for human consumption. In India, the smaller-seeded types are known as Lakhori and 100 seeds weigh approximately 0.18–0.25 oz (5–7 g), the larger ones as lakh, and 100 seeds weigh about 0.25–0.56 oz (7–16 g).

Yield

In India seed yields are reported to average between 250 and 400 lb/ac (280–450 kg/ha), although when grown as a pure crop with efficient cultivation and a seed-rate of 40 lb/ac (45 kg/ha), yields of 900–1 000 lb/ac (1 000–1 120 kg/ha) of seed and 1 200–1 400 lb/ac (1 340–1 570 kg/ha) of hay are reported to be obtained.

Main use

In spite of the problem of lathyrism, discussed in the section *Special features*, the grass pea is used as a foodstuff, particularly in times of food shortages, and especially amongst the rural population of the Indian States of Madhya Pradesh, Bihar, Uttar Pradesh and W. Bengal. The seeds may be boiled and eaten as a pulse, split and used as dhal, or ground into a flour and made into unleavened bread, paste balls or curries.

Subsidiary uses

The seeds may be used for poultry or livestock feeding, they can be used as a source of concentrate in rations supplemented with extra calcium. In India the seed is frequently used to adulterate the more expensive grain legumes, such as chick peas or pigeon peas. The grass pea is often grown for fodder in India; the plants can be reaped and fed green, or the standing crop can be grazed. It cannot be used to produce silage, but can be cured into hay under mild climatic conditions. When fed alone fresh young plants are reported to be harmful to horses, but not to sheep, cattle or rabbits. An analysis of the green plant at flowering stage gave the following results (dry basis): protein 17.3 per cent; fat 4.5 per cent; fibre 36.6 per cent; ash 6.0 per cent; phosphorus 0.5 per cent; calcium 1.1 per cent. The approximate composition of grass pea hay has been reported as: moisture 14.6 per cent; crude protein 9.9 per cent; fat 1.9 per cent; fibre 36.5 per cent; N-free extract 31.0 per cent; ash 6.1 per cent. The crop is also grown as a green manure; at a seed-rate of 60 lb/ac (67 kg/ha), it adds approximately 55 lb/ac (62 kg/ha) of nitrogen to the soil.

Secondary and waste products

The seed is occasionally used in homoeopathic medicine and it has been suggested that it could be utilized as a source of protein for the manufacture of plywood adhesives. The young leaves are sometimes used as a pot herb and occasionally the immature pods are boiled and eaten as a vegetable.

Special features

The approximate composition of the seed, per 100 g of edible portion, has been given as: moisture 10 g; protein 25.0 g; fat 1.0 g; total carbohydrate 61.0 g; fibre 15.0 g; ash 3.0 g; calcium 110 mg; iron 5.6 mg; vitamin A 70 iu; thiamine 0.10 mg; riboflavin 0.40 mg. The seed contains 34.8 per cent starch, which consists of amylose 30.3 per cent and amylopectin 69.7 per cent. In addition it contains sucrose 1.5 per cent, pentosans 6.8 per cent, phytin 3.6 per cent, lignin 1.5 per cent, albumin 6.6 per cent, prolamine 1.5 per cent, globulin 13.3 per cent and glutelin 3.75 per cent. The essential amino acids present (g/16 gN) are: arginine 7.85; histidine 2.51; leucine 6.57; isoleucine 6.59; lysine 6.94; methionine 0.38; phenylalanine 4.14; threonine 2.34; tryptophan 0.40; valine 4.68. The grass pea is very deficient in methionine and tryptophan, and supplementation with methionine would greatly enhance its nutritive value.

The seed also contains a rhamnoside of a flavone derivative, and possibly also lycopene. The presence of a compound yielding hydrogen sulphide has also been reported.

Extraction of the seed with ether yields a viscous oil with a strong stinking odour, characteristic taste, and the following characteristics: SG $^{20°C}$ 0.9285; $N_D^{20°C}$ 1.4768; acid val. 64.99; sap val. 172; RM val. 1.99; unsaponifiable matter 16.2 per cent. Stigmasterol and a yellow pigment are present and the oil is reported to have cathartic properties, but to be toxic.

Lathyrism—it has been known for centuries that consumption of the grass pea for any length of time can cause a serious neurological disease, known as lathyrism, which causes paralysis of the lower limbs in both man and animals. The Indian Government has in fact tried to prohibit the sale of grass peas and their products, but has not been successful because of the suitablility of the crop for cultivation under adverse soil and climatic conditions, particularly in areas liable to serious drought. Moreover, the grass pea is available at comparatively cheaper prices than most other edible grain legumes in India, and is widely used as an adulterant. In view of the continuing production and use of the grass pea, the toxic factors present have been subject to considerable investigation during the last decade or so. A neurotoxic amino acid antagonist, characterized as β–N–oxalyl–L–α–β–diamino-propionic acid (ODAP) has been isolated from the seeds; in addition, a water-soluble aliphatic amino acid glycoside with a nitrile group has also been isolated and designated as N–β–D–glucopyranosyl–N–α–L–arabinosyl–α–β–diamino propionitrile. This compound is thought to act synergistically with ODAP. Two approaches have been made towards reducing the toxin content of the grass pea. The first is that of processing. Steeping the seeds or dhal in water followed by sun-drying, or parboiling similarly to rice, has been found to result in a 90 per cent reduction of the neurotoxin content, but there is also a considerable loss of water soluble nutrients. The second is that of the genetic manipulation of the grass pea. A number of types have been identified with low toxin contents (less than 0.1 per cent, compared with 0.5 per cent of many commonly grown types). Work is now in progress on the multiplication of low-toxin types and provided that they produce satisfactory seed yields their general release to growers could help to solve the problem of lathyrism.

Processing

The grass pea may be processed similarly to other pulse crops to produce dhal. The necessity of steeping the seeds in water to reduce their neurotoxin content has already been discussed.

Production and trade

The SE. Asian sub-continent is the major producing area for grass peas. Indian production is estimated to have amounted to some 840 000 t/a in recent years, and has shown a marked increase since the mid-1960s. Production in Bangladesh averaged 60 600 t/a for the period 1972–75. Figures for Pakistan are not available. The crop is consumed locally and does not enter international trade.

Major influences

Despite the problem of lathyrism, the grass pea continues to be widely grown and utilized amongst the rural poor in countries such as India. Because of its ability to yield a seed crop under severe drought conditions it is of considerable importance in areas, such as Bihar, which are subject to famine. It has been shown that consumption of a diet containing 30–50 per cent of grass peas for 3–6 months can lead to lathyrism, the onset of which is sudden, acute, and often irreversible. For this reason there is need to continue and intensify research on effective methods of detoxifying the seeds at the rural level, without reducing their nutritive value, in addition to developing improved low-toxin containing, high-yielding strains.

Bibliography

ANON. 1967. Simple measures for removing the toxic factors of *Lathyrus sativus*. *Nutr. Rev.*, *25*, 231–233.

ANON. 1968. Renewed interest in lathyrism. *Inf. Bull. Br. Biol. Res. Assoc.*, *7*, 121–124.

ANON. 1970. The unsolved mysteries of lathyrism. *Inf. Bull. Br. Biol. Res. Assoc.*, *9*, 121–124.

BAILEY, E. T. 1952. Agronomic studies of vetches and other large-seed legumes in southern Western Australia. *Aust. Counc. Sci. Ind. Res. Organ., Div. Plant Ind., Tech. Pap.* 1, 21 pp.

CHOPRA, K. AND SWAMY, G. 1975. *Pulses: An analysis of demand and supply in India. Institute for social and economic change; Monograph 2.* New Delhi: Sterling Publishers, PVT, Ltd, 132 pp.

GANAPATHY, S. N., KRISHNAMURTHY, K., SWAMINATHAN, M. AND SUBRAHMANYAN, V. 1958. Studies on the non-protein nitrogenous constituents of the seeds of *Lathyrus sativus*. *Food Sci.*, *7*, 361–362.

GOPALAN, C. 1967. A review of recent studies on toxic factors in *Lathyrus sativus* and the possible modes of their removal. *FAO/WHO/UNICEF, Protein Advis. Group, PAG Bull.* 7, pp. 66–70.

HARTMAN, C. P., DIVAKAR, N. G. AND NAGARAJA RAO, U. N. 1974. A study on *Lathyrus sativus*. *Indian J. Nutr. Diet.*, *11*, 178–191.

HOWARD, G. L. C. AND ABDUR RAHMAN KHAN, K. S. 1928. The Indian types of *Lathyrus sativus* L. (Khesari, lakh, lang, teora). *Mem. Dep. Agric., India, (Bot. Ser.)*, *15*, 51–77.

INDIAN COUNCIL OF AGRICULTURAL RESEARCH. 1969. Chickling vetch (*Lathyrus sativus* L.). *Handbook of agriculture.* 3rd ed. pp. 187–188. New Delhi: Indian Counc. Agric. Res., 911 pp.

JESWANI, L. M. 1975. Varietal improvement of seed legumes in India. *Food protein sources.* (Pirie, N. W. and Swaminathan, M. D. eds.). pp. 9–18. London: Cambridge University Press, 260 pp.

JESWANI, L. M., LAL, B. M. AND PRAKASH, S. 1970. Studies on the development of low neurotoxin (β–N–oxalyl amino alanine) lines in *Lathyrus sativus* (Khesari). *Curr. Sci.*, *39*, 518.

JOHRI, P. N., PRASAD, T. AND KHAN, N. A. 1963. Chemical composition, digestibility and nutritive value of khesari (*Lathyrus sativus*) grains. *Indian J. Dairy Sci.*, *16*, 116–120.

LIENER, I. 1975. Antitryptic and other antinutritional factors in legumes. *Nutritional improvement of food legumes by breeding.* (Milner, M. ed.). pp. 239–258. New York/London: John Wiley and Sons, 399 pp.

MALIK, M. Y., AKHTAR ALI SHEIKH AND SHAH, W. H. 1967. Studies on the use of matri in poultry ration. *Pak. J. Sci.*, *19*, 165–170.

MANN, H. H. 1947. Pulse grain crops in the Middle East. *Emp. J. Exp. Agric.*, *15*, 249–259.

MEHTA, T., HSU, A. AND HASKELL, B. E. 1972. Specificity of the neurotoxin from *Lathyrus sativus* as an amino acid antagonist. *Biochemistry*, *11*, 4053–4063.

MURTI, V. V. S., SESHADRI, T. R. AND VENKITASUBRAMANIAN, T. A. 1964. Neurotoxic compounds of the seeds of *Lathyrus sativus. Phytochemistry*, *3*, 73–78.

NAGARAJAN, V. 1972. Prevention of development of toxins in food: Some approaches for (a) prevention of aflatoxin contamination and (b) reducing the neurotoxin in *Lathyrus sativus. Proc. semin., post-harvest technology of cereals and pulses.* pp. 323–326. New Delhi, Indian Counc. Agric. Res., Counc. Sci. Ind. Res., 354 pp.

NEZAMUDDIN, S. 1970. Khesari: *Lathyrus sativus* Linn. *Pulse crops of India.* (Kachroo, P. and Arif, M. eds.). pp. 314–316. New Delhi: Indian Counc. Agric. Res., 334 pp.

PURSEGLOVE, J. W. 1968. *Lathyrus* L. *Tropical crops: Dicotyledons.* Vol. 1. pp. 276–279. London: Longmans, Green and Co Ltd, 332 pp.

RAMANUJAM, S. 1973. Grain legume improvement in India. *Proc. 1st Int. Inst. Trop. Agric., Grain legume improvement workshop.* pp. 37–41. Ibadan, Nigeria, Int. Inst. Trop. Agric., 325 pp.

RAO, S. L. N., MALATHI, K. AND SARMA, P. S. 1969. Lathyrism. *World Rev. Nutr. Diet.*, *10*, 214–238.

SASTRI, B. N. (ed.). 1962. *Lathyrus sativus* Linn. Chickling vetch, grass pea. *The wealth of India: Raw materials.* Vol. 6 (L–M). pp. 37–41. New Delhi: Indian Counc. Sci. Ind. Res., 483 pp.

SUBRAHMANYAN, V., NARAYANA RAO, M. AND SWAMINATHAN, M. 1957. Lathyrism. *Food Sci.*, *6*, 156–159.

SWAMINATHAN, M. S. AND JAIN, H. K. 1975. Food legumes in Indian agriculture. *Nutritional improvement of food legumes by breeding*. (Milner, M. ed.). pp. 69–82. New York/London: John Wiley and Sons, 399 pp.

WESTPHAL, E. 1974. *Lathyrus sativus* L. Pulses in Ethiopia, their taxonomy and agricultural significance. pp. 104–108. *Wageningen, PUDOC, Cent. Agric. Publ. Doc., Agric. Res. Rep.* 815, 278 pp.

Common names **HARICOT BEAN, Common bean, Field bean[1], French bean[2], Kidney bean, Pole bean[3], Runner bean[4], Snap bean[2], String bean[4].**

Botanical name *Phaseolus vulgaris* L.

Family Leguminosae.

Other names Adagora, Adanguare, Adigura-tsada, Ashanguare (Eth.); Bab (Hun.); Bohne (Ger.); Bonchi (kai) (Sri La.); Boontje (S. Afr.); Boontjis (Indon.); Bo-sa-pè (Burm.); Brazilian bean; Bunchu-kai (Tam.); Bush bean (Rhod.); Butingi (Philipp.); Caraota[5] (S. Am.); Chilemba; Chimbamba (Zam.); Chumbinho opaco (Braz.); Cranberry bean (USA); Edihimba (Ug.); Fagiola, Fagiulo commune (It.); Fajola (Eth.); Fasiolos (Gr.); Fasolia (Cyp., Egy., Eth.); Fasulia (Sud.); Fasûlya (Turk.); Fasûlyah nashef[2] (Ar.); Feijão (Port.); F. ervilha (Ang.); Feijoeiro (Port.); Frash bean (Ind.); Frijol (Lat. Am.); Habichuela[5] (S. Am., Philipp.); Haricot à couper, H. commún (Fr.); H. nain (Zar.); H. pain (Fr.); H. princesse (Zar.); H. vert[2] (Fr.); Icaraota[6] (Venez.); Ingen (mame) (Jpn.); Judia (commun) (Sp.); Kachang bunchis, K. pendek, Katjang mèrah[7] (Malays.); Kunde[7] (Zan.); Kuru fasulya (Turk.)[2]; Maharage[6] (E. Afr.); Manawa (Zam.); Michigan pea

---

[1] Also used for the broad bean, *Vicia faba*, and in India for the hyacinth bean, *Lablab purpureus*.

[2] More correctly used for the types grown for the production of the green immature pods, utilized as a vegetable.

[3] Also used for the lima bean, *Phaseolus lunatus*, for the runner bean, *P. coccineus*, and in India for the hyacinth bean, *Lablab purpureus*.

[4] More commonly used, especially in the UK, for the runner bean, *Phaseolus coccineus*.

[5] Also used for the lima bean, *Phaseolus lunatus*.

[6] Also used for other dry beans, especially the lima bean, *Phaseolus lunatus*, the runner bean, *P. coccineus*, and mung bean *Vigna radiata*.

[7] Also used for the cowpea, *Vigna unguiculata*.

bean; Mula (Philipp.); Navy bean; Ojoo (Ug.); Paszuly (Hun.); Pea bean[1]; Pè-bya-galè, Pè-gya(ni) (Burm.); Pinto bean; Porotillo (Peru); Poroto (comun) (S. Am.); Princess bean; Purutu, Quechua (Peru); Rajmah (Ind.); Salad bean[2]; Salboco bulluc (Som.); Sheuit (hagina) (Is); Teiko (Ug.); Tua kack, Tua phum (Thai.); Vilaiti sem (Ind.); Yeshil fasulya (Turk.); Zada-adagonna (Eth.).

Botany

An annual herb showing considerable variation in habit, vegetative characters, flower colour, and the size, shape and colour of the pods and seeds. Climbing or pole types occur, and also dwarf, determinate bush types; in addition there are intermediate types which develop weak runners. All forms have a well-developed tap-root, which grows rapidly, sometimes reaching a depth of 3 ft (90 cm), or more, but with the lateral roots confined mainly to the top 6 in. (15 cm) of soil, and bearing spherical or irregular shaped nodules, approximately 0.24 in. (6 mm) in diameter. The stems are slender, twisted, angled and ribbed. Climbing types can reach 6.5–10 ft (2–3 m) in height and the dwarf or bush types 8–24 in. (20–60 cm). The leaves are alternate, trifoliolate, often somewhat hairy, with a long petiole, grooved above, and a marked pulvinus at the base. The leaflets are ovate, entire, acuminate, 3.2–5.9 x 2.0–4.0 in. (8–15 x 5–10 cm). The flowers are borne on axillary, few-flowered lax racemes, the pedicels are short, 0.2–0.3 in. (5–8 mm) in length, the corolla may be white, creamy-yellow, pink or violet. The haricot bean is usually self-fertilized, pollination taking place at the time the flower opens. The seed-pods are slender 3–8 x 0.4 x 0.6 in. (7.5–20 x 1.0 x 1.5 cm), often glabrous, straight, or slightly curved, edges rounded or convex, beak prominent. The colour can vary from yellow to dark-green, sometimes with pink or purple blotches. The number of seeds can vary from 1 to 12; they show considerable variation in their colour, shape and size.

P. vulgaris is polymorphic and several attempts have been made to divide it into subspecies based on features such as dwarf, versus tall habit size, shape and colour of the seed. However, the variation within the species can be

[1] Also used for the asparagus bean, *Vigna unguiculata* ssp. *sesquipedalis*.

[2] More correctly used for the types grown for the production of the green immature pods, utilized as a vegetable.

attributed to the existence of several hundreds of cultivars and this variation cannot be met satisfactorily by the creation of subspecies. Some plant breeders have approached the problem from an evolutionary viewpoint and have recognized five races of *P. vulgaris* based on an evolutionary sequence. However, for many purposes the following classification according to use is the most convenient:

(a) *Dry-shell or field beans*, grown extensively in Latin America, Canada, the USA and parts of Africa, for their dry, fully mature seeds. Many types exist showing great variation in the colour of the seed-coat, size and shape, growth period, in addition to resistance to pests and diseases, etc. The more important kinds of dry-shell, or field beans, are:

(i) *Pea or navy beans*, with seeds less than 0.3 in. (8mm) in length, not reniform. They are grown extensively for canning, particularly in the USA and Canada, where numerous cultivars such as Michelite, Monroe, Seaway and Gratiot have been developed. In E. Africa, the best pea bean produced for the canning industry is Mexico 142 (known in Uganda as No 212).

(ii) *Medium haricot beans*, with seeds, 0.4–0.5 in. (10–12 mm) in length and their thickness less than half the length. The pinkish-buff mottled pinto beans and the cranberry beans are typical of this group.

(iii) *Marrow beans*, with medium to largish seeds, 0.4–0.6 in. (10–15 mm) long and their thickness exceeding half the length. Yellow-eye, a large, white, round to oblong bean with a yellow oval spot on the hilum, and Steuben, an improved form selected from yellow-eye, are typical examples.

(iv) *Kidney beans*, with seeds 0.6 in. (15 mm) or more in length, more or less reniform in shape. The colour may be white, various shades of red or purple, sometimes with mottling. The Great Northern group grown in the USA and the Canellini beans of the Mediterranean are typical white kidney beans. Of the coloured types, Redkote is grown extensively in the USA, and the cultivars Diacol Nima, Algarrbo and Higuerillo in Colombia.

(b) *Green-shell haricot beans*, grown mainly in S. America, parts of Africa and Europe for the mature, fully developed beans, which are extracted from mature, but not dry pods. The white, green, red-seeded, 'Flageolet' types of France, such as Michelet and Coco are typical of this group.

(c) *Horticultural haricot beans*—grown mainly for their immature pods for use as a vegetable and generally referred to as; French, snap or string beans. The cultivars Yellow Waxpod, Cherokee Wax, Kentucky Wonder, Tropic Wonder and Tendercrop are typical of this group.

There are many hundreds of cultivars and new ones are constantly being introduced as plant breeders, particularly in the Americas and Europe, produce improved strains to meet the requirements of local growers, especially as regards resistance to endemic pests and diseases.

Origin and distribution

The haricot bean, *Phaseolus vulgaris*, is the best known and the most widely distributed of the *Phaseolus* spp. It is thought to have originated in the western Mexico-Guatemala area, but there is evidence that suggests multiple domestication within C. America, from a widespread and polymorphic ancestral species. It is now widely distributed in many parts of the tropics, subtropics and temperate regions and is the most important food legume throughout Latin America and parts of Africa.

Cultivation conditions

Temperature—for optimum growth average temperatures of between 60° and 75°F (16–24°C) are required, and the haricot bean is grown best in the tropics and subtropics under decreasing temperature conditions. Growth stops at temperatures below 50°F (10°C) and the plant is killed by frost. It can only be grown in areas where there is a frost-free period of 105–120 days. A temperature of 86°F (30°C) would appear to set the upper limit for successful cultivation since above this blossom-drop is very serious, and above 95°F (35°C) 100 per cent failure to set seed has been observed. Where dry, hot conditions prevail the use of plant regulators such as α–naphthyl acetamide, β–naphthoxyacetic acid and chloro-phenoxyacetic acid has been found to increase pod-set significantly. In some cultivars reduction of photosynthetic efficiency has been observed when night temperatures fall between 50° and 64°F (10–18°C).

Haricot bean

Rainfall—the haricot bean is grown as a rain-fed crop in areas with an annual average rainfall ranging from 20–60 in. (500–1 500 mm). An evenly distributed rainfall is required throughout the growing period and during flowering the relative humidity should be preferably above 50 per cent. Hail can cause serious defoliation, sometimes resulting in the complete loss of the crop. The water requirements of the haricot bean have been estimated at between 0.5 and 1.0 in./ac/wk (32–64 mm/ha/wk), depending upon the water-holding capacity of the soils, the run-off and the evaporation rate of the soil and the crop. Since it is a short-season crop it can be grown in summer-rainfall regions in the tropics. In E. Africa the production of dry beans is most successful in those areas where 12–15 in. (300–380 mm) of rainfall occurs over a 10–week growing period, followed by 4 weeks of sunny, cool weather. In Colombia, it has been demonstrated that for high yields of dry beans, a total of 12 to 16 in. (300–400 mm) of rain are required during crop growth, of which 5 to 7 in. (125–175 mm) should be between planting and first flowering, 1 to 3 in. (25–75 mm) during flowering, and about 6 in. (150 mm) from flowering until the pods are well filled, thereafter the drier the better. Hot dry winds during flowering can cause severe blossom-drop, resulting in lower yields and non-uniform ripening of the seed.

A large proportion of the world's haricot bean crop is grown under irrigation. In arid regions, 16–18 in. (400–450 mm) of water are usually applied in seven to eight, or more, irrigations, depending upon the soil type, the climate and the length of growing season. In the USA on average 10–16 in. (250–400 mm) of water are usually applied to irrigated crops, but as little as 6 in. (150 mm) has produced successful crops. Furrow or overhead irrigation is preferable to flood.

The haricot bean is very sensitive to soil-water balance, a slight excess, or deficiency, having a marked effect upon yield. Moisture stress at the flowering period is critical and can reduce yields by as much as 20 per cent. It has been recommended that the available soil moisture should be maintained above 50 per cent during the flowering period.

Soil—the haricot bean can be grown successfully on most soil types, from light sands to heavy clays, but a friable, deep, well-drained soil is preferred. When grown as a rain-fed crop, where the occurrence of rains can be erratic, the soils should be 24–40 in. (60–100 cm) deep; so that the roots may draw on moisture reserves. In badly drained soils germination may be poor and heavy losses can occur. The pH should be preferably between 6.0 and 6.8, although critical problems seldom develop unless it falls below 5.2, or rises above 7.0. Below pH 5.2 manganese toxicity symptoms (stunting, chlorosis and puckering of the leaves) may become apparent, but above pH 6.8–7.0, manganese deficiency can occur, causing retardation of growth and also chlorosis of the leaves. Magnesium deficiency can occur when the crop is grown on sandy, acid soils, and molybdenum deficiency is reported to occur when it is grown on some acid, sandy soils in Australia. The haricot bean has a relatively high zinc requirement and deficiency symptoms (poor pod-set) can occur when it is grown on calcareous soils; foliar sprays can result in a 20 per cent increase in yield. It is also reported to be sensitive to high concentrations of aluminium, boron and sodium, but cultivars show considerable variation in their sensitivity to the last element; French or snap beans tend to be the most sensitive.

In view of the large number of cultivars, the widely differing standards of crop management, and the very varied results that have been reported from fertilizer trials, it is difficult to make any generalized statement regarding fertilizer application. In fact dry beans produced by smallholders in the tropics normally receive very little fertilizer treatment. Where available the application of farmyard manure 10 T/ac (25 t/ha) has been recommended for French beans. As a generalization, with soils high in nitrogen the application of 100–150 lb/ac (112–170 kg/ha) double superphosphate has been suggested, where the nitrogen status is average a 10: 20: 0 NPK mixture 200–300 lb/ac (225–340 kg/ha) and with soils with a marginal deficiency of potassium, 250–300 lb/ac (280–340 kg/ha) of a 8: 16: 8 mixture. However, very variable responses to nitrogen have been reported. Fertilizers should be placed in bands 3 in. (7.5 cm) to the side of the seed and slightly deeper, to prevent injury. With a seed yield of 1 600 lb/ac (1 790 kg/ha) and 1 400 lb/ac (1 570 kg/ha) of straw, the nutrient removal is estimated to be 147 lb/ac (165 kg/ha) nitrogen, 60 lb/ac (67 kg/ha) phosphoric acid and 122 lb/ac (137 kg/ha) potash.

Altitude—in the tropics the haricot bean is normally grown at elevations of between 2 000 and 6 500 ft (600–1 950 m). In Kenya optimum results are obtained at about 3 000–5 000 ft (900–1 500 m) and in Ethiopia at between 5 500 and 6 500 ft (1 650–1 950 m).

Day-length—long-day, short-day and day-neutral types exist; most French or snap beans are day-neutral and many intermediate climbing types are short-day.

Haricot bean

Planting procedure

Material—seed is used and germination is epigeal. For good germination most cultivars require a soil temperature of at least 60°F (16°C), preferably 64° to 68°F (18–20°C). In the UK the cold tolerant pea bean cultivar Purley King, requires a soil temperature, at a depth of 4 in. (10 cm), of 54° to 55°F (12–13°C). For optimum yields certified seed which has passed inspection for freedom from disease, a minimum germination rate of 75 per cent, and varietal purity, should be used, and treatment with fungicide/insecticide seed dressings (frequently thiram, captan or chloranil, combined with dieldrin/aldrin) is often recommended. Seed with 13 per cent moisture in hygroscopic equilibrium stored at ambient temperatures of about 77°F (25°C) may retain its viability and vigour for up to 13 months, and with 10 per cent moisture for up to 3 years. With prolonged storage, however, the germination rate declines markedly, also seedling vigour, while growth abnormalities increase.

Very variable results have been obtained with *Rhizobium* inoculation. The only circumstances when this might be worthwhile would seem to be on land where a bean crop has not been grown previously. Seed treatment with molybdenum combined with inoculation has given a significantly positive result experimentally.

Method—haricot bean production systems are many and varied, ranging from the primitive 'covered bean' system used by smallholders in certain areas of Latin America, to sophisticated, highly mechanized systems employed in countries such as the USA. The seed may be broadcast, drilled in close rows, or planted in wide rows to permit mechanical cultivation. Planting depths range from 1–2 in. (2.5–5 cm) in heavy soils to 2–4 in. (5–10 cm) in light soils. Hand planting is carried out with a dibber, or any sharp pointed tool. Small mechanical planters which can be operated by hand or with small tractors are sometimes used for smallholder production. For large-scale commercial production of dry beans large drill-planters, either manufactured specifically for beans or sugar beet, or those designed for planting maize or cotton, are widely used. These machines can plant both drilled and hill-dropped seeds with considerable precision. Some can also apply herbicides and fertilizer in one operation. It is essential for planters to operate at a relatively slow speed, 2–3 miles/h (3.2–4.8 km/h), at higher speeds seed injury occurs, resulting in low and uneven germination and weak and unproductive plants.

The haricot bean may be grown in monoculture, in a rotation with other crops, (by far the most common in Latin America is a maize-bean rotation), or interplanted with crops such as maize, sweet potatoes, bananas, cotton or coffee. Pole or climbing cultivars require staking and various types of trellis supports are used. In the tropics pole beans are normally grown as a market garden crop under irrigation, or in village compounds, where they are allowed to twine around fences and trees. Planting in furrows is usual if the crop is planted between existing crops or when it is to be irrigated. Hill planting is used for smallholder production, usually with associated or interplanting. Clean shallow cultivation is recommended to control weed growth, but care is required because of the likelihood of injuring the plants by destroying the roots near the surface. Herbicides may be used to control weed growth, but their use requires care as their effects are very much dependent upon the cultivar, the soil type and climatic conditions. The use of MCPB [4–(4–chloro–2–methylphenoxy) butyric acid] has been recommended for the control of annual weeds, but recent work has indicated that it is not entirely satisfactory. Pre-planting application of EPTC [S–ethyl NN–dipropyl–(thiocarbamate)] and trifluralin gives reasonable weed control, but is expensive. Ametryne and linuron have been used successfully in certain areas. Although 2, 4–D [2, 4–dichloro–phenoxyacetic acid] has been used successfully, its use can cause 7–10 days delay in the ripening of the seeds of certain cultivars.

Field-spacing—with pure stands the row width and distance between the spaces usually varies from 21 to 36 in. (52.5–90 cm) and 2 to 9 in. (5–22.5 cm) between the plants, depending upon the cultivar, the rainfall, the soil type, and the method of cultivation practised. In Colombia, the use of a double-row system, with 12 in. (30 cm) between the rows and 24 in. (60 cm) between pairs of rows, and 6 in. (15 cm) within the row, is reported to be satisfactory for mechanized cultivation.

Seed-rate—varies according to the cultivar and the spacing used, etc. In the USA the average seed-rate for the Michigan pea bean is 40 lb/ac (45 kg/ha), for Red Kidney 80–100 lb/ac (90–112 kg/ha) and for Yellow-eye 60 lb/ac (67 kg/ha), but some growers use a considerably lower rate. Bush snap beans are normally planted at rates of from 50 to 150 lb/ac (56–170 kg/ha) depending upon the cultivar, climatic and soil conditions, but the 70 lb/ac (78 kg/ha) is probably about average. The seed-rate for pole snap beans is usually between 20 and 30 lb/ac (22–34 kg/ha), since the rows are normally spaced further apart. In Queensland the seed-rate for bush beans is generally between 50 and 60 lb/ac (56–67 kg/ha) and 18–60 lb/ac (20–67 kg/ha) for pole beans. In E. Africa the seed-rate for the cultivar Canadian Wonder is normally 50–60 lb/ac (56–67 kg/ha).

Pests and diseases
The haricot bean is subject to attack from a large number of pests and diseases and this is considered to be the major factor limiting production in the tropics. The susceptibility of cultivars to attack from specific pests or diseases varies greatly and the development of resistant lines would seem to offer the best prospects of crop protection, particularly if this is combined with the use of healthy, clean seed, efficient weed control and crop rotation. Some of the more important pests and diseases are listed below.

Pests—several species of aphids attack the haricot bean. They are of considerable economic importance not only on account of the damage they cause to the crop, but because they are the vectors of several virus diseases. The black bean aphid, *Aphis fabae*, is the most important of the aphids infesting the haricot bean. It is widely distributed in Europe, parts of Africa, Asia and the Americas. Infestations prevent normal plant growth, sometimes causing yellowing and distortion of the leaves; in severe cases the plants can become desiccated and die. In the UK effective control is reported to be obtained by spraying with Schradan, nicotine sulphate or malathion. In E. Africa the use of Menazon, endosulphan or pyrethrum sprays has been suggested. In the USA seed dressing with Menazon is reported to give up to 4 weeks' control after planting and is recommended where early infestations are likely to occur. Later infestations are usually controlled by the use of Metasystox, Rogor, Menazon or Thiodan. The use of this last insecticide is suggested where bollworm infestation also occurs. Of the beetles, the Mexican bean beetle, *Epilachna varivestis*, is very troublesome in the USA, Canada, Mexico and Guatemala. In the USA it is the most destructive insect pest of beans from the Mississippi river valley eastward in every State, except Minnesota. Both the larvae and the adult beetles feed on the leaves of the haricot and the lima bean. Diazinon, malathion, methoxychlor, dimethoate, parathion or carbaryl sprays or dusts are reported to give effective control. The bean leaf beetle, *Cerotoma trifurcata*, is also widely distributed in the USA and attacks haricot and lima beans, peas, cowpeas and soyabeans. The larvae attack the roots and the adult beetles the leaves and stems. The use of carbaryl, DDT or rotenone is commonly recommended for effectively controlling this pest.

Bean flies, are widely distributed in Africa and Australia and can cause considerable crop losses in these areas. *Ophiomyia phaseoli* (=*Melanagromyza phaseoli*) is the most common species in most regions. The adult bean flies lay their eggs on the bean leaves or in the hypocotyl. The larvae bore into the leaves and move downwards to pupate in the stems at ground level. The bases of the stems become thickened and cracked, and the plants become stunted and yellow and may die eventually. Cultivars vary greatly in their susceptibility to attack, probably due to differences in their ability to produce adventitious roots. Effective control is reported to be obtained by seed dressing with aldrin or dieldrin, particularly if it is combined with early planting, crop rotation and the removal of crop residues and volunteer plants.

White flies, *Bemisia tabaci*, like aphids, are of considerable economic importance as they are the vectors of bean virus diseases, particularly golden yellow mosaic and mottle dwarf virus. They are found in large populations in Central, and to a lesser extent in South American countries, and are responsible for the transmission of these two virus diseases throughout the bean crop of Latin America.

Bean pod weevils, *Apion* spp., are of considerable economic importance in certain bean growing areas, attacking the seeds when they are developing in the pods. The most common species causing damage in Latin America is *Apion godmani*. In El Salvador losses due to this pest have been reported to be as high as 60 per cent in some areas.

The leaf hopper, *Circulifer tenellus*, is of importance in the USA. The potato leaf hopper, *Empoasca fabae*, is troublesome in parts of the USA, Canada, Costa Rica, Puerto Rico, Cuba, the Dominican Republic, Argentina, Brazil, Peru, Venezuela and Belize. The green leaf hopper, *Empoasca kraemeri*, is also widespread in tropical America. This pest is often responsible for heavy crop losses during the dry season in Brazil, and in certain seasons in El Salvador total crop losses, due to stunting injury, have been reported. Effective control is reported to be obtained by the use of the systemic insecticides phosdrin and Thiodan (endosulphan). The use of resistant cultivars has been suggested and Mexican research workers claim to have discovered five resistant ones. The American cotton bollworm, *Heliothis armigera*, is reported to be a serious pest of the haricot bean and other pulse crops in parts of Africa. It is particularly troublesome on dwarf beans in the Arusha area of Tanzania, where it is being controlled by the aerial application of DDT or Thiodan when the plants are in the early pod stage of growth.

Haricot bean

Another pest of some economic importance in E. Africa is the spotted borer, *Maruca testulalis.* Although not so serious as *Heliothis*, it is becoming increasingly troublesome in parts of Tanzania. The larvae bore into the pods and eat the seeds, but because of their habit of remaining inside the pod, control with insecticides which kill on contact is difficult. The use of Thiodan at the early stages of pod development is reported to give reasonably satisfactory control. Other pests in E. Africa which damage the seed are the spiny bugs, *Acanthomia horrida* and *A. tomentosicollis.* They not only cause damage by sucking, but can introduce a fungus disease, *Nematospora* sp. Spraying with dieldrin or Thiodan is reported to give effective control. The cornseed maggot, *Hylemya platura* (=*H.* (*Delia*) *cilicrura*), known as the bean seed fly in the UK, and the onion fly in Australia, is very widespread in most growing areas and can be troublesome in certain seasons. The larvae attack the germinating seed or seedling and in cases of severe infestation can destroy the seedling completely. Effective control is usually obtained by treating the seed with insecticides such as chlorodane, lindane, dieldrin or aldrin. Another insect pest of the haricot bean and other food legume crops in the tropics and subtropics is the green stink bug, or southern green stink bug, *Nezara viridula.* In E. Africa it is successfully controlled by the use of BHC dust. Spider mites, *Tetranychus* spp., and various cut and armyworms are also troublesome in many areas.

The haricot bean is also susceptible to attack from root-knot nematodes, *Meloidogyne* spp. Affected plants become stunted and wilt, despite an adequate supply of water; losses can amount to 50 per cent of the crop in heavily infected soils. Crop rotation and the cultivation of resistant cultivars are the most effective control measures.

Storage pests—the haricot bean is attacked by many insect pests during storage, the most important of these are the bean weevil or bruchid, *Acanthoscelides obtectus*, the cowpea weevil, *Callosobruchus chinensis*, and the southern cowpea beetle, *Callosobruchus maculatus.* All three are widespread throughout the tropics and can move from the field to stores and vice versa. Infestation can therefore start in the field, but usually builds up when the beans are stored, since these insects can breed on stored seed. The life cycle of both *C. maculatus* and *A. obtectus* is completed in approximately 4 weeks at 86°F (30°C) and 70 per cent relative humidity, thus damage is greatest when bean seed is stored in the tropics. In E. Africa, *A. obtectus* is the main storage pest and is a serious problem in the production of haricot beans for canning. Other storage pests of somewhat lesser importance which may attack the haricot bean in certain areas are the Indian meal moth *Plodia interpunctella*, and the tropical warehouse moth, *Ephestia cautella*.

Diseases—the haricot bean is susceptible to a large number of diseases, some of which are limited by climatic conditions and certain insect carriers and are not present in all the haricot bean-producing areas. Moreover, the prevalency and severity of many diseases is often very dependent upon environmental factors, such as temperature and humidity, and can vary from season to season causing marked fluctuations in yield. Bacterial and fungal diseases are generally more important in the lowland humid tropics and subtropics, while virus diseases are usually more severe in drier climates. The most effective control measures are: (i) the development of disease resistant lines; (ii) efficient crop rotation; (iii) good sanitation in the field; (iv) the use of healthy seed; (v) spraying with fungicides.

A destructive fungal disease found in almost all producing areas is anthracnose, *Colletotrichum lindemuthianum*. It is particularly serious in the more temperate humid areas; in New Zealand for example, at one time 30 to 50 per cent infection of the dwarf haricot bean crop was by no means uncommon. It attacks the leaves, stems and pods, and therefore is of considerable economic importance when haricot beans are marketed as green pods. The symptoms of anthracnose show as rust-coloured or purplish lesions which later become dark brown or almost black on the stems, but more obviously on the petiole and the underside of the leaf veins. If the petiole is badly affected the leaf drops and cannot recover its normal position. Anthracnose is most easily recognized by the appearance of infected pods, where it appears first as small, often very numerous, reddish-brown spots, which become more or less circular with a sunken centre and a dark brown or black border. Mature lesions may be 0.24 in. (6 mm) or more in diameter. The causal agent is seed-borne so that the use of disease-free seed, preferably from arid areas, is essential. The fungus is very sensitive to changes in temperature and humidity and requires cool, humid weather for its development, so that it largely disappears under hot, dry conditions. However, the spores can survive for at least 2 years under field conditions so that where the danger of infection exists it is advisable only to grow haricot beans on the same ground about every 3 or 4 years. Regular spraying with copper or dithiocarbamate is reported to give reasonable control in mild outbreaks, but the most effective control is the use of resistant cultivars, and there are several which are virtually immune to all races of this disease.

Haricot bean

Bean rust, *Uromyces phaseoli* var. *typica*, is another fungal disease of considerable economic importance. It has been reported from almost every part of the world and is prevalent throughout Latin America. In Brazil and Peru it is widely considered to be the most troublesome disease of the haricot bean. In E. Africa it is the most important factor limiting the production of white haricot beans for canning and can result in total failure of the crop in certain seasons. The optimum conditions for infection are an average temperature of 62°F (17°C) and the relative humidity maintained above 95 per cent for 8 to 10 hours. The organism attacks the leaves principally. The first symptoms appear on the lower surface as small white spots. Within a few days these develop into rust-coloured lesions or pustules, later the leaf turns yellow and finally dries up. Sulphur dusting at frequent intervals, starting early in the season before any rust spots appear, is reported to give good control. Breeding of resistant cultivars is difficult because of the many races of the organism; more than 35 races have been identified in the USA. Bean rust is not seed-borne, but is disseminated by farm implements, insects, animals and wind, especially wind. Infested bean straw should not be used therefore for feeding or bedding livestock, preferably it should be destroyed. It is also inadvisable to plant beans on land that produced a rust-infected crop in the preceding season.

Ashy stem blight, caused by *Macrophomina phaseoli*, sometimes known as blight, stem blight, root, charcoal, or macrophoma rot, is another troublesome fungal disease. It has been reported from many parts of the world including Africa, India, N. America and the Mediterranean area. High temperatures favour the disease and losses of 60 to 65 per cent have occurred in parts of the USA. The fungus causes conspicuous black, sunken lesions on the seedlings before, or soon after, they emerge. The causal agent is seed-borne so that the use of disease-free seed is the primary method for controlling this disease effectively, particularly if the seed is treated with an organic mercury compound such as Ceresan before planting. The pathogen has over 100 hosts, including maize and sweet potatoes, so that crop rotation, although sometimes recommended, is of little value. Angular, or grey leaf spot, caused by *Isariopsis griseola*, is another widespread fungal disease of the haricot bean, especially in the tropics and subtropics. It is reported to cause serious crop losses in Brazil, Colombia and El Salvador. Spots which originate on the underside of the leaves are delimited by the veins. The lesions are grey initially, but later become brown. In severe cases it can cause complete defoliation of the plant beginning with the lower leaves. Control measures have not been perfected so far, but efforts are being made, notably in Colombia and Australia, to develop resistant cultivars. Where the disease is not serious spraying with Bordeaux mixture has been recommended.

The fungus, *Erysiphe polygoni*, the cause of powdery mildew on the haricot bean, also affects peas (*Pisum sativum*) and other *Phaseolus* spp., and also other genera of economic importance including *Brassica*, *Lupinus*, *Lycopersicon* and *Vicia*. The fungus attacks all parts of the bean plant, except the roots, occurring first on the leaves and spreading later to the stem and pods. The pods may become stunted and malformed and when the white powdery coating is rubbed off a brownish or purple discolouration often referred to as russeting, is revealed. Spraying with Bordeaux mixture or other fungicides will give reasonable control. Different cultivars show a considerable variation in their susceptibility to this disease, probably due to the large number of races which exist. The haricot bean is also susceptible to attack from a number of root rots. Dry root rot caused by *Fusarium solani* f. *phaseoli*, rhizoctonia root rot caused by *Rhizoctonia solani*, black root rot, *Thielaviopsis basicola*, and phythium root rot caused by several species of *Phythium* are widespread. Crop rotation is the most effective control measure for these diseases.

Sclerotinia wilt, also known as white mould or water soft rot, caused by the fungus *Sclerotinia sclerotiorum*, is widely distributed and causes heavy losses in the field and during transportation of haricot beans shipped as pods. Outbreaks of this disease occur after a period of warm humid weather and may result in crop losses of 50 per cent or even higher. The disease appears first as irregularly-shaped water soaked spots on the stems followed by similar spots on the branches and leaves. The organism grows rapidly, causing a soft watery rot of the affected parts including the pods. No completely effective control measures are known for sclerotinia wilt. Efficient field sanitation will help reduce the amount of infection. Planting the beans in wide rows to allow good air circulation in the field will reduce the humidity and help check the development of this fungal rot. A rotation of 2–3 years, or longer, with cereal crops has been suggested as a control measure. Recently workers have reported that spraying with fungicides, particularly Benlate R, gives reasonable control.

Southern blight, southern wilt or crown rot, caused by *Sclerotium rolfsii*, is an important disease of the haricot bean in the tropics and subtropics. The optimum temperature for its growth is 86°F (30°C) and it is reported to be susceptible to injury by low temperatures, hence it is not usually of economic importance in the more temperate haricot bean growing areas. The first symptoms of this disease are a slight yellowing of the lower leaves and a water-soaking of the stem just below the soil line. As the disease progresses, the leaves become yellowed and eventually drop off. The infection of the underground part of the stem and tap-root extends downwards and destroys the cortex; occasionally the organism invades the vascular system of the stem as far as the lower branches, causing a dark discolouration of this tissue. There appear to be no effective measures for the control of this disease, although rotation with cereal crops may be of some assistance.

There are three bacterial diseases which can be troublesome to the haricot bean. Common blight, *Xanthomonas phaseoli*, is widespread, it requires high moisture conditions and the severity of outbreaks varies from season to season, depending upon weather conditions. Losses ranging from 40 to 60 per cent have been reported from the USA, where it was a very serious disease in many eastern, midwestern and southern states. However, the widespread use of disease-free seed produced in more arid areas has effectively controlled this disease in that country. The disease manifests itself initially as small water-soaked spots on the leaves, which later develop into large brown necrotic lesions. On the pods water-soaked spots appear which gradually enlarge to cover much of the pod and become brick-red in colour. The disease is seed-borne, but is also spread by rain and field equipment. The use of disease-free seed is the most effective control measure, although spraying with copper fungicides is reported to give reasonable control.

Halo blight, *Pseudomonas phaseolicola*, has symptoms very similar to those of the common blight, but is favoured by cool, humid temperatures and so is not very widespread in the tropics and subtropics. It is, however, troublesome in some seasons in parts of the USA, and E. Africa. In certain seasons in the UK, it can cause heavy losses to the dwarf French bean crop because the lesions on the pods result in their rejection by the processing industry. In the UK outbreaks of this disease can usually be traced to an infected seed-lot, since the disease does not over-winter in the soil. In some countries halo blight is reported to be controlled successfully by spraying with copper or streptomycin formulations, or by using resistant cultivars. Another seed-borne, bacterial disease infecting the haricot bean is bean wilt, caused by *Corynebacterium flaccumfaciens* (*Bacterium flaccumfaciens*). This wilt often occurs with the common and halo blights and many of its symptoms are similar so that it is sometimes confused with them. The exact distribution of bean wilt is not known, but it has been reported from several areas in the USA, southern Africa, Australia, New Zealand and parts of Europe, notably Bulgaria.

The haricot bean is susceptible to a number of virus diseases which cause considerable crop losses. Bean common mosaic, (BCM), also known as bean virus 1, is the most troublesome. It is probably world-wide in its distribution and has a wide range of host plants in addition to the haricot bean. It causes severe mottling and malformation of the leaves and a serious reduction in yield. This seed-borne disease can be transmitted by aphids so that effective control of aphid infestation is essential. The most satisfactory method of control, however, is the use of resistant cultivars, of which a number, both of field and horticultural types, exist. Bean virus 2, or bean yellow mosaic, (BYMV) has a wide range of host plants, and is also transmitted by aphids. The leaves of infected plants also become mottled and there is pronounced dwarfing of the plant and bunchiness. Losses caused by BYMV can be very severe and it is more virulent than BCM. Control is difficult, but since the most important sources of infection seem to be white sweet clover, (*Melilotus alba*), crimson clover, (*Trifolium incarnatum*), and gladioli (*Gladiolus spp.*), beans should not be planted near these crops. Curly top, although of lesser importance than BCM, is another virus disease which can be troublesome, particularly in the USA, where it is transmitted by the leaf hopper, *Circulifer tenellus*. It produces curling of the leaves which become brittle, and often results in the death of the plant. It can be controlled most satisfactorily by growing resistant cultivars. In general field beans are more resistant to curly top than the horticultural French or snap beans. In many Latin American countries the most important virus disease is golden mosaic, which is transmitted by the sweet potato white fly, *Bemisia tabasci*. In the Pacific coastal plains of C. America this disease produces heavy losses. Infected plants exhibit a severe bright yellow mosaic and show a pronounced reduction of yield. In trials in El Salvador and Costa Rica with over 5 000 cultivars no resistance has been found to this virus. In addition, white fly also transmit mottle dwarf virus disease, which produces symptoms similar to curly top and is sometimes known as pseudo curly top; it produces curling of the leaves, stunting of the plant and a severe reduction in crop yield.

Growth period

Haricot beans take from about 60 to 150 days to reach maturity depending upon the cultivar and local climatic conditions. In Queensland 80 per cent of the haricot bean crop normally reaches maturity 98–112 days after planting. In the USA the pea bean cultivar Sanilac matures in about 88 days and the cultivar Seaway in about 79 days. In E. Africa the crop cycle is approximately 90 days. In the UK the cold tolerant cultivar Purley King produces seed in approximately 120 days.

The green immature pods of the haricot bean are normally ready for harvesting 50 to 56 days after sowing, some 14–28 days after the first flowers have appeared. In Hawaii during the summer months the French bean crop matures from 45 to 60 days, and in 60 to 75 days when grown at higher elevations during the winter months.

Harvesting and handling

Dry or field beans—dry beans are harvested as soon as a high proportion of the pods are fully mature and when the moisture content of the seed is above 40 per cent. At this moisture content most cultivars will have approximately 80 per cent of their pods yellow and ripe. Shattering and uneven ripening are problems with certain cultivars and if the moisture content of the seeds is less than 40 per cent at harvest, losses from shattering and mechanical damage are likely to increase significantly. Some commercial growers, notably in the USA and Canada, apply chemical defoliants, containing magnesium or sodium chlorate and small quantities of boron or magnesium chloride, when the leaves begin to yellow and the pods and seeds are well developed. These accelerate maturity, promote uniform ripening and reduce the possibility of damage due to wet weather. Defoliants must be applied when the temperature is above 60°F (16°C) and no rain is expected for at least 6 hours.

Smallholders usually harvest by uprooting the whole plant. The plants are then tied into bunches and hung to dry on a frame, or spread out in heaps on the bare earth, mats, tarpaulins or corrugated iron, and left to dry. Drying may take from 7 to 10 days, depending upon the weather. When dry, the plants are often beaten with sticks and the haulms and pods removed by hand or by winnowing. With large-scale cultivation the beans are harvested mechanically. Sometimes a two-bladed puller that pulls the plants out of the soil, or cuts them slightly below the surface of the soil, is used. The puller normally harvests two rows at a time and works them into a windrow. In order to minimize shattering the plants are usually harvested in the morning while they are still damp and the pods are turgid. After pulling the crop is left to dry for up to 10 days. The windrow normally incorporates 6–8 rows of plants to prevent too rapid drying. Field drying of haricot beans is a critical operation as some cultivars, particularly the white-skinned types discolour with excessive sun-drying, while in some areas high humidity can cause discolouration of the seed. When dry the seeds (preferably with a moisture content of 18 per cent, or less) are separated from the pods and haulms by threshing. This is often achieved by collecting the plants into large heaps on a hessian sheet supported by a layer of grass or straw and running over the dry plants with a rubber-wheeled tractor. During the operation the seeds drop to the bottom of the pile and are protected from damage by the straw. Commercial grain threshers may also be used provided that they have been specially adapted by the removal of a number of the cylinder teeth and by reducing the speed. The use of a rubber-belt thresher has been recommended in the USA. Throughout threshing and the subsequent cleaning and grading operations haricot beans must be handled with care to avoid damage and the speed of the thresher must be adjusted according to the moisture content of the seeds. (Speeds of between 250 and 450 r/m are generally satisfactory). For safe storage haricot beans should have a maximum moisture content of 15 per cent, preferably below 12 per cent. If pre-storage drying is insufficient they are subject to attack from moulds, including a number of toxicogenic species, such as *Aspergillus flavus*.

In some countries, eg the USA, beans with a moisture content of over 18 per cent are dried artificially, but this operation requires great care in order to prevent wrinkling and splitting of the seed-coat. Moreover, beans dried artificially tend to become mushy when cooked. For drying beans with a moisture content of 25 per cent the initial air temperature should not exceed 80°F (27°C), for beans of a lower moisture content it may be higher, but never above 90°F (32°C). It is not advisable to dry haricot beans artificially to a moisture content of less than 18 per cent.

Before storage the beans are usually sorted and graded, and protected against insect infestation, which can cause serious losses. Build-up of insect pests in the stores can be prevented by treating the beans with an insecticide such as pyrethrum, preferably synergized with piperonyl butoxide, or lindane, or by fumigating. In E. Africa carbon bisulphide, Phostoxin, methyl bromide or Killoptera (3 parts ethyl dichloride to 1 part carbon tetrachloride) are commonly used fumigants.

In Latin America and Africa haricot beans for local use are usually stored in a wide range of storage containers, such as earthenware pots, baskets and metal drums. The latter can be made airtight and provide a simple and effective method for protecting small quantities of beans against insect infestation. Generally it is recommended that the moisture content should be 12 per cent, or less, otherwise off-flavours may develop. However, in Brazil recent investigations have indicated that black beans, moisture content 13 per cent, can be stored satisfactorily in aerated bins with an ambient relative humidity of between 65 and 85 per cent, and the temperature below 64°F (18°C). With large-scale production, particularly the production of canning beans for export, the beans are carefully sorted after fumigation to remove all traces of trash, defective or damaged seed, and other extraneous matter, and are sometimes polished by high speed admixture with sawdust, followed by sorting by specific gravity. The beans are then subjected to a final sorting and grading, either by hand, or mechanically (sometimes electronic sorters are used) before being bagged and stored ready for shipment. The beans are normally treated with an insecticide dressing when bagged and are often fumigated again before shipment.

Haricot beans require care when stored in order to avoid loss of quality, and adverse changes in their processing, organoleptic and culinary characteristics. Experiments have shown that canning beans with a moisture content of 10 per cent, or more, when stored at 77°F (25°C) for 2 years, develop off-flavours. For long-term storage drying slowly to a moisture content of 8–9 per cent has been recommended, although users are liable to have problems with soaking and handling of beans with a low moisture content, due to the development of hardshell. Recent investigations indicate that short-term dry heat treatment prior to storage could have possibilities as a low-cost technology which would reduce deterioration of unprocessed beans during storage. In countries such as the USA and Canada, beans for canning are frequently stored in bulk in bins or silos at 60°F (16°C), or less, preferably 50°F (10°C), and 50 per cent relative humidity, in order to prevent the development of moulds and musty off-flavours. Haricot beans for export markets are often packed in strong 50 lb (22.5 kg) or 100 lb (45 kg) jute (burlap) sacks. The use of plastic (polythene) sack-liners is becoming widespread and is recommended, provided the beans have been dried to a safe moisture content and the sacks are not exposed to direct sunlight. In recent years American exporters have shipped canning beans to the UK in van containers each holding 400 bags. Shipments are normally carried at a temperature below 75°F (24°C) and a relative humidity of less than 80 per cent.

French or snap beans—when grown for the fresh vegetable market the green immature pods are usually harvested by hand at intervals of 3 to 4 days to ensure that only young beans are marketed. They are often belt-graded to sort out any diseased or broken pods before being packed in baskets, shallow trays, wooden boxes or crates. Lining the package with good quality paper is frequently practised in the UK to reduce wilting; with some types of packages lining is also needed to prevent the beans falling out. In the USA over 90 per cent of the French bean crop grown for processing is now harvested mechanically by specially designed harvesters, the stage of maturity at which the crop is harvested being dependent upon the sieve size distribution needs of the processor purchasing the crop. French beans deteriorate rapidly at temperatures above 68°F (20°C) and so are frequently cooled rapidly to 40°F (4°C) after harvest and then shipped or stored under refrigeration to prolong their shelf-life. Storage at a temperature of 40–45°F (4–7°C) and a relative humidity of 90–95 per cent is recommended; when they may be expected to keep for about 7 to 10 days. If stored at below 40°F (4°C) they are subject to chilling injury, evident as surface pitting in storage, and russet discolouration when the beans are removed for marketing. If the beans have been hydro-cooled they must be stored under refrigeration, otherwise they are liable to develop soft rot. Storage in an atmosphere of low oxygen content (2–3 per cent) and 5–10 per cent carbon dioxide has been suggested to prolong the shelf-life of green beans, but the principal benefit of this treatment appears to be retardation of yellowing. Waxing has also been suggested for extending the shelf-life. Bean containers should be stacked to allow abundant air circulation otherwise deterioration can be rapid due to the high incidence of various market diseases such as watery soft rot, *Sclerotinia* spp., anthracnose, *Colletotrichum lindemuthianum*, cotton leak, *Phythium* spp., bacterial soft rot, *Erwinia carotovora*, bacterial bean wilt, *Corynebacterium flaccumfaciens* and grey mould rot, *Botrytis cinerea*. Post-harvest treatment with hot water, 125°F (52°C), with or without dichloran (450 ppm), has been reported to give satisfactory control of *Phythium* and *Sclerotinia* rots.

Primary product

Seed—which show considerable variation in shape, size and colour. The seeds are often oblong, but may be ellipsoid, globular or kidney-shaped, the ratio of length, breadth and thickness is also very variable. The colour can be white, buff, yellow-ochre, brown, red, purple, grey or black, or mottled with a combination of these colours. The hilum is oblong, sometimes elliptic, approximately 0.08–0.14 in. (2–3.5 mm) in length, usually white, but sometimes surrounded by a darker coloured ring. One hundred seeds weigh between 0.7 and 2.1 oz (20–60 g). The following types of dry haricot beans are recognized in international trade:

(i) *Pea beans*—also commonly known as navy, Michigan or small white beans and grown mainly in the USA, Canada, Ethiopia, E. Africa and Europe. Similar, but slightly larger than the navy or pea beans, are the Cristales beans grown in Chile and the Otenashi or Tebo beans grown in Japan.

(ii) *White kidney beans*—this group includes the Great Northern group, grown mainly in the USA, in the states of Nebraska, Idaho and Wyoming, and the large white haricots and Cannellini beans of the Mediterranean region.

(iii) *Red kidney beans*—production and consumption of this group is confined largely to the Americas and E. Africa.

(iv) *Pinto beans*—a buff or light-brown bean grown and consumed mainly in the Americas.

(v) *Cranberry beans*—this group includes the Romano and Borlotti spotted beans (white or light-brown and red) popular in Italy, Spain and parts of Latin America, both as a dry bean and as a mature green bean.

(vi) *Black beans*—this group includes the turtle soup bean. Black beans are the most popular type of haricot bean in Brazil, Venezuela, Central America and the Caribbean.

(vii) *Yellow-eye beans*—these beans are white with a brown spot and are produced and consumed mainly in N. America.

(viii) *Brown beans*—this group includes brown and various shades of yellow and are sometimes known as 'Brown Dutch' beans; Beka is a well-known type. They are grown mainly in the Netherlands, Sweden and Angola, and are popular throughout Scandinavia, the Netherlands and in Surinam.

(ix) *Pink beans*—a pink-coloured bean of minor importance grown in the USA (Idaho, Washington and California) and Latin America.

Yield

The yield of dry haricot beans varies greatly according to the climatic and soil conditions, the level of crop management, the purity of seed used, and the efficiency of pest and disease control. The world average yield is about 445 lb/ac (500 kg/ha), but there are very wide variations from country to country, and sometimes in different areas within a country. In the USA the average yield is about 1 100 lb/ac (1 230 kg/ha), although in 1970 it was reported that the maximum known commercial yield of dry beans was 3 600 lb/ac (4 030 kg/ha), attained under irrigation in Colorado. In Brazil, the average yield is about 530 lb/ac (600 kg/ha), but under good climatic conditions yields of approximately 1 000 lb/ac (1 120 kg/ha) are obtained. In the Constanza valley of the Dominican Republic yields of between 3 560 and 4 450 lb/ac (4 000–5 000 kg/ha) have been reported. In Mexico the average yield of haricot beans grown under irrigation is 1 160 lb/ac (1 300 kg/ha) in the State of Sinaloa, compared with the national average of only 400 lb/ac (450 kg/ha), which is affected by the very low yields obtained when the haricot bean is grown in dry-land areas. In E. Africa the average yield varies from 200–600 lb/ac (225–670 kg/ha), although with better crop management, and more efficient crop protection, a yield of 1 000 lb/ac (1 120 kg/ha) is attainable. The cultivar Canadian Wonder is reported to have given a yield of 2 800 lb/ac (3 140 kg/ha) experimentally.

The yield of immature beans in the pod (French or snap beans) is reported to average 4 600 lb/ac (5 150 kg/ha) in the USA, but with efficient crop management considerably higher yields can be obtained. In the UK yields average about 6 700 lb/ac (7 500 kg/ha) in a good season, but adverse weather conditions can cause a reduction of some 50 per cent in some seasons. In India the yield of green pods normally ranges from 2 200–3 100 lb/ac (2 500–3 500 kg/ha).

Main use

Mature, dry, haricot beans are widely used as a human foodstuff in many parts of the world. The haricot bean is in fact the most widely accepted grain legume in Latin America and is an important protein foodstuff in Asia and parts of Africa. In S. America the beans are often cooked in water for 4–5 hours, usually with the addition of onions and other flavouring ingredients, the softened beans are then mixed with rice, maize, cassava or plantains, or fried with fat, and eaten. The beans may also be utilized for the production of flour. In addition, considerable quantities of pea beans are canned as 'baked beans', especially in the UK and the USA. Attempts are also being made to develop various quick-cooking food preparations such as instant bean powders.

Subsidiary uses
Haricot beans may be utilized for the production of protein concentrates, eg milk. In S. America and parts of Africa the mature green beans are a very popular foodstuff and are eaten as a vegetable. The immature, unshelled green pods of the horticultural types of haricot beans (French or snap beans) are widely used as a vegetable—fresh, canned, dehydrated or quick-frozen.

Secondary and waste products
In parts of Asia the young green leaves are sometimes used as a salad ingredient. Cull dry beans can be used, after cooking, for livestock or poultry feed, in the USA they are sometimes used to fatten sheep. Similarly, discoloured pods, or those otherwise unsuitable for human consumption, and the waste from factories processing French beans, can be used for feeding livestock. The dried haulms or straw can also be used for livestock feeding. An approximate analysis has been given as: moisture 11.0 per cent; protein 6.0 per cent; fat 1.5 per cent; N-free extract 34.0 per cent; fibre 40.0 per cent; ash 7.5 per cent; calcium 1.7 per cent; phosphorus 0.1 per cent; potassium 1.0 per cent; digestible nutrients 45 per cent; nutritive ratio 14:1.

Special features
The seed-coat constitutes approximately 6.6 to 9.2 per cent of the total weight of the haricot bean. An approximate analysis of the dry mature beans is: moisture 11.0 per cent; protein 22.0 per cent; fat 1.6 per cent; carbohydrate 57.8 per cent; fibre 4.0 per cent; ash 3.6 per cent; calcium 137 mg/100 g; iron 6.7 mg/100 g; vitamin A 30 iu/100 g; thiamine 0.54 mg/100 g; riboflavin 0.18 mg/100 g; niacin 2.1 mg/100 g; ascorbic acid 3.0 mg/100 g. The approximate amino acid content (mg/gN) of the seeds has been reported as: isoleucine 262; leucine 476; lysine 450; methionine 66; cystine 53; phenylalanine 326; tyrosine 158; threonine 248; tryptophan 63; valine 287; arginine 355; histidine 177; alanine 262; aspartic acid 784; glutamic acid 924; glycine 237; proline 223; serine 347. There is considerable variation in the composition of different cultivars, and in the same cultivar grown under different environmental conditions. Protein contents ranging from as low as 14.6 per cent to as high as 33.0 per cent have been reported. High protein content tends to be associated with small, spherical, black or white seed, from vining

plants grown in tropical areas; low protein content tends to be associated with elongated seed from bush-type plants grown in more temperate areas. Recently it has been reported that beans with a relatively high total protein content have relatively low levels of the essential amino acids, methionine and cystine, and that the highest levels of these amino acids occur in beans with a protein content of between 20 and 22 per cent.

The major constituents of the seed protein have been reported as phaseolin (20 per cent of the seed dry weight), phaselin (2 per cent), and conphaseolin (0.35–0.4 per cent). Recently, using improved techniques, the globulin fraction of Mexican black beans has been investigated and found to consist of four major components, one of which is a glycoprotein containing 4.95 per cent carbohydrate (as mannose) and 1.19 per cent hexosamine (as glucosamine).

The carbohydrate constituents of the haricot bean have been given as: sugars 1.6 per cent; dextrins 3.7 per cent; starch 35.2 per cent; pentosans 8.4 per cent; galactans 1.3 per cent; cellulose 3.1 per cent. In addition, approximately 0.7 per cent of pectin is present. The haricot bean also contains a golden yellow fatty oil (1–2 per cent) with the following characteristics: $SG^{15.5°C}$ 0.9603; $N_D^{30°C}$ 1.4808; sap. val. 132.6; iod. val. 149.8; acid val. 20.5; RM val. 1.0; Pol. val. 2.0; unsaponifiable matter 7.0 per cent. The oil contains 19 per cent of saturated fatty acids (palmitic, plus a small quantity of carnaubic), and 63.3 per cent of the unsaturated fatty acids (oleic, linoleic and linolenic).

Raw haricot beans contain varying amounts of a heat-labile trypsin inhibitor, four haemagglutinins, in varying proportions according to the cultivar, and a goitrogen, but under normal circumstances cooking destroys these toxic factors. However, there have been cases where toxic haemagglutinins have been present in certain quick-cooking bean preparations.

As has been mentioned in the section *Harvesting and handling*, haricot beans are extremely susceptible to the development of hardshell during prolonged storage, which makes them difficult to cook; in addition they are liable to produce flatulence when eaten. Both these factors handicap their use as a foodstuff. They are also susceptible to the development of moulds during storage, and contamination with aflatoxin has been reported.

Haricot bean

Haricot beans normally have a bland flavour, the following volatile constituents have been identified in raw red haricot beans: oct–1–en–3–ol and hex–*cis*, 3–enol and in the cooked beans: thialdine, p–vinylguaiacol, 2,4–dimethyl–5–ethylthiazole and 2–acetythiazole. They are, however, very susceptible to the development of off-flavours, particularly an earthy, musty off-flavour, recently identified as geosmin. Another off-flavour, nona–2,4,6–trienal, is thought to arise as the result of the oxidative breakdown of the linolenic acid present in the oil.

The immature green pods of French or snap beans consist of approximately 94 per cent edible material with the following approximate composition: moisture 90.5 per cent; protein 2.0 per cent; fat 0.1 per cent; total carbohydrate 6.8 per cent; fibre 1.0 per cent; ash 0.6 per cent; calcium 75 mg/100 g; phosphorus 38 mg/100 g; iron 0.8 mg/100 g; sodium 2.0 mg/100 g; potassium 182 mg/100 g. Average vitamin contents have been reported as: vitamin A, 525 iu/100 g; thiamine 0.07 mg/100 g; riboflavin 0.09 mg/100 g; niacin 0.70 mg/100 g; ascorbic acid 15 mg/100 g. The pods contain pectic substances (as calcium pectate 9–15 per cent dry basis, depending upon the pod length), oxalic acid (0.03 per cent), mannitol and sugars, including glucose and fructose (1.16 per cent).

Processing

Dry beans, canning—dry haricot beans of all types may be canned successfully. White pea beans such as navy or Michigan pea beans and Great Northern types are frequently processed with ingredients such as tomato or other flavoured sauces, sometimes with the addition of pork, bacon or sausages, while various types of red kidney beans are normally canned in brine or spiced sauces.

There is considerable variation in the canning procedures and the ingredients used by individual canners, but the following basic operations are carried out to produce the well-known, popular pack, baked beans in tomato sauce. The beans are first carefully sorted and graded, so that all mis-shapen, discoloured, damaged, or wrong-sized beans are removed, together with extraneous matter. The beans are then left to soak in water until they have become soft and plump. The time of soaking depends upon such factors as the age of the beans, their moisture content, the thickness of the skin and the hardness of the water. Old and dry beans tend to require a longer-soaking period, but as a broad generalization, beans with a moisture content of 15 per cent require approximately 10 hours soaking, those with 12 per cent, 12 hours, and those with 9 per cent, 14 hours. Water with a hardness of 4 to 9 degrees gives the best results. Extremely soft water can result in splitting and matting of the beans in the can. The addition of sodium hexametaphosphate to hard water (0.2 per cent in water of 26–29 degrees hardness) is frequently recommended to improve the texture of the final product. Soaking is usually carried out in shallow non-corrosive metal, glass or enamel-lined tanks, and often the water is changed once or twice during the soaking period to reduce the likelihood of souring occurring. After soaking the beans are usually run over a water riffle to remove any small stones which may be present. The soaked beans are next blanched. Some canners blanch the beans in water at 180° to 200°F (82–93°C) for about 15 minutes, while others use a longer blanching period of up to 40 minutes in water at 170° to 212°F (77–100°C). The advantage of using a longer blanching period is that the soaking period may be reduced or even omitted altogether. However, its use requires efficient quality control, particularly the careful checking at all stages of the moisture content of the beans, the control of can filling weights, the amount of sauce added, and cooking times. The moisture content of the beans after blanching should be preferably between 50 and 60 per cent, but when a longer blanching period is used it may be less than 50 per cent, in this case it is necessary to cook the beans in the can for a short period at 212° to 214°F (100–101°C) to allow them to absorb the maximum amount of moisture, before processing them in the normal way. After blanching the beans are subjected to a final inspection and all broken, shrivelled or otherwise defective beans removed. They are then sometimes baked for a short time before being filled into cans and tomato sauce added. Pieces of pork, bacon, or sausage may also be added at this stage. Filling requires great care since the beans absorb a considerable amount of sauce and it is essential to fill the cans completely before closing, otherwise the final product is liable to be dry and flavourless. Recipes for the tomato sauce vary greatly, but it usually contains, in addition to tomato puree, varying amounts of sugar, salt, and various spices, sometimes with the addition of starch and vinegar. The beans and sauce are normally filled hot at a temperature of at least 170°F (77°C), and after closing, are processed immediately at 240° to 250°F (116–121°C) for 35 to 65 minutes for A 2½ (401 x 411), or smaller cans, and 55 to 100 minutes for A 10 (603 x 700) cans, provided there is no starch in the sauce. In heavy sauces containing starch, processing times normally range from 45 to 70 minutes for picnic size cans (211 x 400) to 175 to 210 minutes for large A 10 (603 x 700) cans. The processing time, however, is dependent upon the filling temperature, the type of pack, the quality of the beans, the blanching time used and the degree of hardness of the water. The pack is very solid so that heat penetration is slow and the use of agitating retorts is advisable, particularly in the production of large catering size packs. Efficient cooling after heat processing is essential and this is normally accomplished by spraying the cans with cold water. Plain cans with lacquered ends are usually used for this pack.

Considerable quantities of red kidney beans are also canned, the operations of cleaning, sorting, blanching, can filling and heat processing being very similar to those described for white pea beans. However, the pack is somewhat susceptible to discolouration and the final texture is also very much dependent upon the hardness of the water used in the canning operations. Water with a total hardness of 5 to 7 degrees is usually considered best, below this the beans usually have a soft mushy texture, unless calcium chloride is added to the canning medium; above 10 to 12 degrees hardness, longer soaking, blanching and processing times are often necessary. Like pea beans, red kidney beans are also susceptible to hardshell. Many canners consider that for successful processing, red kidney beans should have a moisture content of between 12 and 14 per cent, and sometimes condition them for several days in an atmosphere of moderate humidity prior to canning, in order to increase the permeability of the seed-coat and to reduce the incidence of incompletely soaked beans. The addition of 0.25 to 0.50 per cent citric acid to the soaking water has been suggested to improve the colour and texture of the final product. Blanching is usually carried out in water at 200–210°F (93–99°C) for 5–10 minutes. There is considerable variation in the formulae of the brine or sauce used, but many canners use a brine containing approximately 2 per cent salt, plus 3–4.5 per cent sugar, and the spiced sauce normally contains cayenne pepper, cloves, cinnamon and onion salt, in addition to sugar and salt. The addition of sodium EDTA (ethylene diamine tetra-acetate), up to 165 ppm, to the brine or sauce, has been found to improve the texture of the beans significantly. The brine or sauce should be added to the cans at a temperature as near boiling point as possible, otherwise steam-flow closure may be necessary. Suggested processing times are small cans (401 x 411, or smaller) 45 minutes at 240°F (116°C), 30 minutes at 245°F (118°C) and 20 minutes at 250°F (121°C); for A 10 cans (603 x 700) 70 minutes at 240°F (116°C), 55 minutes at 245°F (118°C) and 40 minutes at 250°F (121°C). After processing the cans are cooled immediately, normally under sprays of cold water. If continuous agitators are used for processing starch weeping can be a problem, but this is usually overcome by the addition of a small quantity of calcium chloride to the canning medium. Red kidney beans are also sometimes canned in thick sauces when they are processed similarly to white pea beans.

Pre-cooked, dehydrated, whole haricot beans—this product has been developed in order to increase bean consumption by producing a product of more uniform quality which can be cooked quickly. The processes used for the preparation of pre-cooked, dehydrated beans consist essentially of soaking the beans in water for approximately 8 hours, steam-cooking for 20 minutes at 250°F (121°C), followed by dehydration. There are, however, a number of variations to the basic process. Freezing before or after cooking, or dipping in a sugar solution, are two methods sometimes used to avoid butterflying, ie the development of fissures in the beans during drying. In another process the beans are subjected to an intermittent vacuum treatment for 30 to 60 minutes in a solution of sodium chloride, tripolyphosphate, bicarbonate and carbonate, after which they are left to soak in the salt solution for a further 6 hours. The beans are then thoroughly rinsed and dried slowly to a moisture content of from 9.5 to 10.5 per cent. Pre-cooked dehydrated beans normally require between 10 and 30 minutes cooking.

Pre-cooked bean flour—this product may be prepared from whole beans by soaking, cooking and dehydrating the beans followed by grinding, or by pulping the cooked beans and then drying the resultant puree on single-, or double-, drum driers, or by spray-drying. Discolouration is often a problem, particularly if black beans are used, but bean flour is very suitable for use as a soup ingredient, requiring a cooking period of about 10 to 15 minutes.

Canned French or snap beans—the immature green pods of certain horticultural cultivars are canned and are a very popular processed vegetable. In the processing of French beans it is essential that the pods are harvested at an immature stage before they become tough and fibrous and that there is the minimum of delay between harvesting and processing. On arrival at the cannery the pods are cleaned, usually mechanically, by the use of shakers and air cleaners, and stones, twigs, leaves and pieces of vine are removed. The pods are next washed thoroughly, then inspected and graded, after which they are snipped and sliced, in the case of packs of various styles of cut beans. After this the beans are blanched for 2 to 3 minutes in water at 170° to 180°F (77–82°C), washed in cold water and thoroughly drained, before being packed into cans and covered with a 2 per cent brine solution at a minimum temperature of 200°F (93°C). Some canners also add sugar to the brine, about 1.5 per cent, and sometimes a little calcium chloride and, or, monosodium glutamate, plus a little green colouring matter, if the domestic food regulations permit its use. The cans, which preferably should be coated throughout with a sulphur absorbent lacquer, are exhausted and then processed. Processing times vary according to the type of French bean used, the style of pack, the can size and the filling temperature, but normally range between 17 and 45 minutes at 240°F (116°C). After processing the cans are cooled immediately, usually under sprays of cold water, in order to prevent softening and loss of colour.

Quick-frozen French or snap beans—this product is also a very popular processed vegetable. The initial processes of cleaning, washing, grading and slicing are very similar to those used for the preparation of canned French beans. The beans are normally blanched in steam or hot water for 2 to 3 minutes, then immediately cooled and drained before being sorted, packed and frozen. Whole French beans are often packed in cartons and pass through belt-tunnel freezers through which refrigerated air at –25°F (–32°C) is circulating. Considerable quantities of sliced French beans are now individually frozen by the fluidized-bed method and then packed in bulk and stored at 0°F (–18°C), or lower.

Dehydrated French or snap beans—certain types of horticultural haricot beans, such as the cultivar Tendergreen, can be dehydrated successfully. The initial preparation is the same as for the production of canned French beans. The sliced beans are normally blanched for about 4 minutes in atmospheric steam and then immediately cooled with water sprays. Some processors use boiling water and sometimes a longer blanching period, up to about 10 minutes. Treatment with sulphur dioxide during blanching, or immediately afterwards, to give a sulphur dioxide content of approximately 500 ppm in the finished product, helps preserve the colour and flavour. Freezing the blanched beans in still air at 0°F (–18°C) prior to dehydration results in a much improved product. Sliced beans are often dried in a cabinet drier for approximately 6 hours at 145°F (63°C), and then transferred to a bin drier, where they are held at 115°F (46°C) for about 48 hours. The moisture content of the final product is usually 5 per cent, or less.

Production and trade

Production—haricot beans are widely grown in the tropics, subtropics and warmer temperate regions, for the production of the dry mature beans. They are the dominant grain legume crop of the Americas and are also of major importance in parts of Africa. It is, however, very difficult to assess accurately global production, since not only is there considerable production at subsistence level which is probably never recorded in the official production figures of many countries in Latin America and Africa, but in the *FAO Production yearbooks* haricot beans are included in a composite item 'dry edible beans', which also includes the considerable production of mung beans, moth beans, urd and hyacinth beans in Asia, in addition to lima beans and various other dry beans of lesser importance such as adzuki and rice beans. According to the *FAO Production yearbooks* world production of 'dry edible beans' averaged 11 551 000 t/a for the years 1970–74, compared with 10 674 000 t/a for 1965–69, an increase of 8 per cent, and for 1975 was 12 745 000 t, and for 1976 12 580 000 t. Analysis of the statistical data included under the heading 'dry edible beans' indicates that haricot beans probably constituted about 60 per cent of the total production and it therefore seems likely that global production of dry haricot beans has probably averaged about 7 000 000 t/a since 1970.

Brazil, the USA and Mexico are the leading producers of haricot beans and the figures quoted in the following table for these countries may be assumed to consist very largely of haricot beans. Production of various types of dry haricot beans in the USA, for example, averaged some 715 850 t/a for the period 1970–74, with pea beans amounting to 35 per cent of the total production and pinto beans a further 32 per cent. The production figures for India and the People's Republic of China, however, include considerable quantities of various other dry beans, particularly mung, moth and urd, and in the case of India only a relatively small proportion of the total consists of haricot beans. The figures for Burundi, Tanzania and Uganda, include, in addition to haricot beans, a not inconsiderable production of lima beans, plus lesser quantities of various other beans, such as mung.

Haricot bean

Dry edible beans: Major producing countries

Quantity tonnes

	Annual average 1965–69	Annual average 1970–74	1975	1976
Brazil	2 321 000	2 305 000	2 271 000	1 923 000
India	1 855 000	2 171 000	2 778 000	2 600 000
People's Repub. China	1 395 000	1 659 000	2 125 000	2 174 000
USA	801 000	800 000	790 000	781 000
Mexico	891 000	893 000	1 027 000	1 149 000
Uganda	129 000	169 800	195 000	258 000
Burundi	138 000	229 800	148 000	150 000
Tanzania	104 000	142 200	134 000	146 000

In addition to the dry haricot beans there is also a considerable production of haricot beans for consumption as a vegetable, either as fresh fully developed beans or as the immature pods. These are frequently referred to as French, snap, string or green beans. In the *FAO Production yearbooks* two separate categories are shown: (i) String beans, ie climbing or pole type beans; (ii) Green beans, ie dwarf, bushy types. Both categories include a large proportion of haricot beans, although various other types of beans, such as runner beans, *P. coccineus*, and lima beans, *P. lunatus*, are understood to be also included. Production of string beans averaged 1 231 000 t/a for the years 1970–74, compared with 1 105 000 t/a for 1965–69, an increase of 11 per cent. In 1975 production was estimated to be 1 243 000 t.

String beans: Major producing countries

Quantity tonnes

	Annual average 1965–69	Annual average 1970–74	1975
USA	689 000	732 000	741 000
Turkey	184 000	245 000	250 000
France	201 000	213 000	204 000
Chile	35 000	35 000	n/a
Argentina	26 000	34 000	40 000

n/a Not available.

Production of green beans averaged 1 947 000 t/a for the years 1970–74 compared with 1 678 000 t/a for 1965–69, an increase of 16 per cent, and in 1975 amounted to 2 274 000 t, and in 1976 to 2 208 000 t.

Green beans: Major producing countries

Quantity tonnes

	Annual average 1965–69	Annual average 1970–74	1975	1976
Italy	268 000	264 000	278 000	281 000
Spain	112 000	186 000	204 000	213 000
USA	205 000	170 000	186 000	111 000
Egypt	111 000	146 000	160 000	162 000
UK	83 000	110 000	92 000	105 000

Trade—dry haricot beans are a major item in international trade although many countries do not classify them separately but often include them under a composite item 'dry beans'. It is estimated that haricot beans constitute about 20 per cent of all the grain legumes entering international trade and that during the decade 1965–74 the total import trade averaged some 350 000 t/a.

Haricot beans (dry): Major exporting countries

Quantity tonnes

	Annual average 1965–69	Annual average 1970–74	1975
USA	101 655	117 252	166 037
Canada	23 909	40 486	54 704
Ethiopia	18 340[1]	38 093	39 644
Argentina	14 743	39 648	65 392
Angola	14 816	14 404[2]	n/a
Turkey	3 408	12 350[1]	n/a
Chile	9 319[1]	12 325	n/a

1 Four year average.
2 Three year average.
n/a Not available.

Haricot bean

Haricot beans (dry): Major importing countries

Quantity tonnes

	Annual average 1965–69	Annual average 1970–74	1975	1976
Cuba	71 895	86 349[1]	n/a	n/a
UK	75 349	78 753	93 648	101 372
Japan	60 873	47 791	39 522	n/a
France	30 089	43 609	45 323	n/a
Italy	16 714[2]	30 111[3]	32 433	28 930
Netherlands	22 530	23 757	39 152	63 231

1 Three year average.
2 Four year average
3 Two year average.
n/a Not available.

There is also a limited trade in fresh haricot beans (French, snap, or green beans), chiefly in Western Europe and N. America. The European trade is principally during the production season, with small quantities being imported from outside Europe, during the off-season, mainly between October and May. Unfortunately statistical data are somewhat fragmentary, but the following details are available.

Haricot beans (French, snap, green): Major exporting countries

Quantity tonnes

	Annual average 1965–69	Annual average 1970–74	1975	1976
Italy	27 000	21 600	23 400	25 284
Spain	7 300	14 000	19 100	n/a
Netherlands	2 800	5 800	6 400	n/a
Morocco	2 400	3 600	1 500	n/a

n/a Not available.

Haricot beans (French, snap, green): Major importing countries

Quantity tonnes

	Annual average 1965–69	Annual average 1970–74	1975	1976
France	13 000	19 000	29 000	n/a
German Fed. Repub.	14 000	14 600	13 300	12 143
Netherlands	19 000	11 000	8 000	n/a
Canada[1]	5 600	7 000	7 000	n/a
Belgium/Luxembourg	3 800	5 200	6 000	5 896

1 Obtained mainly from the USA and Mexico.
n/a Not available.

Prices—there has been a steady increase in haricot bean prices over the period 1965–74. There is, however a considerable variation in the prices paid for the various types and grades which are produced. The average price paid to producers for black beans in Brazil was £94.05/t for the period 1970–74, compared with £39.73/t for 1965–69. The average fob value of US exports of various types of haricot beans was as follows:

Navy or pea beans, 1965–69, £65.80/t, 1970–74, £108.60/t, 1975, £173.00/t;

Great Northern beans, 1965–69, £73.20/t, 1970–74, £131.00/t, 1975, £228.00/t;

Red Kidney beans, 1965–69, £100.00/t, 1970–74, £153.60/t, 1975, £287.00/t;

White beans, 1965–69, £68.40/t, 1970–74, £104.20/t, 1975, £208.00;

Black beans, 1965–69, £84.40/t, 1970–74, £156.40/t, 1975, £273.00;

Pinto beans, 1965–69, £76.20/t, 1970–74, £130.20/t, 1975, £279.00.

The average cif price of imports of haricot beans, largely navy or pea beans for canning, into the UK from the three main sources of supply were:

USA, 1965–69, £80.40/t, 1970–74, £154.00/t, 1975, £238.00/t, 1976, £319.00/t;

Tanzania, 1965–69, £85.00/t, 1970–74, £144.00/t, 1976, £221.00/t;

Ethiopia, 1965–69, £62.00/t, 1970–74, £143.00/t, 1976, £248.00/t.

Haricot bean

Prices for fresh French beans fluctuate considerably according to the season and the quantities available at any particular time on the wholesale fresh vegetable markets. The average cif price of imports into the German Federal Republic was £335.76/t in 1976 and £243.24/t in 1975, compared with £155.09/t for the period 1970–74 and £88.78/t for 1966–69.

Major influences

Haricot beans are one of the most widely grown and utilized of the food legumes. They are particularly important as a source of protein for a high percentage of the population of S. America, where they constitute about 80 per cent of the total production of food legumes. Although haricot beans can be grown over a wide range of ecological conditions they are especially suited for cultivation at medium and higher elevations in the tropics, and under drier, irrigated conditions. They are, however, extremely susceptible to attack from pests and diseases, particularly when grown under humid conditions. The susceptibility of haricot beans to attack from pests and diseases is the most important factor limiting their production in the tropics, and there is an urgent need to develop efficient methods for the control of the major pests and diseases, especially the development of resistant cultivars, before their full potential is likely to be realized. Currently there is a very wide variation in the average yields obtained in various growing areas, according to climatic conditions, the level of crop management, the purity of the seed used, and the efficiency of pest and disease control. The world average yield is reported to be approximately 445 lb/ac (500 kg/ha), but in the USA it averages about 1 100 lb/ac (1 230 kg/ha), and yields in excess of 3 000 lb/ac (3 300 kg/ha) are attainable, which indicates the potential productivity of this food legume.

Although regarded as a nutritious foodstuff, the average protein content of the dry beans is 22 per cent and they are deficient in the essential amino acids methionine and cystine, so that the breeding of cultivars with higher protein contents and an improved amino acid balance is desirable. Factors which undoubtedly affect the popularity and more widespread use of haricot beans as a foodstuff are the relatively long cooking period they require, the presence of several anti-nutritional factors and a flatulence factor, and their susceptibility to the development of hard-shell and off-flavours during storage. The improvement of storage and processing methods and the development of new food products could increase consumption significantly in many producing countries.

The countries of Western Europe and Japan provide large import markets for dry haricot beans. The UK dominates the trade in small pea beans, importing on average 78 000 t/a, valued at around £300.00/t in 1975, for the production of canned baked beans in tomato sauce. Attempts to develop domestic production, by the cultivation of cold-tolerant cultivars, have not been very successful as yet and the requirements of the canning industry continue to be supplied largely from N. America. The production of dry haricot beans, particularly pea beans, for export would seem to offer opportunities for a number of countries in the tropics, provided that economic yields and high quality can be attained by efficient crop management, particularly the control of pests and diseases.

Bibliography

ADAMS, J. M. 1977. A bibliography on post-harvest losses in cereals and pulses with particular reference to tropical and subtropical countries. *Rep. Trop. Prod. Inst.*, G 110, iv + 23 pp.

ADAMS, M. W. 1973. Plant architecture and physiological efficiency in the field bean. Potentials of field beans and other food legumes in Latin America. (Wall, D. ed.). pp. 266–278. *Cali, Colombia, CIAT, Ser. Semin.* 2E, 388 pp.

ALLARD, H. A. and ZAUMEYER, W. J. 1944. Responses of beans (*Phaseolus*) and other legumes to length of day. *US Dep. Agric., Tech. Bull.* 867, 24 pp.

ALLEN, A. G. and WALKER, W. F. 1973. Edible dry beans. *Tasman. J. Agric., 44*, 145–150.

ANANDASWAMY, B. and IYENGAR, N. V. R. 1961. Prepackaging studies on fresh snap beans (*Phaseolus vulgaris*). *Food Sci., 10*, 279–283.

ANDERSEN, A. L. 1955. Dry bean production in the Eastern States. *US Dep. Agric., Farmers' Bull.* 2083, 29 pp.

ANDERSEN, A. L. 1965. Dry bean production in the Lake and Northeastern States. *US Dep. Agric., Agric. Res. Serv. Agric. Handbk.* 285, 32 pp.

ANON. 1967. French bean growing in Queensland. *Qd. Agric. J., 93*, 408–415.

ANON. 1974. Bean seed industry in the dry tropics. *Qd. Agric. J., 100*, 289–290.

ANTHONY, J. P. 1973. A cost comparison for four container systems to export dry edible beans to the UK. *US Dep. Agric., Agric. Res. Serv.*, ARS-NE- 31, 20 pp.

Haricot bean

Appadurai, R. R., Rajakaruna, S. B. and Gunasena, H. 1967. Effect of spacing and leaf area on pod yields of kidney bean (*Phaseolus vulgaris* L.). *Indian J. Agric. Sci., 37*, 22–26.

Arnon, I. 1972. Pulses or grain legumes. *Crop production in dry regions.* Vol. 2. pp. 217–260. London: Leonard Hill Books, 683 pp.

Ayonoad, U. W. U. 1974. Races of bean anthracnose in Malawi. *Turrialba, 24*, 311–314.

Bakker-Arkema, F. W., Bedford, C. L., Patterson, R. J., McConnell, B. B., Palnitkar, M. and Hall, C. W. 1967. Drum and spray drying and characteristics of precooked bean powders. *Rep. 8th dry bean res. conf.* (1966). pp. 24–39. *US Dep. Agric., Agric. Res. Serv.*, ARS 74–41, 76 pp.

Bakker-Arkema, F. W. and Patterson, R. J. 1971. Artificial drying of red kidney beans. *Rep. 10th dry bean res. conf.* (1970). pp. 108–116. *US Dep. Agric., Agric. Res. Serv.*, ARS 74–56, 146 pp.

Bakker-Arkema, F. W., Patterson, R. J. and Bedford, C. L. 1969. The manufacturing, utilization and marketing of instant legume powders. *Rep. 9th dry bean res. conf.* (1968). pp. 35–45. *US Dep. Agric., Agric. Res. Serv.*, ARS 74–50, 94 pp.

Ballantyne, B. 1971. Diseases of dry beans. *Agric. Gaz. N.S.W., 82*, 270–279.

Ballantyne, B. 1971. Summer death of beans. *Agric. Gaz. N.S.W., 82*, 295–297.

Barker, R. D. J., Derbyshire, E., Yarwood, A. and Boulter, D. 1976. Purification and characterization of the major storage proteins of *Phaseolus vulgaris* seeds and their intracellular and cotyledonary distribution. *Phytochemistry, 15*, 751–757.

Bird, J. and Lopez Rosa, J. H. 1973. Whitefly and aphid-borne viruses of beans in Puerto Rico. *Proc. 1st Int. Inst. Trop. Agric., Grain legume improvement workshop.* pp. 276–278. Ibadan, Nigeria, Int. Inst. Trop. Agric., 325 pp.

Bird, J. and Maramorosch, K. (eds.). 1975. *Tropical diseases of legumes.* New York/San Francisco/London: Academic Press, 171 pp.

Bourne, M. C. 1967. Size, density and hardshell in dry beans. *Food Technol., 21*, 335–338.

BRESSANI, R., FLORES, M. and ELIÁS, L. G. 1973. Acceptability and value of food legumes in the human diets. Potentials of field beans and other food legumes in Latin America. (Wall, D. ed.). pp. 17–48. *Cali, Colombia, CIAT, Ser. Semin.* 2E, 388 pp.

BRIEN, R. M., CHAMBERLAIN, E. E., COTTIER, W., CRUICKSHANK, I. A. M., DYE, D. W., JACKS, H. and REID, W. D. 1955. Diseases and pests of peas and beans in New Zealand and their control. pp. 34–68. *N.Z. Dep. Sci. Ind. Res., Bull.* 114, 91 pp.

BROUWER, H. M., STEVENS, G. R. and FLETCHER, J. G. 1975. Zinc foliar sprays increase yields of navy beans. *Qd. Agric. J., 101*, 705–707.

BROWN, A. H. and CARLSON, R. A. 1969. Drying characteristics of quick-cooking dry beans. *Rep. 9th dry beans res. conf.* (1968). pp. 23–27. *US Dep. Agric., Agric. Res. Serv.*, ARS 74–50, 94 pp.

BUREN, J. P. VAN. 1968. Adding calcium to snap beans at different stages in processing: calcium uptake and texture of the canned product. *Food Technol., 22*, 790–793.

BUREN, J. P. VAN and DOWNING, D. L. 1969. Can characteristics, metal additives and chelating agents; effect on the colour of canned wax beans. *Food Technol., 23*, 800–802.

BURKE, D. W., HAMPTON, R. O. and ZAUMEYER, W. J. 1965. Results of field experiments on microbiological and cultural methods for control of fusarium root rot. *Rep. 7th dry bean res. conf.* (1964). pp. 65–69. *US Dep. Agric., Agric. Res. Serv.*, ARS 74–32, 94 pp.

BURR, H. K. and KON, S. 1967. Factors influencing the cooking rate of stored dry beans. *Rep. 8th dry bean res. conf.* (1966). pp. 50–56. *US Dep. Agric., Agric. Res. Serv.*, ARS 74–51, 76 pp.

BURR, H. K., KON, S. and MORRIS, H. J. 1968. Cooking rates of dry beans as influenced by moisture content and temperature and time of storage. *Food Technol., 22*, 336–338.

BUTTERY, R. G. 1975. Nona –2, 4, 6–trienal, an unusual component of blended dry beans. *J. Agric. Food Chem., 23*, 1003–1004.

BUTTERY, R. G., GUADAGNI, D. G. and LING, L. C. 1976. Geosmin, a musty off-flavour of dry beans. *J. Agric. Food Chem., 24*, 419–420.

BUTTERY, R. G., SIEFERT, R. M. and LING, L. C. 1975. Characterization of some volatile constituents of dry red beans. *J. Agric. Food Chem.*, *23*, 516–519.

CAMPBELL, J. S. and HODNETT, G. E. 1960. Spacing experiments with dwarf beans (*Phaseolus vulgaris* L.) in Trinidad. *Trop. Agric.*, (*Trinidad*), *37*, 265–270.

CANADA DEPARTMENT OF INDUSTRY, TRADE and COMMERCE. 1974. *World pulses market survey*. pp. 6–7. Ottawa, Dep. Ind. Trade Commer., Agric., Fish., Food Prod. Branch, 123 pp.

CARVALHO JUNQUEIRA, P. DE, CANCEGLIERO, L. F. B., MATSUNAGA, M. and YAMAGUISHI, C. T. 1971. Aspectas econômicos de produção e comercialização do feijão, 1971. [Economic aspects of the production and commercialization of beans, 1971]. *Agric. Sao Paulo, Bol. Tec. Inst. Econ. Agric.*, *17* (7/8), 1–64.

CENTRO INTERNACIONAL DE AGRICULTURA TROPICAL. 1975. Bean production systems program. *Cali, Colombia, CIAT*, *Ser.* FE—5, 38 pp.

CENTRO INTERNACIONAL DE AGRICULTURA TROPICAL. 1976. *Abstracts on field beans* (*Phaseolus vulgaris* L.). Volume 1. Cali., Colombia, Cent. Int. Agric. Trop. (Bean Information Center), 494 pp.

CHOUDHURY, B. 1970. Beans. *Pulse crops of India*. (Kachroo, P. and Arif, M. eds.). pp. 233–255. New Delhi: Indian Counc. Agric. Res., 334 pp.

CHRISTEN, R. and ECHANDI, E. 1967. Razas fisiólogicas más communes de la roya *Uromyces phaseoli* var. *phaseoli* en Costa Rica y evaluación de la resistencia de algunos cultivares de frijol a la roya. [The commonest physiological races of bean rust, *Uromyces phaseoli* var. *phaseoli*, in Costa Rica and the assessment of rust resistance in some bean cultivars]. *Turrialba*, *17*, 7–10.

COMMONWEALTH BUREAU OF PASTURES AND FIELD CROPS. nd. *Phaseolus vulgaris*, general agronomy including fertilizers. Annotated bibliography 1313, 1948–72. *Commonw. Agric. Bur.*, *Bur. Pastures & Field Crops, Hurley, Maidenhead, Berks.*, *England*, 14 pp.

COYNE, D. P. and SCHUSTER, M. L. 1971. Breeding Great Northern dry beans tolerant to the common blight (*Xanthomonas phaseoli*) and wilt bacteria (*Corynebacterium flaccumfaciens*). *Rep. 10th dry bean res. conf.* (1970). pp. 7–11. *US Dep. Agric.*, *Agric. Res. Serv.*, ARS, 74–56, 146 pp.

CRAMP, K. V., EWAN, J. W., HUME, W. G., SHEARD, G. F. and FINCH, C. C. 1962. Beans. pp. 4–11. *Lond.*, *Minist. Agric.*, *Fish. Food Bull. 87*, 24 pp.

DAVIES, J. C. 1959. A note on the control of bean pests in Uganda. *East Afr. Agric. J.*, *24*, 174–178.

DAVIES, J. C. 1962. A note on in-sack storage of beans using 0.04% gamma BHC dust. *East Afr. Agric. For. J.*, *27*, 223–224.

DAVIS, D. R. 1976. Effect of blanching methods and processes on quality of canned dried beans. *Food Prod. Dev.*, *10* (7), 74–76; 78.

DAVIS, J. J. 1969. Beanfly and its control. *Qd. Agric. J.*, *95*, 101–106.

DEAN, L. L. and LA FERRIERE, L. 1958. Diseases of beans in Idaho. *Univ. Idaho, Idaho Coll. Agric., Agric. Exp. Stn. Bull.* 293, 19 pp.

DICKSON, M. H. and HACKLER, L. R. 1975. Protein quantity and quality in high-yielding beans. *Nutritional improvement of food legumes by breeding.* (Milner, M. ed.). pp. 185–192. New York/London: John Wiley and Sons, 399 pp.

DOWNEY, R. L. 1959. The sloughing problem in canned green beans. *Cann. Trade*, *81* (44), 9–10.

ECHANDI, E. 1966. Principales enfermedades del frijol observadas en diferentes zonas écológicas de Costa Rica. [Principal bean diseases observed in different ecological zones of Costa Rica]. *Turrialba*, *16*, 359–363.

ECHANDI, E. 1967. Amarillamiento del frijol (*Phaseolus vulgaris* L.) provocado por *Fusarium oxysporum* f. *phaseoli.* [Yellowing of beans (*Phaseolus vulgaris* L.) caused by *Fusarium oxysporum* f. *phaseoli*]. *Turrialba*, *17*, 409–410.

EDJE, O. T., AYONOADU, U. W. U. and MUGHOGHO, L. K. 1971. Effects of fertilizer on the yield of beans under rain-fed and irrigated conditions. *Univ. Malawi, Bunda Coll. Agric., Res. Bull.*, 2, 20–28.

EDJE, O. T., MUGHOGHO, L. K. and AYONOADU, U. W. U. 1975. Bean yield and yield components as affected by fertilizer and plant population. *Turrialba*, *25*, 79–84.

EIJNATTEN, C. L. M. Van. 1975. Report on a literature review and field study of agriculture in Kirinyaga district with special reference to beans (*Phaseolus vulgaris*). *Kenya, Univ. Nairobi, Facul. Agric., Dep. Crop. Sci., Tech. Commun.* 13, 35 pp.

ELIÁS, L. G., BRESSANI, R. and FLORES, M. 1973. Problems and potentials in storage and processing of food legumes in Latin America. Potential of field beans and other food legumes in Latin America. (Wall, D. ed.). pp. 52–87. *Cali, Colombia, CIAT, Ser. Semin.* 2E, 388 pp.

Haricot bean

Elías, L. G., Cristales, F. R., Bressani, R. and Miranda, H. 1976. Composición quimica y valor nutritivo de algunas legumisosas de grano. [Chemical composition and nutritive value of some grain legumes]. *Turrialba, 26*, 375–380.

Elías, L. G., Hernández, M. and Bressani, R. 1976. The nutritive value of precooked legume flours processed by different methods. *Nutr. Rep. Int., 14*, 385–403.

El-Nadi, A. H. 1975. Water relations of beans: (iii) Pod and seed yield of haricot beans under different irrigation in the Sudan. *Exp. Agric., 11*, 155–158.

El Nahry, F., Darwish, N. M. and Tharwat, S. 1977. Effect of preparation and cooking on the nutritive value of local kidney bean (*Phaseolus vulgaris* var. Giza 3). *Qual. Plant, Plant Food Human Nutr., 27*, 141–150.

Enriquez, G. A. 1976. Effect of temperature and daylength on time of flowering in beans, *Phaseolus vulgaris* L. *New York State Coll. Agric. Life Sci., Cornell Inst. Agric., Diss. Abstr., Ser.* 1, *3*, 8–10.

Enyi, A. C. 1975. Effect of plant population on grain yield, production and distribution of dry matter in beans (*Phaseolus vulgaris*). *Ghana J. Sci., 15*, 159–169.

Erdmann, M. H., Robertson, L. S., Janes, R. L., White, R. G., Adams, M. W. and Andersen, A. L. 1965. Field bean production in Michigan. *Michigan State Univ., Ext. Bull. 513*, (*Farm Sci. Ser.*), 10 pp.

Evans, A. M. 1975. Genetic improvement of *Phaseolus vulgaris. Nutritional improvement of food legumes by breeding.* (Milner, M. ed.). pp. 107–115. New York/London: John Wiley and Sons, 399 pp.

Evans, A. M. 1976. Beans. *Evolution of crop plants.* (Simmonds, N. W. ed.). pp. 168–172. London: Longman Group Ltd, 339 pp.

Fassbender, H. W. 1967. La fertilización de frijol (*Phaseolus* sp.). [The manuring of beans (*Phaseolus* sp.)]. *Turrialba, 17*, 46–52.

Feinberg, B. 1973. Vegetables. *Food dehydration.* (Arsdel, W. B. Van, Copley, M. J. and Morgan, A. I. eds.). 2nd ed. Vol. 2. pp. 1–82. Westport, Connecticut: Avi Publishing Co Inc, 529 pp.

FEINBERG, B., WINTER, F. and ROTH, T. L. 1968. The preparation for freezing and freezing of vegetables. *The freezing preservation of foods*. (Tressler, D. K., Arsdel, W. B. Van and Copley, M. J. eds.). 4th ed. Vol. 3. pp. 150–194. Westport, Connecticut: Avi Publishing Co Inc, 486 pp.

FENNEMA, O. and WECKEL, K. G. 1961. Factors influencing the physical and chemical properties of dehydrated green snap beans. *Univ. Wisconsin Res. Bull.* 224, 24 pp.

FERNANDEZ, F. and FRANKLIN, D. L. 1973. Bean production systems. Potential of field beans and other food legumes in Latin America. (Wall, D. ed.). pp. 188–217. *Cali, Colombia, CIAT, Ser. Semin.* 2E, 388 pp.

FODA, Y. H., EL-WARAKI, A. and ZAID, M. A. 1967. Effect of dehydration, freeze-drying and packaging on the quality of green beans. *Food Technol.*, *21*, 1021–1024.

FOOD AND AGRICULTURE ORGANIZATION OF THE UNITED NATIONS. 1970. Amino acid content of foods and biological data on proteins. *FAO Nutr. Stud.* 24, pp. 50–51. Rome: FAO, 285 pp.

FOOTE, L. E. and CHURCHILL, B. R. 1962. A study of chemical and cultural weed control treatments in navy, cranberry and kidney beans. *Michigan State Univ. Q. Bull.*, (45), pp. 318–324.

FOSTER, G. H. and FISCUS, D. E. 1969. Physical damage from mechanical handling of pea beans. *Rep. 9th dry bean res. conf.* (1968). pp. 68–72. *US Dep. Agric., Agric. Res. Serv.*, ARS 74–50, 94 pp.

FRANCIS, C. A. (ed.). 1974. Field beans. *Guide for field crops in the tropics and the subtropics*. (Litzenberger, S. C. ed.). pp. 101–108. Washington, Off. Agric., Tech. Assist. Bur., Agency Int. Dev., 321 pp.

FRANCIS, C. A. and HERNÁNDEZ-BRAVO, G. 1974. The Latin American program for increasing yields of dry beans. *1st ASEAN workshop on grain legumes. Bogor, Indonesia, Minist. Agric., ASEAN 74 FA/Wrks., GL1/Sdo-02, Pap.* 35, 17 pp.

FROUSSIOS, G. 1970. Genetic diversity and agricultural potential in *Phaseolus vulgaris* L. *Exp. Agric.*, *6*, 129–141.

GABRIAL, G. N., HUSSEIN, L. and MORCOS, S. R. 1975. Some nutritional studies on kidney bean proteins (*Phaseolus vulgaris* var. Giza 3). *Qual. Plant., Plant Food Hum. Nutr.*, *24*, 61–70.

Haricot bean

GALLACHER, E. C. 1972. Navy beans. *Qd. Agric. J.*, *98*, 562–567.

GANE, A. J., KING, J. M., GENT, G. P., BIDDLE, A. J., HANDLEY, R. D. and BINGHAM, R. J. B. 1975. *Pea and bean growing handbook*. Vol. 2. *Beans*. Peterborough, England, Processors and Growers Res. Organ., 7 sections, (loose leaf).

GENTRY, H. S. 1969. Origin of the common bean, *Phaseolus vulgaris*. *Econ. Bot.*, *23*, 55–69.

GIUDICE, P. M. DEL, ALVARENGA, S. C. DE, CONDE, A. and DOS, R. 1973. Estudo comparativo de diferentes processos de armazenagem de feijão preto. [A comparison of various black bean storage methods]. *Experientiae*, *13*, 273–313. (*Hortic. Abstr.*, *43*, 2036).

GRIEG, J. K. and GWIN, R. E. (JR.). 1966. Dry bean production in Kansas. *Kansas State Univ., Agric. Exp. Stn. Bull.* 486, 19 pp.

GROESCHEL, E. C., NELSON, A. I. and STEINBERG, M. P. 1966. Changes in colour and other characteristics of green beans stored in controlled refrigerated atmospheres. *J. Food Sci.*, *31*, 488–496.

GUTHRIE, J. W., DEAN, L. L., BUTCHER, C. L., FENWICK, H. S. and FINLEY, A. M. 1975. The epidemiology and control of halo blight in Idaho. *Univ. Idaho Coll. Agric., Agric. Exp. Stn. Bull.* 550, 11 pp.

HABISH, H. A. 1972. Aflatoxin in haricot bean and other pulses. *Exp. Agric.*, *8*, 135–137.

HABISH, H. A. and ISHAG, H. M. 1974. Nodulation of legumes in the Sudan: (iii) Responses of haricot bean to inoculation. *Exp. Agric.*, *10*, 45–50.

HACKLER, L. R. and DICKSON, M. H. 1973. A comparison of the amino acid and nitrogen content of pods and seeds of beans (*Phaseolus vulgaris* L.). *New York State Agric. Exp. Stn. Geneva, Search Agric.*, *3* (5), 1–6.

HANSSEN, K. B. 1970. Production of seed beans for export. *Rhod. Agric. J.*, *67*, 45–50.

HARTMAN, R. W. 1969. Photoperiod responses of *Phaseolus* plant introduction in Hawaii. *J. Am. Soc. Hortic. Sci.*, *94*, 437–440.

HELLENDOORN, E. W. 1974. Digestibility and flatulence activity of beans. *1st ASEAN workshop on grain legumes. Bogor, Indonesia, Minist. Agric., ASEAN 74 FA/Wrks., GL1/Wop-08, Pap.* 23, 17 pp.

HERNÁNDEZ-BRAVO, G. 1973. Potentials and problems of production of dry beans in the lowland tropics. Potential of field beans and other food legumes in Latin America. (Wall, D. ed.). pp. 144–156. *Cali, Colombia, CIAT Ser. Semin.* 2E, 388 pp.

HILLS, W. A., DARBY, J. F., THAMES, W. H. (JR.) and FORSEE, W. T. (JR.). 1953. Bush snap bean production on the sandy soils of Florida. *Univ. Florida Agric. Exp. Stn. Bull.* 530, 23 pp.

HOFF, J. E. and NELSON, P. E. 1967. Methods of accelerating the processing of dry beans. *Rep. 8th dry bean res. conf.* (1966). pp. 39–49. *US Dep. Agric., Agric. Res. Serv.*, ARS 74–41, 76 pp.

HOLMAN, L. E. and SEFCOVIC, M. S. 1961. Aeration of bulk-stored pea beans. *US Dep. Agric., Agric. Mark. Res. Rep.* 481, 43 pp.

HOWLAND, A. K. 1966. East African bean rust studies. *East Afr. Agric. For. J.*, *32*, 208–210.

HUGHES, P. A. and SANDSTED, R. F. 1975. Effect of temperature, relative humidity and light on the colour of California light red kidney bean seed during storage. *Hortscience*, *10*, 421–423.

INNES, N. L. and HARDWICK, R. C. 1974. Possibilities for the genetic improvement of *Phaseolus* beans in the UK. *Outlook on Agric.*, *8*, 126–132.

INSTITUTO INTERAMERICANO DE CIENCIAS AGRICOLAS DE LA OEA. 1972. Bibliografia frijol (*Phaseolus* spp.). [Beans (*Phaseolus* spp.), bibliography]. *Turrialba, Costa Rica, Cent. Interam. Doc. Inf. Agric., Bibliogr.* 4, 299 pp.

ISHINO, K., ORTEGA, D. M. L. 1975. Fractionation and characterization of major reserve proteins from seeds of *Phaseolus vulgaris*. *J. Agric. Food Chem.*, *23*, 529–533.

JAFFÉ, W. G. 1973. Factors affecting the nutritional value of beans. *Nutritional improvement of food legumes by breeding.* (Milner, M. ed.). pp. 43–48. New York/London: John Wiley and Sons, 399 pp.

JAFFÉ, W. G. and GOMEZ, M. J. 1975. Beans of high or low toxicity. *Qual. Plant., Plant Food Hum. Nutr.*, *24*, 359–365.

JARVIS, C. D. 1908. American varieties of beans. *New York State Coll. Agric. Cornell Univ., Agric. Exp. Stn. Bull.* 260, 110 pp.

JOHNSON, R. M. 1965. Maintenance of dry bean quality during transportation, handling and storage. *Rep. 7th dry bean res. conf.* (1964). pp. 20–23. *US Dep. Agric., Agric. Res. Serv.*, ARS 74–32, 94 pp.

Haricot bean

KAKADE, M. L. and EVANS, R. J. 1965. Growth depression of rats fed fractions of raw navy beans. *Rep. 7th dry bean res. conf.* (1964). pp. 61–63. *US Dep. Agric., Agric. Res. Serv.*, ARS 74–32, 94 pp.

KAKADE, M. L. and EVANS, R. J. 1965. Nutritive value of different varieties of navy beans. *Michigan State Univ., Agric. Exp. Stn. Q. Bull.*, (48), pp. 89–93.

KAPLAN, L. 1965. Archaeology and domestication in American *Phaseolus* (beans). *Econ. Bot.*, *19*, 358–368.

KASASIAN, L. 1971. Vegetable crops. *Weed control in the tropics.* pp. 163–175. London: Leonard Hill Books, 307 pp.

KELLY, J. F. and RHODES, B. B. 1975. The potential for improving the nutrient composition of horticultural crops. *Food Technol.*, *29*, 134–140.

KERR, W. E. 1962. Bean seed production. *Rhod. Agric. J.*, *59*, 159–164.

KING, P. 1975. The market for French beans in selected Western European countries. *Rep. Trop. Prod. Inst.* G 92, 56 pp.

KON, S., WAGNER, J. R. and BECKER, R. 1971. Inactivation of bean enzymes by control of pH to produce highly nutritious food products. *Rep. 10th dry bean res. conf.* (1970). pp. 91–107. *US Dep. Agric., Agric. Res. Serv.*, ARS 74–56, 146 pp.

KOOISTRA, E. 1971. Germinability of beans (*Phaseolus vulgaris*) at low temperatures. *Euphytica*, *20*, 208–213.

LABELLE, R. L., HACKLER, L. R. and DANIEWSKI, M. H. 1969. Processing pre-cooked dehydrated beans for suitable balance of quality and nutritional value. *Rep. 9th dry bean research conf.* (1968). pp. 53–64. *US Dep. Agric., Agric. Res. Serv.*, ARS 74–50, 94 pp.

LANTZ, E. M., GOUGH, H. W. and CAMPBELL, A. M. 1958. Effect of variety, location and years on the protein and amino acid content of dried beans. *J. Agric. Food Chem.*, *6*, 58–60.

LEAKEY, C. L. A. 1970. The improvement of beans (*Phaseolus vulgaris*) in East Africa. Crop improvement in East Africa. (Leakey, C. L. A. ed.). pp. 99–128. *Commonw. Agric. Bur., Bur. Plant Breed. Genet., Tech. Commun.* 19, 280 pp.

LEAKEY, C. L. A. and SIMBWA-BUNNYA, M. 1972. Races of *Colletotrichum lindemuthianum* and implications for bean breeding in Uganda. *Ann. Appl. Biol.*, *70*, 25–34.

LE BARON, M., DEAN, L. L. and PORTMAN, R. 1969. Bean production in Idaho. *Univ. Idaho Coll. Agric. Exp. Stn. Bull.* 282, (revised), 27 pp.

LEGGETT, E. K. 1973. Growing edible dry beans. *Agric. Gaz. N.S.W.*, *84*, 198–202.

LEPIGRE, M. 1965. Étude sur les possibilités d'amélioration de la conservation des haricots au Togo en milieu rural. [A study of the possibilities of improving the storage of beans in rural areas of Togo]. *Agron. Trop.*, *20*, 388–430.

LEWIS, W. E. 1958. Refrigeration and handling of two vegetables at retail. Green snap beans and southern yellow summer squashes. *US Dep. Agric., Agric. Mark. Res. Rep.* 276, 6 pp.

LIENER, I. E. 1967. Phytohemagglutinins in dry beans. *Rep. 8th dry bean res. conf.* (1966). pp. 61–67. *US Dep. Agric., Agric. Res. Serv.*, ARS 74–41, 76 pp.

LIENER, I. E. 1973. Toxic factors associated with legume proteins. *Indian J. Nutr. Diet.*, *10*, 303–322.

LIENER, I. E., YADAV, N. K. and BROWN, C. 1977. The nutritive value of dry roasted bean flour. *League Int. Food Educ. Newsl.*, January, pp. 1–3.

LOCK, A. 1969. Canning of vegetables: (i) Green beans; (ii) Beans in tomato sauce. *Practical canning*. (i) pp. 225–242; (ii) pp. 242–256. London: Food Trade Press Ltd, 415 pp.

LOPEZ, A. 1975. (i) Beans, green and wax; (ii) Pork and beans (beans with pork); (iii) Red kidney beans. *A complete course in canning*. 10th ed. (i) pp. 348–355; (ii) pp. 573–580; (iii) pp. 580–587. Baltimore, Maryland: The Canning Trade Inc, 755 pp.

MACARTNEY, J. C. 1966. The selection of haricot bean varieties suitable for canning. *East Afr. Agric. For. J.*, *32*, 214–219.

MACARTNEY, J. C. and WATSON, D. R. W. nd. Beans. *Tanzania, Min. Agric. For. Wildl. Bull.* 20, (*Tengeru Rep.* 80), 28 pp.

MARSHALL, J. J. and LAUDA, C. M. 1975. Purification and properties of phaseolamin, an inhibitor of α-amylose, from the kidney bean *Phaseolus vulgaris*. *J. Biol. Chem.*, *250*, 8030–8037.

MCFARLANE, J. A. 1970. Pulses. *Food storage manual*. pp. 335–346. Rome: FAO, World food programme, 799 pp., plus indexes, (loose leaf).

Haricot bean

McGregor, W. G. and Wallen, V. R. 1966. Field beans in Canada. *Can., Dep. Agric. Publ.* 843, 15 pp.

McMillan, R. T. (Jr.). 1969. Post-harvest control of *Sclerotinia sclerotiorum* of pole beans. *Proc. Florida State Hortic. Soc., 82*, 139–140.

Miller, C. F., Guadagni, D. G. and Kon, S. 1973. Vitamin retention in bean products: cooked, canned and instant bean powders. *J. Food Sci., 38*, 493–495.

Mislivec, P. B., Dieter, C. T. and Bruce, V. R. 1975. Mycotoxin producing potential of mold flora of dried beans. *Appl. Microbiol., 29*, 522–526.

Molina, M. R., Baten, M. A., Gomez-Brenes, R. A., King, K. W. and Bressani, R. 1976. Heat treatment: A process to control the development of the hard-to-cook phenomenon in black beans (*Phaseolus vulgaris*). *J. Food Sci., 41*, 661–666.

Molina, M. R., Fuente, G. de la, and Bressani, R. 1975. Inter-relationships between storage, soaking time, cooking time, nutritive value and other characteristics of the black bean (*Phaseolus vulgaris*). *J. Food Sci., 40*, 587–591.

Molina, M. R., Gudiel, H., Fuente, G. de la, and Bressani, R. 1974. Use of *Phaseolus vulgaris* in high protein quality products. *Madrid, 4th Int. Congr. Food Sci. & Technol., Work Doc. 8b, Exploiting local food raw materials, Pap. 5*, pp. 12–14.

Moraes, R. M. de and Angelucci, E. 1971. Chemical composition and amino acid contents of Brazilian beans (*Phaseolus vulgaris*). *J. Food Sci., 36*, 493–494.

Moreira, M. A., Brune, W. and Batista, C. M. 1976. Avaliação do teor de metionine em sementes de feijão (*Phaseolus vulgaris* L.). [Methionine content in beans (*Phaseolus vulgaris* L.)]. *Turrialba, 26*, 225–231.

Morris, H. J. 1965. Changes in cooking qualities of raw beans as influenced by moisture content and storage time. *Rep. 7th dry bean res. conf.* (1964). pp. 37–44. *US Dep. Agric., Agric. Res. Serv.*, ARS 74–32, 94 pp.

Morris, H. J. and Seifert, R. M. 1962. Constituents and treatments affecting cooking of dry beans. *Rep. 5th dry bean res. conf.* (1961). pp. 42–46. *US Dep. Agric., Agric. Res. Serv.*, 62 pp.

Morris, H. J. and Wood, E. R. 1956. Influence of moisture content on keeping quality of dry beans. *Food Technol., 10*, 225–229.

MORRISON, K. J. and BURKE, D. W. 1962. Growing field beans in the Columbia basin. *Washington State Univ., Agric. Sci. Ext. Bull.* 497, 12 pp.

MUCKLE, T. B. and STIRLING, H. G. 1971. Review of the drying of cereals and legumes in the tropics. *Trop. Stored Prod. Inf.*, (22), pp. 11–30.

MUNETA, P. 1964. The cooking time of dry beans after extended storage. *Food Technol.*, *18*, 1240–1241.

MURPHY, E. L., KON, S. and SEIFERT, R. M. 1965. The preparation of bland, colourless, nonflatulent high-protein concentrates from dry beans. *Rep. 7th dry bean res. conf.* (1964). pp. 63–65. *US Dep. Agric., Agric. Res. Serv.*, ARS 74–32, 94 pp.

NAKAGAWA, Y. 1957. Snap bean growing in Hawaii. *Univ. Hawaii Ext. Circ.* 383, 11 pp.

NANCARROW, J. 1976. Dry-edible beans: A high return, but high risk crop. *J. Agric., Vict. Dep. Agric.*, *74*, 267–272.

NATTI, J. J. 1965. Control of halo blight with streptomycin. *Rep. 7th dry bean res. conf.* (1964). pp. 23–24. *US Dep. Agric., Agric. Res. Serv.*, ARS 74–32, 94 pp.

NATTI, J. J. 1971. Control of halo blight of bean by foliage sprays. *New York State, Agric. Exp. Stn. Geneva, Search Agric.*, *1* (11), 1–9.

NGUNDO, B. W. and TAYLOR, D. P. 1975. Comparative development of *Meloidogyne incognita* and *M. javanica* in six bean cultivars. *East Afr. Agric. For. J.*, *41*, 72–75.

PALMER, R., MCINTOSH, A. and PUSZTAI, A. 1973. The nutritional evaluation of kidney beans (*Phaseolus vulgaris*): The effect on nutritional value of seed germination and changes in trypsin inhibitor content. *J. Sci. Food Agric.*, *24*, 937–944.

PASSLOW, T. 1969. Pest control for quality navy beans. *Qd. Agric. J.*, *95*, 711–712.

PAYNE, H. W. 1972. Improving the yield of red peas in Jamaica. *Cajanus*, *5*, 277–284.

PINCHINAT, A. M. 1966. El cultivo del frijol en Centro America. [The cultivation of the bean in Central America]. *Ext. Am.*, *11* (2), 27–32.

PORRITT, S. W. 1974. Beans green or snap. Commercial storage of fruits and vegetables. p. 27. *Can. Dep. Agric. Publ.* 1532, 56 pp.

PURSEGLOVE, J. W. 1968. *Phaseolus vulgaris* L. *Tropical crops: Dicotyledons.* Vol. 1. pp. 304–310. London: Longmans, Green and Co Ltd, 332 pp.

PUSZTAI, A. and PALMER, R. 1977. Nutritional evaluation of kidney beans (*Phaseolus vulgaris*): The toxic principle. *J. Sci. Food Agric.*, *28*, 620–623.

RAWSON, J. E. 1969. Use of pre-emergence herbicides in navy beans. *Qd. J. Agric. Anim. Sci.*, *26*, 231–234.

ROBERTSON, J. K. 1955. The growing of canning beans in Tanganyika. *World Crops*, 7, 23–25.

ROBINS, J. S. and HOWE, O. W. 1961. Irrigating dry beans in the west. *US Dep. Agric.*, *Leafl.* 499, 6 pp.

ROCKLAND, L. B., HEINRICH, J. D. and DORNBACK, K. J. 1971. Recent progress on the development of new and improved quick-cooking products from lima and other dry beans. *Rep. 10th dry bean res. conf.* (1970). pp. 121–131. *US Dep. Agric.*, *Agric. Res. Serv.*, ARS 74–56, 146 pp.

RUTGER, J. N. 1971. Variation in protein content and its relation to other characters in beans (*Phaseolus vulgaris* L.). *Rep. 10th dry bean res. conf.* (1970). pp. 59–83. *US Dep. Agric.*, *Agric. Res. Serv.*, ARS 74–56, 146 pp.

RYALL, A. L. and LIPTON, W. J. 1972. Commodity requirements of unripe fruits and miscellaneous structures. *Handling, transportation and storage of fruits and vegetables.* Vol. 1. pp. 122–141. Westport, Connecticut, Avi Publishing Co Inc, 473 pp.

SAETTLER, A. W. and POTTER, H. S. 1970. Chemical control of halo bacterial blight in field beans. *Michigan State Univ.*, *Agric. Exp. Stn. East Lansing Res. Rep.* 98, (*Farm Sci.*), 8 pp.

SANDSTED, R. F. 1966. Commercial snap bean production in New York State. *New York State Coll. Agric.*, *Cornell Univ.*, *Ext. Bull.* 1163, 30 pp.

SANDSTED, R. F., HOW, R. B., MUKA, A. A. and SHERF, A. F. 1971. Growing dry beans in New York State. *New York State Coll. Agric.*, *Cornell Univ.*, *Inf. Bull.* 2, (*Plant Sci.*; *Veg. Crops* 1), 22 pp.

SANTOS, H. P. (Filho) and WAITE, B. H. 1969. Algunas doenças importantes do feijão na Bahia. [Some important diseases of beans in Bahia]. *Bahia, Inst. Pesqui. Exp. Agropecu. Leste, Circ.* 15, 35 pp.

SATTERLEE, L. D., BEMBERS, M. and KENDRICK, J. G. 1975. Functional properties of the Great Northern bean (*Phaseolus vulgaris*) protein isolate. *J. Food Sci.*, *40*, 81–84.

Scarisbrick, D. H. 1976. Navy beans seem to face an uncertain future. *Commer. Grow.*, (4175), pp. 97–99.

Scarisbrick, D. H., Carr, M. K. V. and Wilkes, J. M. 1976. The effect of sowing date and season on the development and yield of navy beans (*Phaseolus vulgaris*) in south-east England. *J. Agric. Sci.*, (*Camb.*), *86*, 65–76.

Scarisbrick, D. H., Wilkes, J. M. and Kempson, R. 1977. The effect of varying plant population density on the seed yield of navy beans (*Phaseolus vulgaris*) in south-east England. *J. Agric. Sci.* (*Camb.*), *88*, 567–577.

Schieber, E. 1970. Enfermedades del frijol (*Phaseolus vulgaris*) en la Republica Dominica. [Diseases of the bean (*Phaseolus vulgaris*) in the Dominican Republic]. *Turrialba*, *20*, 20–23.

Sevilla, U. L. and Luh, B. S. 1974. Several factors influencing color and texture of canned red kidney beans. *Madrid, 4th Int. Congr. Food Sci. & Technol., Work Doc. 1a, Chemical constituents of foods in relation to flavour, color and texture. Pap.* 20, pp. 52–54.

Siegel, A. and Fawcett, B. 1976. *Food legume processing and utilization* (*with special emphasis on application in developing countries*). Ottawa, Int. Dev. Res. Cent., IDRC–TS1, 88 pp.

Silbernagel, M. J. 1971. Bean protein improvement work by USDA: Bean and pea investigations. *Rep. 10th dry bean res. conf.* (1970). pp. 70–83. *US Dep. Agric., Agric. Res. Serv.*, ARS 74–56, 146 pp.

Silbernagel, M. J. and Burke, D. W. 1973. Harvesting high quality bean seed with a rubber-belt thresher. *Washington State Univ., Agric. Exp. Stn. Coll. Bull*, 777, 7 pp.

Simbwa-Bunnya, M. 1972. Fungicidal control of bean diseases at Kawanda, Uganda. *Plant Dis. Rep.*, *56*, 901–903.

Simpson, I. H. 1969. Bean waste a potential sheep feed. *Agric. Gaz. N.S.W.*, *80*, 30.

Singh, B. and Linvill, D. E. 1977. Field drying of navy beans in the harvesting period. *Trans. Am. Soc. Agric. Engin.*, *20*, 228–231.

Singh, R. P., Buelow, F. H. and Lund, D. B. 1973. Storage behaviour of artificially waxed snap beans. *J. Food Sci.*, *38*, 542–543.

Smartt, J. 1976. *Tropical pulses*. pp. 70–74; 102; 106; 112–114; 246–260. London: Longman Group Ltd, 348 pp.

Haricot bean

SMITH, M. A., MCCOLLOCH, L. P. and FRIEDMAN, B. A. 1966. Snap beans. Market diseases of asparagus, onions, beans, peas, carrots, celery and related vegetables. pp. 21–31. *US Dep. Agric., Agric. Res. Serv., Agric. Handb.* 303, 85 pp.

SMITTLE, D. A. and WILLIAMSON, R. E. 1976. Potential for pea bean (navy bean) production in Georgia. *Tifton, Georgia, Univ. Georgia Coll. Agric., Agric. Exp. Stn. Res. Rep.* 222, 14 pp.

SPURLING, A. T. 1973. Field trials with Canadian Wonder beans in Malawi. *Exp. Agric., 9*, 97–105.

STEINKRAUS, K. H., BUREN, J. P. VAN, LABELLE, R. L. and HAND, D. B. 1964. Some studies on the production of precooked dehydrated beans. *Food Technol., 18*, 1945.

STEPHENS, D. 1967. The effects of ammonium sulphate and other fertilizer and inoculation treatments on beans (*Phaseolus vulgaris*). *East Afr. Agric. For. J., 32*, 411–417.

STOBBE, E. H., ORMROD, D. P. and WOOLLEY, C. J. 1965. Blossoming and fruit set patterns in *Phaseolus vulgaris* L. as influenced by temperature. *Can. J. Bot., 44*, 813–819.

THOMPSON, J. A., SEFCOVIC, M. S. and KINGSOLIVER, C. H. 1962. Maintaining quality of pea beans during shipment overseas. *US Dep. Agric., Agric. Mark. Res. Rep.* 519, 43 pp.

TOOLE, E. H., TOOLE, K. V., LAY, B. J. and CROWDER, J. T. 1951. Injury to seed beans during threshing and processing. *US Dep. Agric. Circ.* 874, 10 pp.

TORREY, M. 1974. Vegetables: Beans. *Food technology review 13: Dehydration of fruits and vegetables.* pp. 145–156. New Jersey/London: Noyes Data Corp, 286 pp.

TRIM, L. G. 1957. Liming and trace element problems in the bean crop. *Qd. Agric. J., 83*, 327–330.

UNITED STATES DEPARTMENT OF AGRICULTURE. 1966. Control of insects that attack dry beans and peas in storage. *US Dep. Agric., Agric. Res. Serv., Agric. Inf. Bull.* 303, 8 pp.

UNITED STATES DEPARTMENT OF AGRICULTURE. 1968. Controlling the Mexican bean beetle. *US Dep. Agric. Leafl.* 548, 8 pp.

VIEIRA, C. 1966. Effect of seed age on germination and yield of field bean (*Phaseolus vulgaris* L.). *Turrialba, 16*, 396–398.

WELLS, J. M. and COOLEY, T. N. 1973. Control of *Phythium* and *Sclerotinia* rots of snap beans with postharvest hot water and chemical dips. *Plant Dis. Rep., 57*, 234–236.

WESTPHAL, E. 1974. *Phaseolus vulgaris* L. Pulses in Ethiopia, their taxonomy and agricultural significance. pp. 159–176. *Wageningen, PUDOC, Cent. Agric. Publ. Docum, Agric. Res. Rep.* 815, 278 pp.

WILLMAN, J. P., JOHNSON, G. R. and BRANNON, W. F. 1961. Cull beans for fattening lambs. *New York State Coll. Agric., Cornell Univ., Agric. Exp. Stn. Bull.* 959, 16 pp.

WOODROOF, J. G., HEATON, E. K. and ELLIS, C. 1962. Freezing green snap beans. *Univ. Georgia, Agric. Exp. Stn. Bull. NS* 90, 43 pp.

XUAN, N. V. 1969. Structure microscopique et conditions de congélation et d'entreposage des haricots verts. [The microscopic structure and the conditions of freezing and storing French beans]. *Bull. Inst. Int. Froid, Suppl. 6, Aliment. Congéles*, pp. 41–46.

YARNELL, S. H. 1965. Cytogenetics of the vegetable crops: (iv) Legumes, garden bean, *Phaseolus vulgaris* L. *Bot. Rev., 31*, 250–292.

ZAUMEYER, W. J. 1954. Snap beans for marketing, canning and freezing. *US Dep. Agric., Farmers' Bull.* 1915, 16 pp.

ZAUMEYER, W. J. 1965. Visits to dry bean growing areas in Brazil and El Salvador. *Rep. 7th dry bean res. conf.* (1964). pp. 83–89. *US Dep. Agric., Agric. Res. Serv.*, ARS 74–32, 94 pp.

ZAUMEYER, W. J. 1973. Goals and means for protecting *Phaseolus vulgaris* in the tropics. Potentials of field beans and other food legumes in Latin America. (Wall, D. ed.). pp. 218–228. *Cali, Colombia, CIAT*, *Ser. Semin.* 2E, 388 pp.

ZAUMEYER, W. J. and MEINERS, J. P. 1975. Disease resistance in beans. *Ann. Rev. Phytopathol., 13*, 313–334.

ZAUMEYER, W. J. and THOMAS, H. R. 1957. A monographic study on bean diseases and methods for their control. *US Dep. Agric., Tech., Bull.* 868, 255 pp.

Haricot bean

ZAUMEYER, W. J. and THOMAS, H. R. 1962. Diseases of snap and dry beans. *US Dep. Agric., Agric. Res. Serv. Agric. Handb.* 225, 39 pp.

ZINK, E. and ALMEIDA, L. D'A DE. 1970. Estudios sôbre a conservação de sementes de feijoeiro. [Studies on the storage of dry bean seed]. *Bragantia, 29* (1), Nota 10, XLV–XLVII.

Common names HORSE GRAM[1], Horse grain, Kulthi (bean), Madras gram.

Botanical name *Macrotyloma uniflorum* (Lam.) Verdc., syn. *Dolichos uniflorus* Lam. *D. biflorus* auct. *non* L.

Family Leguminosae.

Other names Dolico cavallino (It.); Grain de cheval (Fr.); Hurali (Ind.); Kalai[2] (Assam); Kallu (Ind.); Kollu (Sri La.); Kulat, Kulatha, Kurtikalai, Muthera, Muthira, Muthiva (Ind.); Pè-bi-zât (Burm.); Pferdekorn (Ger.); Ulavalu (Ind.).

Botany

A sub-erect, trailing annual, 12–20 in. (30–50 cm) high, with many slender branches arising from the base of the plant. All parts are covered with soft, grey hairs. The leaves are trifoliolate with a conspicuous stipule. The petioles are about 2 in. (5 cm) long and the leaflets are membranous, yellowish-green in colour, about 1–2 in. (2.5–5 cm) long. The flowers (1–3), are borne on axillary racemes clustered at node-like thickenings of the peduncle. They are creamy-yellow in colour and about 0.4 in. (1.0 cm) in length. The pods are linear, about 1.6–2 in. (4–5 cm) long, recurved, beaked, downy and dehiscent. They contain 5–7 small seeds which may be self-coloured or mottled.

The botanical names applied to the horse gram are confusing. Until recently it was considered to belong to the genus *Dolichos* and it continues to be referred to as *Dolichos uniflorus* in much of the literature. However, Verdcourt (1970) has shown that it belongs to a quite distinct genus *Macrotyloma*, although in 1977 evidence was brought together to show that *Kerstingiella* is also congeneric.

There are a number of different cultivars of horse gram grown in India, which differ in photoperiodic response, period of maturity and seed colour, but very little breeding work has been carried out to improve the crop. A short-duration, high yielding cultivar Co-1 has been released to farmers. Other cultivars which have shown promise include BG–M1–1, BR-10, BR–5, No 35 and DB–7. Recently work has been undertaken on the development of day-neutral cultivars with a high yield potential.

---

[1] Sometimes used for the chick pea, *Cicer arietinum*.
[2] Sometimes used for urd, *Vigna mungo*.

Horse gram

Origin and distribution

The horse gram probably originated in the SE. Asian sub-continent, and is an important crop in S. India. It is also cultivated on a more limited scale in other parts of Asia, particularly Burma, in tropical Africa, the W.I. and Queensland, Australia.

Cultivation conditions

Temperature—the horse gram is essentially a crop for the dry tropics requiring an average temperature of between 68° and 86°F (20–30°C).

Rainfall—it is a drought resistant crop and is grown in areas of moderate rainfall, less than 35 in./a (900 mm/a), or in more humid areas as a dry season crop. It can also be grown successfully under irrigation.

Soil—the horse gram will grow on a wide range of soil types provided that they are well drained and not highly alkaline. It cannot stand waterlogging. In India it is found thriving on light, sandy soils, red loams, black cotton soils and gravels. Little is known of its manurial requirements, although the application of superphosphate is frequently recommended. In Australia up to 336 lb/ac (376 kg/ha) is reported to be used. In India the suggested rate is 45 lb/ac (50 kg/ha), however, recent trials indicate that the application of 27 lb/ac (30 kg/ha) of nitrogen plus 53 lb/ac (60 kg/ha) phosphoric acid is likely to give the most economic return.

Altitude—in India the crop is grown from sea level up to elevations of approximately 5 000 ft (1 500 m).

Day-length—most cultivars require short days.

Planting procedure

Material—seed is used, inoculation, particularly if accompanied by the application of phosphate, has been found to increase seed yields and seed protein content significantly.

Method—the horse gram may be broadcast, or sown in rows about 0.5 to to 1 in. (1.25–2.5 cm) deep. It usually receives little attention. It is sometimes grown in pure stands on new land before cereal crops. In Madras, it is frequently grown as a catch crop, after sesame or cereals, or mixed with a crop of niger, which is sown simultaneously in rows about 3–6 ft (0.9–1.8 m) apart. Sometimes it is drilled in a standing crop of castor after the inter-cultivation has finished.

Field-spacing—when grown as a pure crop it is frequently sown in plough furrows about 9 in. (22.5 cm) apart.

Seed-rate—in S. India the average seed-rate is 40 lb/ac (45 kg/ha); in N. India the rate is lower, approximately 20 to 25 lb/ac, (22–28 kg/ha). In Queensland, when drilled, the average seed-rate is between 5 and 8 lb/ac (5.6–9 kg/ha), and when broadcast between 10 and 12 lb/ac (11–13 kg/ha).

Pests and diseases

The horse gram is reported to be comparatively free from attack from pests and diseases. In India various hairy caterpillars and grasshoppers are common pests. Sometimes the gram leaf caterpillar, *Azazia rubricans*, can be very destructive and occasionally the green pod boring caterpillar, *Etiella zinckenella* is troublesome. Like most other grain legumes the seed is subject to attack from a number of insects when stored.

In India the following diseases have been reported: a root rot caused by *Rhizoctonia* sp. which can sometimes result in serious crop losses, especially when the soil becomes waterlogged; a rust, *Uromyces appendiculatus*, which affects the leaves; anthracnose, *Colletotrichum lindemuthianum*, which affects the stem, leaves and seeds; a dieback caused by *Vermicularia capsici*, which affects the flowers, and a bacterial leaf spot caused by *Xanthomonas phaseoli* var. *sojense*.

Growth period

Mature seed is produced between 120 and 180 days after sowing; black-seeded types are normally of shorter duration than others. The improved cultivar Co-1 is of short duration, producing seed in about 120 days.

Harvesting and handling

The plants are usually uprooted when the pods become light-brown and begin to shrivel and the leaves begin to dry and fall off. They are stacked and left to dry for about 7 days, and then threshed similarly to other pulse crops. In India the seed is very susceptible to contamination with weed seed and thorough cleaning is necessary before it can be stored similarly to other grain legumes. The seed of certain types of horse gram is liable to lose its colour during storage.

Horse gram

Primary product

Seed—horse gram seed is small, flattened, ellipsoid, about 0.12–0.24 in. (3–6 mm) in length, with a shiny seed-coat and an inconspicuous hilum. It may be black, red, brown, grey, or mottled, depending upon the cultivar.

Yield

In India under favourable conditions the horse gram can yield on average 600 lb/ac (670 kg/ha) of dry seed; average yields of the improved cultivar Co-1 are reported to range from 700 to 800 lb/ac (780–900 kg/ha). However, in many areas the crop receives little attention and is frequently sown late in the season so that yields as low as 150 to 300 lb/ac (170–340 kg/ha) are quite common. In Australia the average yield of seed is reported to range between 1 000 and 2 000 lb/ac (1 120–2 240 kg/ha).

Main use

The seed is eaten both boiled and fried by the poorer section of the population in India. Unlike most other Asian grain legumes it is not usually split to produce dhal. In Burma the seed is boiled, pounded with salt and fermented to produce a sauce similar to soy sauce.

Subsidiary uses

The seed may be used as a concentrate for cattle or horses, usually after being boiled or slightly parched.

Secondary and waste products

The seeds are sometimes used medicinally as a diuretic. The stems, leaves and pods left after harvesting may be used for animal feed. The horse gram may be grown for forage, as a green manure, or anti-erosion crop. When grown for forage it is usually harvested about 40–45 days after sowing and in Madras recorded yields are 2 000–5 600 lb/ac (2 240–6 270 kg/ha) for the rain-fed crop and 8 100–12 700 lb/ac (9 070–14 220 kg/ha) when grown under irrigation. Analysis of horse gram hay on a dry weight basis gave the following values: crude protein 10.6 per cent; crude fibre 16.2 per cent; fat 1.8 per cent; N-free extract 58.3 per cent; ash 13.1 per cent; calcium 1.81 per cent; phosphorus 1.18 per cent.

Special features

The seed (red type) is reported to have the following proximate composition: moisture 8.4 per cent; carbohydrate 59.3 per cent; fat 4.8 per cent; protein

24.7 per cent; ash 2.8 per cent; iron 0.02 per cent; calcium 0.34 per cent; phosphorus 0.27 per cent; vitamin A 40–119 iu/100 g; nicotinic acid 1.5 mg/100 g; thiamine 0.40 mg/100 g; riboflavin 0.15 mg/100 g. Globulins account for approximately 80 per cent of the total nitrogen present. The amino acid composition (mg/gN) for red type seeds has been reported as: arginine 519; threonine 212; leucine 506; isoleucine 344; valine 356; histidine 156; phenylalanine 419; lysine 537; methionine 106; tryptophan 60; cystine 62; tyrosine 331. The presence of a trypsin inhibitor has been reported.
In addition, there is evidence that horse gram seeds contain growth inhibiting factors which can be inactivated by autoclaving.

Production and trade

Only scanty statistical data relating to the horse gram are available. Production in India, by far the major producer, has been estimated to be of the order of 377 000 t/a for the period 1968–72. Most of the crop is produced for domestic consumption, particularly in Madras and Hyderabad, which together account for about 90 per cent of the total Indian area under this crop. There is reported to be a limited inter-state trade in horse gram in India, but the seed is of negligible importance in international trade.

Major influences

The horse gram is a cheap source of protein food and with efficient cultivation is capable of reasonable seed yields. It is tolerant of drought and poor soils and could be a potentially valuable food crop in the more arid areas of the tropics. However, little work has as yet been undertaken on breeding improved, short-duration, high-yielding cultivars. Although accepted as a nutritious foodstuff the nutritive value of the seed requires further investigation, particularly the nature of the growth inhibiting factors.

Bibliography

Aykroyd, W. R. and Doughty, J. 1964. Legumes in human nutrition. *FAO Nutr. Stud.* 19, pp. 104; 115; 117. Rome: FAO, 138 pp.

Ayyangar, G. N. R. and Narayanan, T. R. 1940. Fodder crops in the Madras Presidency—a review: Horsegram (*Dolichos biflorus*). *Madras Agric. J., 28*, 58–59.

Chopra, K. and Swamy, G. 1975. *Pulses: An analysis of demand and supply in India. Institute for social and economic change; Monograph 2.* New Delhi: Sterling Publishers, PVT, Ltd, 132 pp.

Horse gram

COMMONWEALTH BUREAU OF PASTURES AND FIELD CROPS. nd. *Dolichos* spp. Annotated bibliography 1220, 1960–69. *Commonw. Agric. Bur., Bur. Pastures & Field Crops. Hurley, Maidenhead, Berks., England*, 2 pp.

JESWANI, L. M. 1975. Varietal improvement of seed legumes in India. *Food protein sources*. (Pirie, N. W. and Swaminathan, M.S. eds.). pp. 9–18. London: Cambridge University Press, 260 pp.

MAHAPATRA, I. C., BAPAT, S. R., SARDANA, M. G. AND BHENDIA, M. L. 1973. Fertilizer response of wheat, gram and horse gram under rainfed conditions. *Fert. News*, *18*, 57–63.

MANAGE, L. AND SOHONIE, K. 1972. Proximate composition, amino acid make up and *in vitro* proteolytic digestibility of horse gram (*Dolichos biflorus* Linn.). *J. Food Sci. Technol.*, *9*, 35–36.

NARAYANAN, T. R. AND DABADGHAO, P. M. 1972. Horse gram (*Dolichos biflorus* Linn.). *Forage crops of India*. pp. 69–71. New Delhi: Indian Counc. Agric. Res.. 373 pp.

NEZAMUDDIN, S. 1970. Horsegram, *Dolichos biflorus* L. *Pulse crops of India*. (Kachroo, P. and Arif, M. eds.). pp. 321–324. New Delhi: Indian Counc. Agric. Res., 334 pp.

PURSELOVE, J. W. 1968. *Dolichos uniflorus* Lam. *Tropical crops: Dicotyledons*. Vol. 1. pp. 263–264. London: Longmans, Green and Co Ltd, 332 pp.

RACHIE, K. O. AND ROBERTS, L. M. 1974. Grain legumes of the lowland tropics. *Adv. Agron.*, *26*, 1–132.

RANGASWAMI, G. N., KRISHNA RAO, P. AND SESHADRI SARMA, P. 1934. Preliminary studies in horse gram (*Dolichos biflorus* L.). *Madras Agric. J.*, *22*, 200–204.

RAY, P. K. 1968. A comparison of growth of rats fed the raw seeds of *Dolichos biflorus* (horse gram). *Sci. Cult.*, *34*, 350–352.

RAY, P. K. 1969. Toxic factor(s) in raw horse gram (*Dolichos biflorus*). *J. Food Sci. Technol.*, *6*, 207–208.

RAY, P. K. 1969. Growth inhibition of rats fed raw seeds of horse gram (*Dolichos biflorus*). *Sci. Cult.*, *35*, 519–524.

RAY, P. K. 1970. Nutritive value of horse gram (*Dolichos biflorus*): Determination of biological value, digestibility and net protein utilization. *Indian J. Nutr. Diet.*, *7*, 71–73.

SAHU, S. K. 1973. Effect of rhizobium inoculation and phosphate application on black gram (*Phaseolus mungo*) and horse gram (*Dolichos biflorus*). *Madras Agric. J.*, *60*, 989–993.

SASTRI, B. N. (ed.). 1952. *Dolichos biflorus* Linn. Horse gram. *The wealth of India: Raw materials*. Vol. 3 (D–E). pp. 101–104. Delhi: Indian Counc. Sci. Ind. Res., 236 pp.

SMARTT, J. 1976. *Dolichos* Lam. *Tropical pulses*. pp. 58–59. London: Longman Group, 348 pp.

STAPLES, I. B. 1966. A promising new legume. *Qd. Agric. J.*, *92*, 388–392.

VEERASWAMY, R. 1960. Madras gets a new horse gram. *Indian Farming*, *10*, 8–9.

VERDCOURT, B. 1970. A new genus of Leguminosae—Phaseoleae. *Kew Bull.*, *24*, 322; 400.

YEGNA NARAYAN AIYER, A. K. 1966. Horse gram (*Dolichos biflorus*). *Field crops of India*. 6th ed. pp. 115–117. Bangalore City: Bangalore Printing and Publishing Co Ltd, 564 pp.

Common names **HYACINTH BEAN, Bonavist(a) bean, Dolichos (bean), Egyptian (kidney) bean, Indian (butter) bean, Lablab (bean).**

Botanical name *Lablab purpureus* (L.) Sweet, syn. *Dolichos lablab* L., *Lablab niger* Medik.

Family Leguminosae.

Other names Agaya (Philipp.); Agni guango ahrua (Iv. C.); Amora-guaya (Eth.); Anataque (Re.); Anumulu (Ind.); Apikak (Philipp.); Ataque (Fr.); Australian pea; Avarai, Avare, Ballar (Ind.); Bannabees (Guy.); Batao, Batau, Beglau (Philipp.); Bonavis (pea) (Trin.); Bunabis (Gren.); Butter bean[1] (Bah., Dom., Guy.); Caraota chivata (Venez.); Carpet legume, Chapprada (avare) (Ind.); Chaucha japonese (Sp.); Chichaso (P.R.); Chikkudu (Ind.); Chimbolo verde (C. Rica); Cumana tupi (Par.); Cumandiata (Braz.); Dâu van (Viet.); Dolic (d'Egypte) (Fr.); Dolicho lablab (Sp.); Dolique d'Egypte, D. du Soudan, Fève d'Egypte (Fr.); Field bean[2] (Ind.); Fiwi (bean) (E. Afr.); Frijol bocon (Peru); F. caballero (Salv.); F. chileno (Peru); F. de la tierra (Cuba); Fuji-mame (Jpn.); Gallinazo blanco (Venez.); Gallinita (Mex.); Gerenga (Eth.); Gueshrangaig (Egy.); Haricot cutelinho (Port.); Helmbohne (Ger.); Itab (Philipp.); Kachang kara, Kara-kara (Malays.); Kashrengeig (Sud.); Kekara (Malays.); Kerara (Indon.); Kikuyu bean (E. Afr.); Labe-labe (Braz.); Louvia (Cyp.); Lubia (h) bean (Eth., Sud.); Macape (Malag.); Macululu (Ang.); Mochai (Ind.); O-cala (Eth.); Ossangue (Zar.); Papaya bean (As.);

[1] More commonly used for the lima bean, *Phaseolus lunatus*, and also used in Kenya to denote a white-seeded cultivar of the runner bean, *Phaseolus coccineus*.

[2] More commonly used for the broad bean, *Vicia faba*, or the haricot bean, *Phaseolus vulgaris*.

Parda (Philipp.); Pavta (Ind.); Pè-gyi (Burm.); Pois boucoussou (Ant.); P. contour, P. coolis, P. d'un sou, P. en tout temps, P. indien (Mart.); Pole bean[3] (Ind.); Poor man's bean (Aust.); Popat (Ind.); Poroto bombero (Chile); P. de Egipto (Arg.); P. japonese (Sp.); Saeme (Carib.); Seim bean, Sem (Trin.); Shim (bi) (Beng.); Sim bean[4] (Assam); Tonga bean[5] (Aust.); Tua nang, T. pab, T. pep (Thai.); Twin flowered bean; Urahi[6], Uri[6], Val (Ind.); Waby bean; Wal (papdi) (Ind.).

Botany

A herbaceous, perennial herb, frequently grown as an annual, usually twining to reach 5 to 20 ft (1.6–6 m), but bushy, semi-erect, and prostrate forms exist. Probably no other legume shows such variation in form and habit. The tap-root is well developed with many laterals and well-developed adventitious roots. The leaves are alternate trifoliolate, the leaflets ovate, the lateral leaflets are oblique, entire 2–6 x 1.6–6 in. (5–15 x 4–15 cm), sometimes covered with soft hairs. The inflorescences are axillary erect, and may be long-stalked 6–18 in. (15–45 cm), with showy flowers, or short-stalked, 2 in. (5 cm) or less, with insignificant flowers. The flowers may be white, pink, red or purple, and usually arise from prominent tubercular thickenings on the peduncle. The pods are very variable in shape and colour; they may be flat or inflated, and can vary from 2 to 8 in. (5–20 cm) in length and are usually 0.4 to 2 in. (1–5 cm) in breadth, sometimes curved, with a curved beak and persistent style. They can vary in colour and may be septate, or non-septate, and usually contain 3 to 6 seeds of varying colours and size.

There has been some altercation over the generic name. *Dolichos* in the old sense has been divided into several genera and the question concerns which one should retain the original name. Verdcourt (1970) and Westphal (1974) have put opposing views, but the 1975 International Botanical Congress ruled in favour of using *Lablab* for the hyacinth bean. Westphal (1975) also casts doubt on the use of the epithet '*purpureus*'.

---

[3] More frequently used in other countries for the lima bean, *Phaseolus lunatus*, or for climbing types of the haricot bean, *P. vulgaris*, or for the runner bean, *P. coccineus*.

[4] Also used for the rice bean, *Vigna umbellata*, and the runner bean, *P. coccineus*.

[5] Also frequently used for the lima bean, *P. lunatus*.

[6] Also used for the runner bean, *P. coccineus*.

Hyacinth bean

In view of the considerable variation in the form and habit of the hyacinth bean several authorities have also attempted to distinguish subspecies based mainly upon the characteristics of the pods and the seeds, but since there is considerable variation within these subspecies and crossing frequently occurs, the distinguishing of subspecies is of doubtful value.

There are numerous cultivars of the hyacinth bean adapted to local conditions and the purpose for which the crop is grown. In India attempts are being made to develop improved high-yielding types. The quick-maturing, cultivar Lab-lab Co-1, which is suitable for the production of green pods and dry seed was released to growers in the late 1960s. More recently three improved cultivars Co-6, Co-7 and Co-8 have been developed from Co-1. These are short, erect, high-yielding, dual purpose types suitable for all the year round cultivation in India.

Origin and distribution

The hyacinth bean is generally considered to have originated in the SE. Asian sub-continent and was taken to Africa in the eighth century, but some authorities consider that it is endemic to Africa. It is now found growing throughout the tropics and subtropics.

Cultivation conditions

Temperature—for optimum results a warm, equable climate is required, with average temperatures between 64° and 86°F (18–30°C). Most cultivars are fairly tolerant of high temperatures and many can withstand frost for a limited period, although night frosts are liable to cause some leaf damage. Cold weather, however, has an adverse effect upon pollination and seed-set.

Rainfall—the hyacinth bean is a hardy, drought resistant crop, suitable for growing as a rain-fed crop in semi-arid tropical areas with an average annual rainfall of 24–36 in. (600–900 mm). In India it is successfully grown, with supplementary irrigation, in areas with a rainfall as low as 16 in./a (400 mm). It requires adequate moisture during the early stages of growth, after which its deep root-system enables it to sustain growth on residual soil moisture. When grown as a market garden crop for the production of the immature pods it requires frequent irrigation throughout its growing period.

Soil—it can be grown on a wide variety of soil types, provided that they are well drained. It does particularly well on sandy loams, pH 6.5, and in Brazil thrives on heavy clays, pH 5.0. It cannot tolerate water-logging, or brackish soils. Little is known of its fertilizer requirements, but it is reported to respond to phosphate and potash. The application of ammonium sulphate, 89 lb/ac (100 kg/ha) and superphosphate 220 lb/ac (250 kg/ha) has been recommended when the hyacinth bean is grown as a field crop in India. When grown for pod production in India the plants normally receive heavy applications of farmyard manure. In addition, the application of a general fertilizer is sometimes recommended, but a recent study indicates that probably this is not necessary. When grown for grazing in Australia the application of molybdenized superphosphate 220 lb/ac (250 kg/ha) has been recommended on all but the most fertile soils, and up to 445 lb/ac (500 kg/ha) on poor coastal soils.

Altitude—normally grown from sea level up to elevations of between 6 000 and 7 000 ft (1 800–2 100 m) in Asia.

Day-length—short-day, long-day and day-neutral cultivars exist.

Planting procedure

Material—seed is used, germination is epigeal and normally takes about 5 days. The seed is reported to remain viable for two or three years and to have an average germination rate of 85–95 per cent. Inoculation is not usually considered to be necessary, if the crop is grown where cowpeas have been grown.

Method—when grown as a field crop, the hyacinth bean is usually sown in rows, either pure, or mixed with a cereal crop such as maize; it is sometimes grown as a cover crop in rotation with sorghum and cotton. When grown as a field crop it frequently receives very little attention although weed control during the early stages of growth may be necessary. Recently the use of pre-emergence herbicides such as chloramben, chlorthal, diphenamid, trifluralin and dinoseb, has been suggested. In India it is invariably grown as a mixed crop with ragi, *Eleusine coracana*, as the main crop. Then the hyacinth bean shares the same interculture weeding and thinning operations which are carried out for the ragi crop. It makes only moderate growth until the ragi crop is harvested, after which it normally makes rapid growth and commences to flower. When grown as a market garden crop the seeds are often planted in well-prepared pits, 36 x 36 in. (90 x 90 cm); normally 6–10 seeds are planted in each pit, and supports, eg bamboo canes, are provided for the beans to climb up. After about 30 days the plants are thinned to four per pit.

Hyacinth bean

Field-spacing—when grown as a pure crop planting distances usually vary from approximately 4 to 10 in. (10–25 cm) between the plants and 12 to 40 in. (30–100 cm) between the rows, according to the cultivar and the purpose for which the crop is grown. In India Lab-lab Co-1, the short-duration, dual purpose cultivar is often planted 4 x 12 in. (10 x 30 cm). In Brazil, when grown as a green manure crop a spacing of 8 x 20 in. (20 x 50 cm) is reported to be used.

Seed-rate—in Asia, when grown as a pure crop for seed, the average seed-rate is 50–60 lb/ac (56–67 kg/ha), when mixed with a cereal crop, such as ragi, it varies from 5 to 15 lb/ac (5.6–17 kg/ha). In E. Africa the seed-rate normally averages 20–30 lb/ac (22–34 kg/ha), and in Brazil, when grown as a green manure, the average seed-rate is 21 lb/ac (24 kg/ha).

Pests and diseases

In India, pod boring larvae are the most serious pests of the hyacinth bean, reducing the crop both in quantity and quality. The larvae of *Adisura atkinsoni* are particularly troublesome, but this pest has been controlled experimentally by a strain HE–111 of *Bacillus cereus* var. *thuringiensis*. In addition, the gram caterpillar, *Heliothis armigera*, the plume moth, *Exelastis atomosa*, and the spotted pod borer, *Maruca testulalis*, are of considerable economic importance.

In Malawi the flowers are reported to suffer attack from *Mylabris* beetles and in Puerto Rico, the bean leaf beetle, *Cerotoma ruficormis*, is a major pest. In the Sudan, cock-shaver larvae, *Schizonycha* sp., are of considerable economic importance; the use of mercurized seed-dressings has been suggested as an effective control measure. Like most grain legumes the mature seeds of the hyacinth bean are susceptible to insect infestation during storage, particularly bruchid beetles, *Callosobruchus* spp., which also attack the crop in the field.

Although relatively free from diseases when grown in the savanna zone, the hyacinth bean can be seriously affected by ashy stem blight, *Macrophomia phaseoli*, and bacterial blight, *Xanthomonas phaseoli*, when grown under humid conditions. In India, anthracnose, *Colletotrichum lindemuthianum*, can cause serious crop losses; spraying with zineb or captan is reported to give reasonable control. Leaf spot, *Cercosphora dolichi*, and powdery mildew, *Leveillula taurica* var. *macrospora*, may also be troublesome and are controlled by spraying with Bordeaux mixture. The crop can also be affected by rust, *Uromyces appendiculatus*, which is controlled by spraying with sulphur dust.

In addition, a virus disease, known as dolichos enation mosaic, which resembles tobacco mosaic, has been reported; it is transmitted by the white fly, *Bemisia tabaci.* In China, a scab caused by the fungus *Elsinoe dolichi* has been recorded and in Pakistan a fungal disease, *Choanephora* sp., which causes dry rot and shedding of the leaves has been reported.

Growth period

The growth period can vary from approximately 75 to 300 days. In India, the improved cultivar Lab-lab Co-1, begins to bear pods approximately 60–65 days after sowing and continues for 90 to 100 days. Other improved quick-maturing cultivars such as Co-6, Co-7, and Co-8, which can be grown all the year round, produce pods 60 days after sowing and continue up to 120 days. Mature seeds are normally harvested 150 to 210 days after sowing, but this can be very dependent upon the cultivar and the time of sowing. In India short-day cultivars can take from 42 to 330 days to produce flowers, according to the sowing date.

The hyacinth bean is sometimes treated as a perennial and allowed to crop for a second season.

Harvesting and handling

The green pods are picked by hand when they have reached a reasonable size, usually when the seeds are about three-quarters ripe. The plants are generally picked over at intervals of 3–4 days, cleaned and graded for size, before being packed in baskets for the market. Both the pods and the immature beans have a relatively short shelf-life and are susceptible to infection by the fungal diseases, anthracnose, *Colletotrichum lindemuthianum*, and black rot, *Rhizopus nigricans*, during storage. Storage at 32–35°F (0–2°C) and a relative humidity of 85–90 per cent is reported to extend the shelf-life of the pods to a maximum of 21 days, and of the shelled beans up to 7 days.

In many cultivars the pods mature in succession up the stem and are liable to shatter once they are ripe. For seed production the pods are frequently picked by hand as soon as they ripen, until the plants reach full maturity and the major proportion of the remaining pods have ripened, at this stage the entire plant is cut close to the ground with a sickle, the vines left to dry for a few days, before being threshed to separate out the seeds. After drying and cleaning the seeds are stored; usually earthenware or metallic containers are used in Asia, and the seeds are covered with a 2 in.

(5 cm) protective layer of sand. The dry seeds are susceptible to infestation by bruchid beetles, *Callosobruchus* spp., and weevils, *Sitophilus* spp., during storage. Harvesting the individual pods as soon as the seed is ripe has been found to reduce infestation, but more important is to ensure that the seed is dried to a moisture content of 10 per cent or less.

Primary product

Pod—these may be flat, inflated, white, green, or purplish in colour, and can vary in length from approximately 2 to 8 in. (5–20 cm) and from 0.4–2 in. (1–5 cm) in width. They may be crescent-shaped to more or less straight and oblong, or sometimes dorsally straight and ventrally deeply curving. Cultivars grown as a vegetable have pods with thick, fleshy skins, with practically no fibre. The pods may be septate, or non-septate; in the former each seed occupies a separate compartment in the pod, while in the latter the pods have a bloated appearance. Each pod normally contains 3–6 seeds. These are generally less than 0.5 in. (1.25 cm) in length, and may be rounded, or oval, and rather flattened, white, cream, beige, red, brown, or black, self-coloured, or variously speckled. The hilum is white, prominent and oblong, usually extending one-third around the seed. One hundred seeds weigh 0.7–1.8 oz. (20–50 g).

Yield

In India, the average yield of green pods varies from 2 300 to 4 000 lb/ac (2 600–4 500 kg/ha), while the average yield of seed is about 400 lb/ac (450 kg/ha) when the hyacinth bean is grown as a mixed crop, and up to 1 300 lb/ac (1 460 kg/ha) in pure stands.

Main use

The hyacinth bean is a nutritious foodstuff. The culinary types are popular in Asia, where they are widely consumed as a vegetable, being eaten boiled similarly to French beans, or used in curries. Sometimes the immature green seeds are extracted from the pods and eaten as a vegetable, either boiled or roasted. In Egypt the hyacinth bean is sometimes used as a substitute for broad beans in the preparation of fried bean cake, 'tanniah'. In Asia the mature seeds are utilized as a pulse, often as dhal. The seeds are sometimes soaked in water overnight and when germination starts they are sun-dried and stored for future use.

Subsidiary uses

The pods and the seeds can also be used for livestock feeding. The hyacinth bean may also be grown as green manure, or as a forage crop, the vines being palatable to cattle and staying green during a prolonged dry period. The average composition of the forage has been given as: fibre 28.1 per cent; fat 3.5 per cent; crude protein 14.2 per cent; carbohydrate 39.4 per cent; ash 14.8 per cent; calcium 1.98 per cent; phosphorus 0.26 per cent.

Secondary and waste products

The young leaves are sometimes boiled and eaten as a vegetable in Asia. After harvesting, the haulms and pods may be used as a nutritious animal feedingstuff, as may be the seed husks and other residues left after the preparation of dhal.

Special features

There is considerable variation in the composition of the pods and the seeds of the hyacinth bean according to cultivar, climatic conditions, and the standard of crop management. The proportion of seed to pod-husk is approximately 1.1:1. The approximate composition of the immature pods has been given as: moisture 82.4 per cent; protein 4.5 per cent; fat 0.1 per cent; fibre 2.0 per cent; carbohydrate 10.0 per cent; ash 1.0 per cent; calcium 0.05 per cent; phosphorus 0.06 per cent; iron 10 mg/100 g; nicotinic acid 0.8 mg/100 g; vitamin C, uncooked samples 0.77–1.12 mg/100 g, cooked samples, 7.33–10.26 mg/100 g. (The increase in vitamin C content on cooking is attributed to the softening of the internal tissue which facilitates extraction). The protein content of the mature seeds normally varies between 21 and 29 per cent. It has recently been shown that American cultivars when grown in India have a high protein content, but are less vigorous and lower-yielding than local cultivars. An approximate analysis of dry, fully ripe Indian seeds have been given as: moisture 9.6 per cent, protein 24.9 per cent; fat 0.8 per cent; fibre 1.4 per cent; carbohydrate 60.1 per cent; ash 3.2 per cent; calcium 0.06 per cent; phosphorus 0.45 per cent; iron 2.0 mg/100 g; nicotinic acid 1.8 mg/100 g. The chief protein is a globulin, dolichosin. The amino acid content (mg/gN) has been reported as: isoleucine 256; leucine 436; lysine 36; methionine 36; cystine 57; phenylalanine 299; tyrosine 197; threonine 207; valine 294; arginine 393; histidine 186; alanine 266; aspartic acid 727; glutamic acid 978; glycine 240; proline 288; serine 323. Infestation with bruchid beetles has been found to decrease the thiamine and methionine content significantly. The seed contains from

38.6 to 46.0 per cent of starch, on a dry weight basis. It has an iodine value of 6.05, corresponding to approximately 30 per cent amylose and is similar to that contained in the chick pea, *Cicer arietinum*, and the haricot bean, *Phaseolus vulgaris*.

The hyacinth bean is a rich source of catechol oxidase. The presence of a cyanogenic glycoside has been reported in certain cultivars. Two haemagglutinins have been isolated from meal prepared from the beans and one of these has been found to cause zonal necrosis of the liver when fed to rats.

Processing

In Asia, the immature pods are sometimes preserved by salting, steaming, followed by drying in the sun. The resulting dried product is reported to have a storage-life of approximately one year and is usually eaten after being fried.

Production and trade

No statistical data have been traced.

Major influences

The hyacinth bean is an important food legume in parts of Asia, notably in southern India, where it supplies a considerable proportion of the protein in the daily diet of the rural population. For this reason there is need to develop improved high-yielding cultivars, with a high protein content. In most other parts of the tropics and subtropics the hyacinth bean does not appear to be being fully exploited for food, or as a forage crop. It is in fact reported to be declining in importance in Kenya, owing to the increasing popularity of the haricot bean. Nevertheless, it could be a useful crop for the semi-arid areas of the tropics and subtropics. However, there is need to study systematically the behaviour of the many cultivars under different soil and climatic conditions and to improve the protein content of existing cultivars.

Bibliography

Acland, J. D. 1971. Bonavist beans. *East African crops*. p. 26. London: Longman Group Ltd, 252 pp.

Ayyangar, G. N. R. and Nambiar, K. K. K. 1941. Lablab—the garden bean. *Indian Farming*, *2*, 469–472.

BACON, G. H. 1948. Crops of the Sudan: *Dolichos lablab* L. 'Lubia', 'Lubia afin'. *Agriculture in the Sudan.* (Tothill, J. D. ed.). pp. 354–356. London: Oxford University Press, 974 pp.

CASSIDY, G. J. 1975. Lablab bean for autumn grazing. *Qd. Agric. J., 101,* 37–40.

COMMONWEALTH BUREAU OF PASTURES AND FIELD CROPS. nd. *Lablab niger* (*Dolichos lablab, Lablab vulgaris.*) Annotated bibliography 1221, 1960–69. *Commonw. Agric. Bur., Bur. Pastures & Field Crops, Hurley, Maidenhead, Berks, England,* 7 pp.

FOOD AND AGRICULTURE ORGANIZATION OF THE UNITED NATIONS. 1970. Amino acid content of foods and biological data on proteins. *FAO Nutr. Stud.* 24, pp. 50–51. Rome: FAO, 286 pp.

GÖHL, B. 1975. *Dolichos lablab* L. (*Lablab vulgaris* Medic., *L. niger* Medic.). *Tropical feeds: Feeds information summaries and nutritive values.* pp. 186; 512; 528. Rome: FAO, 661 pp.

HABISH, H. A. AND MAHDI, A. A. 1976. Effect of soil moisture on nodulation of cowpea and hyacinth bean. *J. Agric. Sci.,* (*Camb.*), *86,* 553–560.

HERKLOTS, G. A. C. 1972. Hyacinth bean. *Vegetables in South-east Asia.* pp. 241–244. London: George Allen and Unwin Ltd, 525 pp.

KASASIAN, L. 1971. Vegetable crops. *Weed control in the tropics.* London: Leonard Hill Books, 307 pp.

KUNJAMMA HRISHI, V. K. AND RAJASEKARAN, V. P. A. 1968. Co.1 Lab-lab: A new dual purpose hybrid strain. *Madras Agric. J., 55,* 411–413.

KUNJAMMA HRISHI, V. K. AND RAJASEKARAN, V. P. A. 1968. A double-purpose pulse for Madras State, Co.1 Lab-lab (Ottu Mochai). *Indian Farming 18,* (8), 18–21.

KUNJAMMA HRISHI, V. K., VEERASWAMY, R., RAJASEKARAN, V. P. A. AND PALANISAMY, G. A. 1970. Preliminary studies on the effect of sowing season on *Lablab niger* Co.1. *Madras Agric. J., 57,* 506–507.

MAHDI, A. A. AND HABISH, H. A. 1975. Effects of light and temperature on nodulation of cowpea and hyacinth bean. *J. Agric. Sci.,* (*Camb.*), *85,* 417–425.

NARAYANA RAO, D., HARIHARAN, K. AND RAJAGOPAL RAO, D. 1976. Purification and properties of a phytohaemagglutinin from *Dolichos lablab* (Field bean). *Lebensm.-Wiss.u-, Technol., 9,* 246–250.

Hyacinth bean

PATIL, J. G. 1957. Flowering and pod-setting in *Dolichos lablab* Roxb. *Poona Agric. Coll. Mag.*, *48* (1), 41–46.

PIPER, C. V. 1915. The bonavist, lablab or hyacinth bean. *US Dep Agric., Bull.* 318, 15 pp.

PURSEGLOVE, J. W. 1968. *Lablab niger* Medik. Hyacinth bean. *Tropical crops: Dicotyledons.* Vol. 1. pp. 273–276. London: Longmans, Green and Co Ltd, 332 pp.

RACHIE, K. O. AND ROBERTS, L. M. 1974. Grain legumes of the lowland tropics. *Adv. Agron.*, *26*, 1–132.

RAMA RAO, G., AND KADKOL, S. B. 1957. Analysis of field beans (*Dolichos lablab*) at different stages of maturity. *Food Sci.*, *6*, 153–154.

REGUPATHY, A., PALANISWAMY, G. A. AND KRISHNAN, R. H. 1970. Assessment of loss in seed yield by pod borers in certain varieties of field bean. *Madras Agric. J.*, *57*, 274–278.

RIVALS, P. 1953. Le dolique d'Egypte ou lablab (*Dolichos lablab* L.). [Egyptian kidney bean or lablab (*Dolichos lablab* L.]. *Rev. Int. Bot. Appl., Agric. Trop.*, *33*, 314–322; 518–537.

RIVALS, P. 1960. Le dolique d'Egypte ou lablab *Dolichos lablab* L. (ii) Notes complémentaires. [Egyptian kidney bean or lablab *Dolichos lablab* L. (ii) Complementary notes]. *J. Agric. Trop. Bot. Appl.*, *7*, 447–450.

ROSENTHAL, F. R. T., ESPINDOLA, L., SERAPIÃO, M. I. S., SILVA, S. M. O. 1971. Lablab bean starch: Preparation of its granules and pastes. *Die Stärke*, *23*, 18–23.

SALGARKAR, S. AND SOHONIE, K. 1965. Haemagglutinins of field bean (*Dolichos lablab*): (i) Isolation, purification and properties of haemagglutinins; (ii) Effect of feeding field bean haemagglutinin A on rat growth. *Indian J. Biochem.*, *2*, 193; 197. (*J. Food Sci. Technol.*, *3*, 79).

SASTRI, B. N. (ed). 1952. *Dolichos lablab* Linn. *The wealth of India: Raw materials.* Vol. 3 (D–E). pp. 104–106. New Delhi: Indian Counc. Sci. Ind. Res., 236 pp.

SCHAAFFHAUSEN, R. VON. 1963. *Dolichos lablab* or hyacinth bean: Its uses for feed, food and soil improvement. *Econ. Bot.*, *17*, 146–153.

SCHAAFFHAUSEN, R. VON. 1963. Economical methods for using the legume *Dolichos lablab* for soil improvement, food and feed. *Turrialba*, *13*, 172–179.

SHEHNAZ, A. AND THEOPHILUS, F. 1975. Effect of insect infestation on the chemical and nutritive value of Bengal gram (*Cicer arietinum*) and field bean (*Dolichos lablab*). *J. Food Sci. Technol.*, *12*, 299–302.

SINGH, A. 1970. Carpet legume for sheep in eastern Rajasthan. *Indian Farming*, *19* (12), 27–28.

SRIVASTAVA, H. C. AND MATHUR, P. B. 1954. Cold storage of field beans (*Dolichos lablab*), *Indian J. Hortic.*, *11*, 125–128.

STANTON, W. R., DOUGHTY, J., ORRACA-TETTEH, R. AND STEELE, W. 1966. *Dolichos* L., hyacinth or dolichos beans. *Grain legumes in Africa.* pp. 95–99. Rome: FAO, 183 pp.

SUBBIAH, K. K. AND MORACHAN, Y. B. 1973. A note on the response of lab-lab (*Dolichos lablab niger* var. *typicus* Linn.) to N. P. K. *Madras Agric. J.*, *60*, 351–352.

TARDIEU, M. 1962. Le haricot dolique au Sénégal: Recherches sur cette espèce au CRA Bambey. [The dolichos bean in Sénegal: Research on this species at CRA Bambey]. *Agron. Trop. 17*, 33–66.

TARR, S. A. J. 1960. Crop protection by seed treatment: (iii) Fodder bean in the Sudan. *World Crops*, *12* (1), 27.

THWIN, A. 1969. Preliminary studies on the infestation of lablab beans (*Dolichos lablab* Linn. var. *lignosus* Prain) by bruchids in the field and in storage. *Union of Burma*, *J. Life Sci.*, *2* (1), 7–9.

VEERASWAMY, R., KUNJAMMA-HRISHI, V. K., PALANISWAMI, G. A. AND RAJASEKARAN, V. P. A. 1973. Development of Lab-lab Co.6, Co.7 and Co.8—Three short term varieties for all seasons. *Madras Agric. J.*, *60*, 1339–1341.

VENKATA RAO, B. V. AND RAMASWAMY, K. 1970. *Dolichos lablab* pod-husk as a cattle feed: a study on its nutritive quality. *Mysore J. Agric. Sci.*, *4*, 223–225.

VENKATRAO, S., NUGGEHALLI, R. N., PINGALE, S. V., SWAMINATHAN, M. AND SUBRAHMANYAN, V. 1960. Effect of insect infestation on stored field bean (*Dolichos lablab*) and black gram (*Phaseolus mungo*). *Food Sci.*, *9* (3), 79–82.

VERDCOURT, B. 1970. Studies in the *Leguminosae–Papilionoidae* for the 'Flora of tropical East Africa': *iv. Kew Bull.*, *24*, 409–411.

Hyacinth bean

VISWANATHA, S. R., SIDDARAMAPPA, R., SHIVASHANKAR, G. AND SURESH, P. 1972. Study of variability for protein content in *Dolichos lab-lab* L.

WESTPHAL, E. 1974. *Dolichos lablab* L. Pulses in Ethiopia, their taxonomy and agricultural significance. pp. 91–104. *Wageningen, PUDOC, Cent. Agric. Publ. Doc., Agric. Res. Rep.* 815, 278 pp.

WESTPHAL, E. 1975. The proposed retypification of *Dolichos* L. a review. *Taxon, 24*, 189–192.

YARAGUNTIAH, R. C. AND GOVINDU, H. C. 1964. A virus disease of *Dolichos lablab* var. *typicum* from Mysore. *Curr. Sci., 33*, 721–722.

YEGNA NARAYAN AIYER, A. K. 1966. Avare (*Dolichos lab-lab*). *Field crops of India*. 6th ed: pp. 107–110. Bangalore City: Bangalore Printing and Publishing Co Ltd, 564 pp.

Common names	**JACK BEAN, Overlook bean.**
Botanical name	*Canavalia ensiformis* (L.) DC.
Family	Leguminosae.
Other names	Abai[1] (Ind.); Awara (Sri La.); Bara sem (Ind.); Baran chaki (Nig.); Chickasaw Lima; Cut-eye bean[1]; Dir-daguer (Som.); Feijão de porco (Braz.); Fève Jack[1] (Haiti); Gisima (Gab.); Goa bean[2] (Indon.); Gotani bean[1] (Rhod.); Grudge pea (Gren.); Haba blanca (Sp.); H. criolla (Venez.); H. de burro[1] (Sp.); Habas[1] (Philipp.); Haricot sabre[1] (Fr.); Horse bean[3] (Ind., Jam.); Kachang parang puteh (Malays.); Lubia elfil[1] (Sud.); Magtámbókan (Philipp.); Makhan s(h)im[1] (Beng.); Maljoe (Gren., Trin.); Marutong (Philipp.); Mbwanda (Zan.); Musémana, Musimana (Gab.); One-eye bean[1]; Patagonian bean; Pataning-espada, P. españa[1] (Philipp.); Pè-dalet (Burm.); Pois gagne (Guad.); Pois sabre[1] (Fr.); Popondo, Poponla (Nig.); Puakani (Haw.); Saar-sar (Som.); Sabre bean[1]; Snake bean[4]; Sunshine bean (S. Afr.); Sword bean[5]; Vella tamma, Vellai tambattai (Ind.).

Botany

A bushy, semi-erect annual, capable of becoming a perennial climber, and usually 24–48 in. (60–120 cm) in height, with a deeply penetrating, well-nodulated root-system. The stems are slightly ribbed, sometimes hollow and glabrous, often becoming woody with age. They have a tendency to twine at the ends, especially if grown in the shade. The leaves are alternate, trifoliolate, on long petioles, 4.3–6.7 in. (11–17 cm); the leaflets are entire, thick and leathery, about 3–5 in. (7.5–12.5 cm) in length, elliptic to ovate. The infloresence

[1] Also frequently used for the sword bean, *Canavalia gladiata*.
[2] More usually used for the winged bean, *Psophocarpus tetragonolobus*.
[3] More commonly used for the broad bean, *Vicia faba*, especially in Europe, also used for the sword bean, *Canavalia gladiata* in Jamaica.
[4] Also used for the asparagus bean, *Vigna unguiculata* ssp. *sesquipedalis*.
[5] More correctly used for the closely related species *Canavalia gladiata*.

is an axillary raceme with up to 50 flowers borne in groups of three to five, from knob-like projections. The flowers are reddish-purple in colour, the standard petal having recurved edges and a white throat. They are normally self-pollinated. The pods are sword-shaped, pendant, hard and firm, with a beak, straw-coloured when ripe, about 9 to 14 in. (22.5–35 cm) in length, just less than 1.0 in. (2.5 cm) in breadth, and containing 10–20 somewhat flattened white or light-brown seeds.

The sword bean, *Canavalia gladiata*, is a very closely related species and there is considerable confusion in the literature between jack and sword beans. The seeds of the two species can be readily distinguished by the length of the hilum, which in the true sword bean is nearly as long as the seed, but less than half its length in the jack bean. Some authorities in fact have listed *C. ensiformis* as synonymous with *C. gladiata*, which is sometimes considered to be a derivative of the closely related wild species *C. virosa*. According to Westphal (1974), it is probable that the three taxa belong to one species *C. ensiformis* which includes wild as well as cultivated types.

Origin and distribution

The jack bean is native to the W.I. and C. America, but is now found scattered throughout the tropics and subtropics, although in most countries it is not cultivated on a large-scale.

Cultivation conditions

Temperature—a fairly long growing season, with moderately high temperatures is required; most cultivars cannot tolerate frost, but many can be grown successfully in shade.

Rainfall—for optimum yields a well-distributed rainfall of approximately 36–48 in./a (900–1 200 mm/a) is required. However, the jack bean, when once established, is fairly drought resistant and can be grown successfully in areas with a rainfall as low as 25 to 30 in./a (650–750 mm/a) provided there is adequate subsoil moisture, or supplementary irrigation is available.

Soil—the jack bean is tolerant of a wide range of soil types provided the pH is between 5 and 6, with more acid soils liming is necessary. Many cultivars are fairly tolerant of waterlogging and salinity. Little appears to be known about the crop's manurial requirements, the application of nitrogen is reported to depress the yield. When grown for green manure the application of superphosphate, 200 lb/ac (225 kg/ha), has been recommended.

Altitude—although grown mainly in the tropical lowlands the jack bean can be found at elevations up to 6 000 ft (1 800 m).

Day-length—the jack bean responds as a short-day plant.

Planting procedure

Material—sound seed is used and inoculation has been found to be beneficial in Africa. Germination is epigeal and normally occurs very quickly, 48–72 hours, after planting.

Method—when grown as a foodstuff, often in village compounds, the seeds are sown 1–2 in. (2.5–5 cm) deep, usually on raised beds or ridges. When grown as a green manure, or cover crop, the seeds are normally broadcast.

Field-spacing—12–16 in. (30–40 cm) square, or in rows, 24–36 in. (60–90 cm) apart, with 6–12 in. (15–30 cm) between the plants is frequently used. For some bushy types a wider spacing of 40–60 x 40 in. (100–150 x 100 cm) is reported to be used.

Seed-rate—when grown for seed 25–28 lb/ac (28–31 kg/ha) is the average seed-rate, and when sown broadcast as a green manure the seed-rate normally ranges from 48 to 62 lb/ac (54–70 kg/ha).

Pests and diseases

The jack bean is relatively free from serious attacks by insect pests and diseases. In Brazil the moth *Laphygma frugiperda* and the pod weevil *Sternechus tuberculatus* are reported to be the major pests attacking the crop. In Malaysia and the Philippines jack beans are sometimes attacked by a scab, *Elsinoe canavaliae*, and in certain areas, eg Jamaica, by root rot, *Colletotrichum lindemuthianum*. They are also reported to be susceptible to a mosaic virus.

Growth period

Normally a crop of mature seed is produced in 180 to 300 days, depending upon the cultivar and local climatic conditions. The green immature pods may be harvested about 90–120 days after planting.

Harvesting and handling

The mature seed pods are usually harvested by hand as soon as they are ripe, since they are liable to shatter. The pods are dried and threshed by hand and the seed cleaned and stored in traditional earthenware storage units or sacks. The seeds have a tough, thick seed-coat and are fairly resistant to insect infestation when stored.

Primary product

Seed—the seed is white, or more rarely buff coloured, somewhat flattened, smooth and glossy, approximately 1 x 0.5 in. (2.5 x 1.25 cm). One hundred seeds weigh approximately 6 oz (172 g). The hilum, situated centrally, is narrow, approximately 0.3 in. (8 mm) in length, brownish-grey in colour with an orange-brown margin. The cotyledons are pale-yellow in colour.

Yield

The average yield of dry beans is reported to be in the order of 1 200 lb/ac (1 340 kg/ha), but in Puerto Rico average yields are considerably higher, ranging from 3 840 to 4 800 lb/ac (4 300–5 400 kg/ha), while in Brazil they range between 710 and 1 070 lb/ac (800–1 200 kg/ha). When grown for forage or green manure the yield is usually between 10 and 20 T/ac (25–50 t/ha).

Main use

The mature dry seeds can be used as a foodstuff, but are not popular because of their unattractive flavour and texture, and the fact that they require soaking and boiling in salt water for several hours to remove the toxic constituents and to soften them. In Indonesia they are often boiled twice, left in running water for 2 days after the removal of the seed-coat, then fermented for 3–4 days, and finally cooked once more.

Subsidiary uses

The dried seeds are sometimes used for livestock feeding, but are not very palatable and can cause outbreaks of poisoning, unless cooked, or limited to less than 30 per cent of the total feed. In Central America the possibility of utilizing the jack bean as a substitute for the black types of haricot beans, *Phaseolus vulgaris*, in the manufacture of pre-cooked bean flours, or for the production of protein concentrates, is being investigated.

Secondary and waste products

The green immature pods are sometimes boiled and eaten as a vegetable. The mature seeds may be roasted and used as a coffee substitute. In Indonesia the flowers and young leaves are sometimes steamed and used as a flavouring. The jack bean is a valuable green manure crop and is often grown between rows of sugar cane, coffee, cocoa, etc. In some countries it is grown for forage, but is palatable only when dried and usually has to be introduced into the diet gradually. The pods, etc, remaining after the seeds have been harvested may also be used for animal feeding in limited amounts. An approximate analysis of the pod, as a percentage of dry matter, has been given as: crude protein 4.5 per cent; fat 1.5 per cent; N-free extract 42.1 per cent; crude fibre 48.1 per cent; ash 3.8 per cent; calcium 0.3 per cent; phosphorus 0.01 per cent.

Special features

The dried beans have the following approximate composition: moisture 11–15.5 per cent; protein 23.8–27.6 per cent; fat 2.3–3.9 per cent; carbohydrate 45.2–56.9 per cent; fibre 4.9–8.0 per cent; ash 2.7–4.2 per cent; calcium 30–158 mg/100 g; phosphorus 54–298 mg/100 g; potassium 141 mg/100 g; magnesium 19 mg/100 g; iron 7 mg/100 g; niacin 2 mg/100 g; pantothenic acid 0.4 mg/100 g; riboflavin 0.4 mg/100 g; thiamine 8.5 mg/100 g. The seed-coat constitutes about 13 per cent of the total weight of the seed.

Jack beans have a relatively high protein content and the amino acid composition (mg/gN) has been reported as: glutamic acid 644; threonine 275; serine 316; alanine 275; glycine 241, valine 288; methionine 85; isoleucine 250; leucine 453; tyrosine 219; phenylalanine 322; lysine 344; histidine 169; arginine 294; tryptophan 75.

The starch present in jack bean seeds is composed of large and small granules, slightly-oval shaped. The large granules measure up to 37 microns. The starch has an amylose content of approximately 28.7 per cent, a gelatinization temperature ranging from 67.5° to 78°C, a swelling power at 95°C of 18.17, and a solubility of 17.77. It is similar to pigeon pea starch and would be suitable for use whenever high viscosity and high stability are desired during relatively long periods of heating.

Four globulins have been isolated from the seed, one showing considerable urease activity. The others are canavalin and concanavalin A and B. Concanavalin A is a phytohaemagglutinin and occurs at levels of between 2.5 and 3.0 per cent by weight. It has been the subject of considerable interest in recent years because of its affinity for agglutinating cells transformed by DNA tumour viruses and carcinogens. The presence of a crystalline di-amino acid canavanine has also been reported, which it is claimed is a powerful inhibitor of the growth of certain strains of the mould *Neurospora*. In addition, L-α–amino–δ-hydroxyvaleric acid, a possible precursor of proline, has been identified as a constituent of the seed, and L-homoserine. Although concanavalin A has been regarded as the toxic principle of the seeds, recently the presence of a non-haemagglutinating toxic factor has been reported. In addition, Tihon (1946), reports a hydrocyanic acid content of 0.0108 per cent.

Production and trade

No reliable statistical data are available relating to the production of, or trade in, the jack bean.

Major influences

The jack bean is one of the minor grain legume crops of the tropics and subtropics. However, it could have considerable future potential because of its suitability for cultivation in areas of low altitude, high temperature and relative humidity, unsuitable for other legumes such as the haricot bean, *Phaseolus vulgaris*. It is relatively free from serious pest and disease problems and has a high yield potential. Relatively little work has as yet been undertaken on improving the species yet in certain areas, eg Puerto Rico, average yields of dry seed of about 4 000 lb/ac (4 500 kg/ha) are being obtained. Recent research suggests that, although deficient in methionine, it could be utilized successfully for the production of protein concentrates. In addition, it could become of commercial value for medicinal purposes.

Bibliography

ADDISON, K. B. 1957. The effect of fertilizing espacement and date of planting on the yield of jack bean (*Canavalia ensiformis*). *Rhod. Agric. J.*, *54*, 521–532.

AFFLECK, H. 1961. Jack bean poisoning in cattle. *Rhod. Agric. J.*, *58*, 21.

BUSSON, F. 1965. *Canavalia* D. C. *Plantes alimentaires de l'ouest Africain: Étude botanique, biologique et chimique*. pp. 236; 238–239; 252–253. Marseille: Leconte, 568 pp.

CHARAVANAPAVAN, C. 1943. The utilization of the sword bean and jack bean as food. *Trop. Agric.*, (*Ceylon*), *99*, 157–159.

COBLEY, L. S. 1956. The leguminous crops: Sword bean, jack bean—*Canavalia ensiformis. An introduction to the botany of tropical crops.* pp. 154–156. London: Longmans, Green and Co Ltd, 357 pp.

DENNISON, C., STEAD, R. H. AND QUICKE, G. V. 1971. A non-haemagglutinating toxic factor from the jack bean (*Canavalia ensiformis*). *Agroplantae, 3* (2), 27–29. (*Field Crop Abstr., 25*, 5735).

FOOD AND AGRICULTURAL ORGANIZATION OF THE UNITED NATIONS. 1959. *Tabulated information on tropical and subtropical grain legumes.* pp. 63–75. Rome: FAO, Plant Prod. Prot. Div., 367 pp.

GÖHL, B. 1975. *Canavalia ensiformis* (L.) DC. *Tropical feeds: Feeds information summaries and nutritive values.* pp. 167; 511; 527; 545. Rome: FAO, 661 pp.

HOROWITZ, N. H. AND SRB, A. M. 1948. Growth inhibition of *Neurospora* by canavanine and its reversal. *J. Biol. Chem., 174*, 371–378. (*Hortic. Abstr., 18*, 1756).

JENKINS, A. E. 1931. Scab of *Canavalia* caused by *Elsinoe canavaliae. J. Agric. Res., 42*, 1–12.

KALB, A. J. AND LUSTIG, A. 1968. The molecular weight of concanavalin A. *Biochim. Biophys. Acta, 168*, 366–367.

LIENER, I. E. 1964. Seed haemagglutinins. *Econ. Bot., 18*, 27–33.

LUCIE-SMITH, M. N. 1933. Photography as a help in the examination of cattle foods: Structure of the pod and seeds of *Canavalia* spp. *J. Southeast Coll. Agric.*, (*Wye*), (32), pp. 42–48.

MOLINA, M. R., ARGUETA, C. E. AND BRESSANI, R. 1974. Extraction of nitrogenous constituents from the jack bean *Canavalia ensiformis. J. Agric. Food Chem., 22*, 309–312.

MULLER, H. M. AND SELLSCHOP, J. 1953. The sword or jack bean. *Farming South Afr., 28* (326), 175.

PIPER, C. V. 1920. The jack bean. *US Dep. Agric., Dep. Circ.* 92, 12 pp.

PURSEGLOVE, J. W. 1968. *Canavalia ensiformis* (L.) DC. *Tropical crops: Dicotyledons.* Vol. 1. pp. 242–244. London: Longmans, Green and Co Ltd, 332 pp.

QUISUMBING, E. 1965. *Canavalia* in the Philippines. *Araneta J. Agric., 12*, 1–7.

RACHIE, K. O. AND ROBERTS, L. M. 1974. Grain legumes of the lowland tropics. *Adv. Agron.*, *26*, 1–132.

ROSENTHAL, F. R. T., ESPINDOLA, L. AND OLIVERIA, S. M. G. DE. 1970. Jack bean starch: (i) Properties of the granules and of the pastes. *Die Stärke*, *22*, 126–129.

SASTRI, B. N. (ed.). 1950. *Canavalia* DC. *The wealth of India: Raw materials.* Vol. 2 (C). pp. 55–56. Delhi: Indian Counc. Sci. Ind. Res., 427 pp.

SAUER, J. AND KAPLAN, L. 1969. Canavalia beans in American prehistory. *Am. Antiq.*, *34*, 417–423. (*Plant Breed. Abstr.*, *41*, 4301).

SHARON, N. AND LIS, H. 1972. Lectins: Cell-agglutinating and sugar specific proteins. *Science*, *177*, 949–959.

SHONE, D. K. 1961. Toxicity of the jack bean. *Rhod. Agric. J.*, *58*, 18–20.

SMARTT, J. 1976. *Canavalia* DC. The sword or jack beans. *Tropical pulses.* pp. 56–58. London: Longman Group Ltd, 348 pp.

STEHLÉ H. 1953. Étude botanique et agronomique des légumineuses autochtones et exotiques des genres: *Canavalia*, *Clitoria* et *Crotolaria* aux Antilles françaises. [A botanical and agronomic study of native and introduced legumes of the genera *Canavalia*, *Clitoria* and *Crotolaria* in the French Antilles]. *Rev. Int. Bot. Appl.*, *Trop. Agric.*, *33*, 490–517.

THOMPSON, J. F., MORRIS, C. J. AND HUNT, G. E. 1964. The identification of L α–amino–δ–hydroxyvaleric acid and L–homoserine in jack bean seeds (*Canavalia ensiformis*). *J. Biol. Chem.*, *239*, 1122–1125.

TIHON, L. 1946. Un propos de deux *Canavalia* rencontrés au Congo Belge. [An account of two *Canavalia* spp. encountered in the Belgian Congo]. *Bull. Agric. Congo Belge*, *37*, 156–162.

VANETTI, F. 1958. Ensaio preliminar de contrôle ao gorgulho das vagens do 'feijão de porco', *Sternechus tuberculatus* Boheman, 1836. [Preliminary trial in controlling the pod weevil, *Sternechus tuberculatus* Boheman, 1836, on the sword bean]. *Ceres, Minas Gerais*, *10*, 317–323. (*Field Crop Abstr.*, *12*, 196).

WESTPHAL, E. 1974. *Canavalia ensiformis* (L.) DC. Pulses in Ethiopia, their taxonomy and agricultural significance. pp. 72–84. *Wageningen*, *PUDOC*, *Cent. Agric. Publ. Doc.*, *Agric. Res. Rep.* 815, 278 pp.

WHITE, C. T. 1943. The sword bean or scimitar bean and the jack bean. *Qd. Agric. J.*, *57*, 25–27.

Common names	**KERSTING'S GROUNDNUT, Kerstingiella (groundnut), Geocarpa (groundnut), Ground bean[1].**
Botanical name	*Kerstingiella geocarpa* Harms, syn. *Voandzeia geocarpa* (Harms) A. Chev.
Family	Leguminosae.
Other names	Bendi, Bindi (Mali); Diéguem tenguéré[2] (Mossi); Dougoufulo (Niger); Doyi (Togo); Fève de kandela (Fr.); Haricot de behanzin, H. royal (Togo, Benin); Hausa groundnut (Nig.); Kandela (W. Afr.); Kandelabohne (Ger.); Kouarourou (Nig.); Kwaruru (W. Afr.); Lentille de terre (Fr.); Pararu[1] (W. Afr.).

Botany

A prostrate, herbaceous annual. The main stem is 2–3.5 in. (5–9 cm) long, hirsute-pubescent, or nearly glabrous, depending upon the cultivar, and has numerous thin, short stolons which spread out on the soil, or are partly buried in it. The leaves are trifoliolate, the leaflets membranous, more or less rounded at both ends. The terminal leaflet is approximately 2.5–3 in. (6–7.5 cm) long and 1.6–2 in. (4–5 cm) broad. The flowers are small, usually in pairs, white, or greenish-white in colour, sometimes tinged with purple. After fertilization the ovary is pushed out of the calyx into the soil where the seed-bearing pods develop in a manner similar to groundnuts. The mature pod is indehiscent, approximately 0.4–0.8 in. (1–2 cm) long and 0.28–0.4 in. (0.7–1.0 cm) wide, with a buff coloured paper-like husk, usually divided by a constriction and a corresponding thin septum into two, or sometimes three, joints. In some forms it may be simple, slightly curved and glabrous. The pods contain 1–3, (commonly 2), seeds.

There are many different forms of Kersting's groundnut and although widely reported to be found only in the cultivated state, Hepper (1963) differentiates between a robust cultivated form, *Kerstingiella geocarpa* var. *geocarpa*, and a wild form, *K. geocarpa* var. *tisseranti*. The Kersting's groundnut is frequently

---

[1] Also used for the bambara groundnut, *Voandzeia subterranea*.

[2] Also used for the African yam bean, *Sphenostylis stenocarpa*, see *Crop and product digest No. 2: Root crops*.

confused with the bambara groundnut, *Voandzeia subterranea*, but can be distinguished by its flowers which have a deeply divided calyx with narrow lobes. In fact *Kerstingiella* now proves to be congeneric with *Macrotyloma*, while *Voandzeia* now proves to be congeneric with *Vigna*, though these very recent ideas have not yet been taken into the nomenclatural framework.

Origin and distribution

The Kersting's groundnut originated in the savanna areas of W. Africa and has a very restricted range of cultivation, being confined largely to Mali, Upper Volta, Nigeria, Togo and Benin.

Cultivation conditions

Temperature—a tropical legume requiring an average shade temperature from 64° to 93°F (18–34°C), with plenty of bright sunshine.

Rainfall—a crop suitable for semi-arid regions, it is grown successfully in areas with an annual rainfall of between 20 and 24 in. (500–600 mm), but is also found on the fringes of the humid tropics.

Soil—a sandy loam, rich in lime, is required for optimum yields, but the Kersting's groundnut can be grown successfully on poor sandy soils.

Planting procedure

Material—sound, plump seeds, free from insect infestation should be used.

Method—the seeds are frequently planted at the beginning of the rainy season in small patches, either as the sole crop, which may constitute the first crop in a rotation, to be followed by other crops such as cassava, or in mixed cropping systems with taller plants. Frequently, seeds remain in the ground after harvesting and these sprout with the first rains. Thus, the crop can persist in a semi-wild state until the land reverts to bush.

Field spacing—usually planted in rows 12–16 in. (30–40 cm) apart, with 6 in. (15 cm) between the plants.

Pests and diseases

The Kersting's groundnut is not subject to serious attack from pests or diseases, apart from weevils, *Piezotrachelus* spp. and pulse beetles, *Bruchidae*.

Growth period

The crop reaches maturity 90 to 150 days after planting, usually at the end of the rainy season in semi-arid regions.

Harvesting and handling

Harvesting takes place when the leaves begin to yellow and wither. It cannot be delayed, as once mature the seeds are liable to germinate or rot, if left in the soil. The pods and leaves form a tangled mass, which is usually dug up with a spade and left to dry for a few days, before the pods are picked out by hand. The seeds are separated by beating the pods with sticks and then winnowing. They are usually left to dry thoroughly in the sun before being dusted with an insecticide and stored, often in earth granaries, or airtight jars.

Primary product

Seed—this is normally kidney-shaped, approximately 0.35 x 0.24 in. (9 x 6 mm). It varies in colour and can be white, brown, reddish-black, self-coloured or mottled, with a white hilum, and a relatively thick seed-coat. Internally the seed is creamy-white in colour.

Yield

It is difficult to obtain accurate information regarding the yield of seed because the crop is grown in scattered small plots in village compounds. However, it has been estimated that the average yield in W. Africa is between 400 and 445 lb/ac (450–500 kg/ha).

Main use

The Kersting's groundnut has a high nutritional value and a pleasant taste and the mature seeds can be boiled or ground into a paste similar to other grain legumes, such as cowpeas. White-seeded types are reported to be the most popular and to fetch premium prices. In Benin, consumption is confined to tribal chiefs and the heads of families, and is forbidden to women. In many parts of W. Africa the Kersting's groundnut is regarded as a speciality food item and is often eaten boiled and mixed with shea butter and salt.

Secondary and waste products

The leaves are sometimes eaten as a vegetable, or in soups.

Special features

The Kersting's groundnut has an approximate moisture content of between 9.3 and 12.5 per cent. The seed-coat constitutes approximately 12 per cent of the total weight of the seed. The average composition, in terms of 100g of edible portion, is as follows: moisture 9.3–12.5 g; protein 19.0–20.4 g; fat 0.9–2.3 g; total carbohydrate 66.6 g; fibre 4.6–5.6 g; ash 2.8–3.3 g;

calcium 85–227 mg; phosphorus 385–425 mg; thiamine 0.67–0.77 mg. The fatty acids present in the fat, expressed as a percentage of the total fatty acids, are: palmitic 31.5; stearic 5.0; oleic 7.0; linoleic 31.5; linolenic 21.3; arachidic 3.1; behenic 0.6. The carbohydrate consists mainly of starch; an early analysis of seed from Mossi reports a starch content of 48.9 per cent and non-reducing sugars 0.4 per cent. The amino acid content (mg/gN) has been reported as follows: arginine 404; cystine 63; histidine 173; isoleucine 281; leucine 479; lysine 413; methionine 86; phenylalanine 363; threonine 238; tyrosine 219; valine 390; aspartic acid 717; glutamic acid 1 088; alanine 279; glycine 279; proline 340; serine 367; tryptophan 50. The seeds appear to be free of alkaloids or cyanogenic compounds.

Production and trade

Production of the Kersting's groundnut is confined mainly to W. Africa, where it is grown at a subsistence level, notably in parts of Benin, Upper Volta and Mali. No statistical data relating to production or trade have been traced.

Major influences

The Kersting's groundnut is a little known grain legume crop with a restricted range of cultivation. Although nutritious, the relative low yield of seed obtainable has undoubtedly handicapped its development and it appears to be declining in importance in most areas in W. Africa.

Bibliography

ANON. 1913. Some new or little-known leguminous feedingstuffs: *Kerstingiella geocarpa* seeds from northern Nigeria. *Bull. Imp. Inst.*, *11*, 230–243.

BUSSON, F. 1965. *Kerstingiella* Harms. *Plantes alimentaires de l'ouest Africain: Étude botanique, biologique et chimique*. pp. 241; 252–253. Marseille: Leconte, 568 pp.

CHEVALIER, A. 1933. Monographie de l'arachide: (iii) Les légumineuses à fruits souterrains autres que l'arachide—Le kerstingiella ou lentille de terre. [A monograph on the groundnut: (iii) The legumes with underground fruits other than the groundnut—The kerstingiella or earth lentil]. *Rev. Int. Bot. Appl. Agric. Trop.*, *13*, 705–711.

HEPPER, F. N. 1963. Plants of the 1957–58 West African expedition: The bambara groundnut (*Voandzeia subterranea*) and Kersting's groundnut (*Kerstingiella geocarpa*) wild in West Africa. *Kew Bull.*, *16*, 395–407.

HOLLAND, J. H. 1922. *Voandzeia* Thouars. The useful plants of Nigeria. p. 231. *Kew Bull. Misc. Inf., Addit. Ser.* 9, 963 pp.

IRVINE, F. R. 1969. Other edible legumes: Kerstingiella, geocarpa or hausa groundnut (*Kerstingiella geocarpa*). *West African agriculture.* 3rd ed. Vol. 2. *West African crops.* pp. 204–205. London: Oxford University Press, 272 pp.

LAMB, P. H. 1913. *Kerstingiella geocarpa. Kew Bull. Misc. Inf.*, (2), pp. 93–94.

LEPIGRE, M. 1965. Étude sur les possibilités d'amélioration de la conservation des haricots au Togo en milieu rural. [A study on the possibilities of improving bean storage under farming conditions in Togo]. *Agron. Trop.*, *20*, 388–430.

OKIGBO, B. N. 1973. Grain legumes in the farming systems of the humid lowland tropics. *Proc. 1st Int. Inst. Trop. Agric., Grain legume improvement workshop.* pp. 211–223. Ibadan, Nigeria, Int. Inst. Trop. Agric., 325 pp.

SCHNELL, R. 1957. *Plantes alimentaires et vie agricole de l'Afrique noire.* p. 168. Paris: Larose, 223 pp.

SMARTT, J. 1976. *Kerstingiella* Harms. *Tropical pulses.* pp. 60–61. London: Longman Group Ltd, 348 pp.

STAPF, O. 1912. A new ground bean (*Kerstingiella geocarpa* Harms) *Kew Bull. Misc. Inf.*, (5), pp. 209–213.

RACHIE, K. O. AND ROBERTS, L. M. 1974. Grain legumes of the lowland tropics. *Adv. Agron.*, *26*, 1–132.

VUILLET, J. 1934. Culture de *Kerstingiella geocarpa* au Soudan français. [Cultivation of *Kerstingiella geocarpa* in the French Sudan]. *Rev. Int. Bot. Appl. Agric. Trop.*, *14*, 210–211.

Common names — **LENTIL, Red dhal[1], Split pea.**

Botanical name — *Lens culinaris* Medik. syn. L. *esculenta* Moench.

Family — Leguminosae.

Other names — Adas (Ar.); Adasha tarbutit, Adashim (Is.); Adesi (E. Afr.); Ads masri (Sud.); Bersem, Birssin, Burssum (Eth.); Chaunangi, Chirisanagalu (Ind.); Echte linse (Ger.); Faki (Cyp.); Hiramame (Jpn.); Leca (Yug.); Lencse (Hun.); Lensie (S. Afr.); Lenteja (Sp.); Lenticchia (It.); Lentilha (Port.); Lentille (Fr.); Linse (Ger.); Linze (Dut.); Massar, Masser, Mas(s)ur (Ind.); Mercimek (Turk.); Messer (Eth.); Misurpappu, Musiripappu, Thulukkappayar (Ind.).

Botany

A much-branched, sub-erect, slightly pubescent annual; usually 6–30 in. (15–75 cm) high, showing considerable variation in form. Three types of root-system may occur: (i) a much-branched shallow root system with numerous tap-root nodules; (ii) a slender deep tap-root; (iii) an intermediate form. The first type of root-system is found on light alluvial soils, and is associated with profuse branching and small seed; the second, is found on quick-drying soils liable to crack, and is associated with rather sparse branching and bold seed, the third on intermediate soil types. The stem is square and ribbed with several basal branches. The leaves are alternate, with 4–7 pairs of opposite, or alternate, leaflets, approximately 0.5 in. (1.25 cm) long. The flowers may be solitary, or in racemes of 2–4 flowers, and may be white, pink, red or violet in colour, according to the cultivar. Although usually self-fertilized, cross-pollination can occur sometimes. The seed-pods are smooth, compressed, usually 0.5–0.8 in. (1.25–2.0 cm) long and nearly as broad, with a short beak. They contain two smooth lens-shaped seeds, which show considerable variation in size and colour.

Lentils are usually divided into two vast geographic groups, morphologically well delimited, each with a definite geographic area: ssp. *macrosperma*, with large, flattened seeds, and ssp. *microsperma*, with small or medium-sized seeds.

---

[1] Applied to the red coloured types of lentil seed.

The latter is more polymorphous and within it, six narrower geographic groups, or varieties, may be designated, including var. *afghanica* and var. *abyssinica*.

The subspecies *macrosperma* includes the so-called Chilean or yellow cotyledon types of lentil, such as the cultivar Large Blond. While the subspecies *microsperma* includes the small-seeded Persian lentils or red cotyledon types; the cultivar Anicia belongs to this subspecies. As a result of varietal improvement programmes, the improved large-seeded cultivar Tekoa has recently been released in the USA, and in India a number of improved, early-maturing types suitable for late sowing under soil moisture stress have been developed including: Pusa–1, Pusa–4, Pusa–6, L–9–12, T–6, T–36, WB-81 and WB–94.

Origin and distribution

Lentils are one of the oldest of the grain legumes and are thought to have originated in Asia Minor, but quickly spread to Egypt, central and southern Europe, the Mediterranean basin, Ethiopia, Afghanistan, northern India and Pakistan. They were successfully introduced into the New World and are now cultivated in the USA, Mexico, Chile, Peru, Argentina and Colombia.

Cultivation conditions

Temperature—lentils are a crop of the warm temperate and subtropical regions, but are grown in the tropics at high elevations, or as a cool season crop. In India they are planted from mid-October to early November, after which yields decline to uneconomic levels. The seed requires a minimum temperature of 59°F (15°C) for germination, the optimum temperature is 64–70°F (18–21°C). Intense or prolonged frost is reported to affect growth seriously, as do temperatures much in excess of 80°F (27°C), and for optimum yield an average temperature of about 75°F (24°C) is required. There is, however, considerable variation in the response of cultivars to temperature. Recent investigations indicate that the optimum temperature range for the cultivar Large Blond is 66–84°F (19–29°C) and for Anicia 70–77°F (21–25°C).

Rainfall—a moderate rainfall of approximately 30 in./a (750 mm/a) is required; dry conditions must prevail just prior to, and at, harvest. In India lentils are frequently grown during the winter as a dry or unirrigated crop, depending upon subsoil moisture and the heavy dews prevalent during this season. They are moderately tolerant of drought, but the degree of tolerance varies according to the cultivar. They may be grown under irrigation but over-watering can damage the crop; in Egypt the application of 3–4 irrigations at 30 day intervals has been recommended. In Bulgaria it has been observed that small-seeded cultivars are more drought-resistant than the large-seeded ones.

Soil—lentils can be grown on a wide range of soil types, ranging from fairly heavy clay soils to low-lying sandy soils. In India good yields are obtained on light loams and alluvial soils and on moderately deep black cotton soils. Fairly heavy clays which are retentive of moisture are very satisfactory. On soils of high natural fertility vegetative growth is liable to be excessive and the yield of seed low. Lentils can tolerate moderately alkaline soils. Little is known of their precise manurial requirements, but there is a significant response to applications of phosphorus. On soils of low fertility in Pakistan, the application of farmyard manure at the rate of 4–6 T/ac (10–15 t/ha), superphosphate 82 lb/ac (92 kg/ha) and potassium muriate 82 lb/ac (92 kg/ha) has been recommended. On sandy loam soils of average fertility the application of nitrogen 20 lb/ac (22 kg/ha), phosphoric acid 40 lb/ac (45 kg/ha) and potash 20 lb/ac (22 kg/ha) has been suggested. Excessive applications of potash can cause hard seeds. Recent investigations in Egypt suggest that there the most economic fertilizer treatment is nitrogen 16 lb/ac (18 kg/ha), phosphoric acid 64 lb/ac (72 kg/ha). Molybdenum is essential for maximum yields and is often applied as molybdated gypsum 50 lb/ac (56 kg/ha), provided that the seed has been inoculated with the appropriate bacterium. The use of foliar sprays has also been suggested, 6.25 gal/ac (60 l/ha) of 0.01–0.04 per cent solution of ammonium molybdate.

Altitude—in the SE. Asian sub-continent lentils are grown up to elevations of 11 500 ft (3 450 m), in Kenya the crop is grown between 5 000 and 7 000 ft (1 500–2 100 m).

Day-length—lentils are quantitative long-day plants, with some cultivars tending to be day-neutral. The cultivar Large Blond for example, requires long days of 14–16 hours in order to flower, whereas Anicia will flower under photoperiods ranging from 9 to 16 hours, with day-lengths of 15–16 hours promoting earlier flowering.

Planting procedure

Material—seed is used and germination is hypogeal. Inoculation increases yields significantly and the use of fungicide seed dressings as a protectorant against seed-borne diseases has been suggested.

Method—lentils are grown as a pure crop, or as in India, mixed with barley, mustard or castor. The seed may be broadcast or planted in rows, when the seed is normally planted at a depth of from 0.5 to 2.5 in. (1.25–6.25 cm), depending on its size and the available soil moisture. The seed-bed should be cultivated to a depth of about 7 in. (17.5 cm) and be completely free from weeds. Competition from weeds, particularly during the early stages of growth, can adversely affect yields. In countries such as the USA and France the pre-planting application of di-allate followed by the post-emergence application of dinoseb is reported to give effective weed control. The pre-emergence application of prometryne followed by barban post-emergence, has also given promising results, as have linuron, metobromuron and methoprotryne in Bulgaria.

Field-spacing—in India when grown as a pure crop a field-spacing of 9 x 12 in. (22.5 x 30 cm) is frequently used. In the USA lentils are commonly planted with small grain equipment, 7–12 in. (17.5–30 cm) between the plants and 24–36 in (60–90 cm) between the rows.

Seed-rate—when grown as a pure crop in India the average seed-rate is 20–25 lb/ac (22–28 kg/ha) depending on the size of the seed, when sown as a mixed crop the seed-rate averages 10–12 lb/ac (11–13 kg/ha). In the USA the seed-rate ranges between 70 and 80 lb/ac (78–90 kg/ha) for the cultivar Tekoa, compared with 60 to 70 lb/ac (67–78 kg/ha) for Chilean type seed.

Pests and diseases

The main pests reported to attack lentils in the SE. Asian sub-continent are: the gram caterpillar, *Heliothis obsoleta*, white ants, *Clotermes* sp., the gram cutworm, *Ochropleura flammatra*, and the weevil, *Callosobruchus analis*, the last is also a serious pest of the seed in store. In the United States aphids, *Aphis* spp., and the vetch bruchid, *Bruchus brachialis*, can sometimes be troublesome.

Two diseases which are of considerable economic importance, particularly, in the SE. Asian sub-continent, are rust, *Uromyces viciae-fabae* and wilt, *Fusarium oxysporum* f. sp. *lentis*. The leaves and stems of plants affected by rust lose their green colour and become purple, and in the case of serious infection may die. High humidity and moderate temperatures, 63–77°F (17–25°C), favour the spread of this disease. Wilt causes curling of the leaves and affects the development of the root-system. It is favoured by light dry soils, soil moisture about 25 per cent. Crop rotation, destruction of diseased plant debris, the fungicidal treatment of seed, and the growing of

resistant cultivars have been suggested as control measures.
Other diseases which affect lentils include: powdery mildew caused by *Erysiphe polygoni*, downy mildew, *Peronospora lentis*, leaf blight, *Alternaria alternata*, and several root or stem rots including: *Corticium rolfsii*, *Ascochyta pinodella*, *Botrytis cinerea*, *Rhizoctonia* sp., *Sclerotinia sclerotiorum* and *Verticillium albo-atrum*. Of these *B. cinerea* is the most devastating in the USA. Lentils are also susceptible to several virus diseases including: alfalfa mosaic virus, bean yellow mosaic virus, cucumber mosaic virus, and pea leaf roll virus, all of which are transmitted by aphids and can seriously affect yields.

Growth period

Early-maturing cultivars have a crop cycle ranging from 80 to 110 days and late-maturing from 125 to 130 days. Teoka, for example, has a growth period of 91 days, Pusa 1–180 days, Pusa–6 90 days and L–9–12 110 days.

Harvesting and handling

Harvesting takes place when the pods are a golden-yellow colour and the lower ones are still firm. It cannot be delayed as the lower seed-pods become over-ripe and shatter, resulting in a considerable crop loss. In many areas the plant is cut down to ground level manually and left to dry for about 10 days, before being threshed and winnowed. In the USA and the USSR lentils are harvested mechanically. In the former country specially adapted cereal harvesters are used. In the latter, problems of mechanical harvesting have been partially overcome by growing lentils mixed with a fibre plant, *Camelina* spp., which acts as a support. Lentil seed is often badly contaminated with weed seed, particularly vetch seed, and requires careful cleaning. In the USA and certain other countries, the seed is usually cleaned by air and sieves to remove all foreign matter, cracked seeds, etc, and to produce a graded product. A common commercial standard for red lentils is for at least 50 per cent of the sample to remain on a 4.35 mm round hole perforated screen and 100 per cent to remain on a 3 mm screen. Yellow lentil seed is usually graded according to the percentages which remain on 7 mm, 6 mm, and 5 mm round hole perforated screens. The premium grade is the 7 mm size.

The seed should be dried to a moisture content of between 11 and 14 per cent; below 11 per cent breakage and hardshell are likely to be troublesome.

In the SE. Asian sub-continent lentils are frequently stored in bins made of mud or bamboo and covered with dried leaves or straw. Storage losses can amount to as much as 10 per cent due to the development of moulds, insect infestation and rodent damage.

Primary product

Seed—which is flat and lens shaped and may vary in size from approximately 0.12–0.35 in. (3–9 mm) in diameter, and may vary in colour from a pale-pink or buff, to a dark reddish-brown, and various shades of greenish-yellow, sometimes mottled with grey or dark-green. The hilum is small and very narrow. In international trade two main types of lentil seed are recognized, the so-called Chilean, yellow or light-green types, produced mainly in the USA and S. America, and the small-seeded Persian, or red types, cultivated in the Mediterranean area and Asia. The 1 000-seed weight of Chilean type lentils ranges between approximately 1.4 and 2.9 oz (40–82 g) and that of the small-seeded types between 0.4 and 1.4 oz (13–40 g). For example, the 1 000-seed weight of the cultivar Large Blond is about 2.9 oz (82 g) and that of Anicia, 1 oz (28 g).

Yield

In the SE. Asian sub-continent seed yields average between 300 and 600 lb/ac (340–670 kg/ha) with mixed cultivation or dry conditions, and between 800 and 1 000 lb/ac (900–1 120 kg/ha) for pure crops. With efficient cultivation and improved cultivars yields of the order of 2 000 lb/ac (2 240 kg/ha) are reported to be attainable. In Ethiopia yields are reported to average 530 lb/ac (600 kg/ha). In Egypt the average yield of the standard cultivar Giza 9 in 1973 was reported to range from 2 220 to 2 700 lb/ac (2 500–3 020 kg/ha). In the USA yields are reported to average about 1 100 lb/ac (1 230 kg/ha), but under favourable conditions can reach between 1 500 and 1 600 lb/ac (1 680–1 790 kg/ha).

Main use

Lentils are a nutritious foodstuff and are used mainly in the form of dhal as an ingredient in soups. Flour prepared from the ground seeds can also be used mixed with cereal flours in cakes or bread and in the preparation of invalid and baby foods. In parts of India, eg Uttar Pradesh, the whole seed is often eaten salted and fried. The canning of lentils and the production of instant lentil flour is being developed in the USA.

Subsidiary uses

The seeds are sometimes used as a source of commercial starch for use in the textile and printing industries. The yield of starch from the seeds is approximately 28.5 per cent. The young immature pods may be eaten as a vegetable.

Secondary and waste products

Lentils are also grown sometimes as a green manure crop, or for forage. Lentil hay has the following approximate composition: moisture 10.2 per cent; fat 1.8 per cent; protein 4.4 per cent; carbohydrate 50.0 per cent; fibre 21.4 per cent; ash 12.2 per cent. The seed-coat and bran left after the production of dhal is used as an animal feedingstuff. The approximate composition of the seed-coat (dry basis) is: fat 0.8 per cent; protein 12.6 per cent; N–free extract 54.1 per cent; fibre 29.0 per cent; ash 3.5 per cent.

The residue left after starch extraction can also be used as an animal feedingstuff, the approximate composition of a sample (air-dried) has been given as: moisture 8.3 per cent; protein 38.9 per cent; starch 21.9 per cent; crude fibre 26.7 per cent; ash 4.2 per cent.

Special features

The average composition of lentil seed is: moisture 12.4 per cent; fat 0.7 per cent; carbohydrate 59.7 per cent; protein 25.1 per cent; ash 2.1 per cent; calcium 38.6 mg/100 g; phosphorus 242 mg/100 g; iron 7.62 mg/100 g; sodium 36.0 mg/100 g; magnesium 76.5 mg/100 g. The vitamin content has been reported as: thiamine 0.26 mg/100 g; riboflavin 0.21 mg/100 g; nicotinic acid 1.70 mg/100 g; choline 223.00 mg/100 g; folic acid 107.00 mg/100 g; inositol 130.00 mg/100 g; pantothenic acid 1.60 mg/100 g; biotin 1.30 mg/100 g; pyridoxine 0.49 mg/100 g; carotene 1.60 mg/100 g; ascorbic acid 4.20 mg/100g; vitamin K 0.25 mg/100 g; tocopherol 2.00 mg/100 g.

There is considerable variation in the protein content of lentils from various sources, certain Indian cultivars have protein contents of about 30 per cent, but values as low as 17 per cent have been reported in some Pakistani cultivars. The proteins are similar to those of beans and peas and consist of: globulin 44.0 per cent; glutelin 20.6 per cent; prolamin 1.8 per cent and a water-soluble fraction 25.9 per cent. The amino acid composition (mg/gN) has been given as: isoleucine 270; leucine 477; lysine 449; methionine 50; cystine 57; phenylalanine 327; tyrosine 204; threonine 248; valine 313; arginine 543; histidine 171; alanine 269; aspartic acid 723; glutamic acid 1 037; glycine 264; proline 267; serine 329. Recently the presence of the amino acids γ–hydroxyornithine, γ–hydroxyarginine and homoarginine has been reported. The starch content is about 40.4 per cent; total sugars 2.7 per cent; reducing sugars 1.8 per cent. The following sugars are reported to be present: verbascose, stachyose, sucrose, glucose and fructose. The starch has a grain size of 3.5–15.7 x 9.8 microns; its viscosity remains virtually unchanged over a wide range of temperatures.

Lentils are considered to be the most easily digested of the grain legumes. The presence of a heat labile trypsin inhibitor and a low molecular-weight phytohaemagglutinin has been reported. The seed also contains amylase, phosphatase and phytase, and a saponin, esculenin, (MP 173–175°C), has also been isolated.

Processing

In India lentils are processed into dhal by a dry method. The seeds are partially crushed in 'chakkies' or dhal rollers, and the whole husked grain separated from the seed-coat. The bruised grain frequently has to pass 2–3 times over the rollers before the process is completed. The husked grain or dhal is generally polished with magnesia powder to improve its appearance and to remove any particles of seed-coat which may be still adhering to it. The improved method developed by the Central Food Technological Research Institute, Mysore, (CFTRI), for the more economic milling of grain legumes, as described in the section on the chick pea, can be used successfully for lentils. Using the CFTRI method the yield of dhal averages 83 per cent, compared with the maximum theoretical yield of 88 per cent and yields of between 65 and 80 per cent obtained by the traditional dry method.

Production and trade

Production—lentils constitute about 2.4 per cent of the total world output of grain legumes. World production averaged 1 094 400 t/a for the period 1970–74, compared with 1 018 000 t/a for 1965–69, an increase of just over 7 per cent. Production in 1975 amounted to 1 120 000 t, and in 1976 to 1 236 000 t.

Lentils: Major producing countries

Quantity tonnes

	Annual average 1965–69	Annual average 1970–74	1975	1976
India	368 200	387 600	463 000	466 000
Ethiopia	102 000	111 000	61 000	55 000
Turkey	100 000	98 000	135 000	225 000
USSR	52 000	79 000	50 000	90 000
Syria	64 000	67 400	67 000	59 000

Lentil

Trade—there is a considerable international trade in lentils. Large light-green types are favoured by European markets and the small red types by markets in the Middle East and Asia. It is difficult to estimate the quantities of lentils entering international trade since many countries do not classify them separately in their trade returns. However, from the available statistical data, it is estimated that for the period 1970–74 the total quantity of lentils entering international trade was approximately 166 000 t/a, compared with some 121 000 t/a for 1965–69, an increase of 40 per cent.

Lentils: Major exporting countries

Quantity tonnes

	Annual average 1965–69	Annual average 1970–74	1975
USA	24 095	28 718	37 435
Lebanon	17 649	21 186	n/a
Ethiopia	19 148	22 979	34 650
Syria	n/a	18 284[1]	n/a
Morocco	5 229	13 774[1]	13 603
Turkey	14 516	12 619[1]	n/a
Jordan	12 059[2]	10 889	3 942

1 Four year average.
2 1969 only.
n/a Not available.

Lentils: Major importing countries

Quantity tonnes

	Annual average 1965–69	Annual average 1970–74	1975	1976
Ger. Fed. Repub.	23 751	19 590	17 692	16 694
Lebanon	19 245[1]	19 780[2]	n/a	n/a
France	13 688	14 347	20 886	n/a
Italy	8 336	10 678	8 885	11 173
UK	11 404	10 029	9 311	12 630
Malaysia	6 635	6 410	n/a	n/a

1 1969 only.
2 Four year average.
n/a Not available.

Prices—lentil prices depend largely upon the size and colour of the seed, light-green lentils tend to fetch premium prices in international trade. Prices have risen in recent years. In Uruguay, for example, the producer price averaged £79.50/t in 1969, but by 1973 had risen to £140.07/t, and in 1974 was £271.40/t, while in Pakistan wholesale prices averaged £68.67/t for the years 1965–69, but by 1971 had reached £96.19/t. The average fob value of exports from the USA was £202.00/t in 1975 and £112.40/t for the years 1970–74, compared with £77.80/t for 1965–69, while the average value of Indian exports was £302.00/t in 1975 and £100.25/t for the years 1970–74, compared with £81.75/t for 1965–69.

Major influences

Lentils are an important food legume crop because of their relatively high protein content, which can reach 30 per cent, and the ease with which they are digested. They are widely cultivated in temperate and subtropical regions and are an important crop at higher elevations in the tropics, or during the cool season. Their susceptibility to disease, particularly rust and wilt, has tended to handicap their development as a food legume crop and if production is to be increased there is need to develop improved, high-yielding, short-duration, disease-resistant cultivars.

Bibliography

ANON. 1954. Note sur la lentille large verte d'Algérie. [Note on the Algerian large-green lentil]. *Terre Maroc.*, *28*, 225–229.

ANON. 1971. Tekoa lentil and its culture. *Washington State Univ. Coll. Agric., Ext. Cir.* 375, (leafl. unpaginated).

ANON. 1975. US dry peas and lentils—a thriving industry. *Int. Fruit World*, *34*, 71–78.

CHOPRA, K. AND SWAMY, G. 1975. *Pulses: An analysis of demand and supply in India. Institute for social and economic change; Monograph 2.* New Delhi: Sterling Publishers, PVT, Ltd, 132 pp.

COMMONWEALTH BUREAU OF PASTURES AND FIELD CROPS. nd. Lentils (*Lens esculenta*). Annotated bibliography 1297, 1950–72. *Commonw. Agric. Bur., Bur. Pastures & Field Crops, Hurley, Maidenhead, Berks, England*, 12 pp.

DIXIT, R. K. 1974. Inter-relationship among some agronomic traits in lentil (*Lens esculenta* Moench.). *Madras Agric. J.*, *61*, 588–591.

Lentil

El-Bagoury, O. H. 1976. Effect of different fertilizers on the germination and hard seed percentage of lentil seeds (*Lens culinaris* Med.). *Seed. Sci. Technol.*, *2*, 427–434. (*Field Crop Abstr.*, *29*, 321).

Entenman, F. M., Morrison, K. J. and Youngman, V. E. 1968. Growing lentils in Washington. *Washington State Univ. Coll. Agric.*, *Ext. Bull.* 590, 6 pp.

Food and Agriculture Organization of the United Nations. 1970. Amino acid content of foods and biological data on proteins. *FAO Nutr. Stud.* 24, pp. 54–55. Rome: FAO, 286 pp.

Fotidar, M. R. 1949. Lentil as green manure. *Indian Farming*, *10* (3), 111–112.

Golubev, V. D. and Lugovskikh, M. A. 1974. [Pre-sowing treatment of lentil seeds with ammonium molybdate]. *Khimiya v Sel'skom Khozyaistve*, *12* (3), 24–25. (*Field Crop Abstr.*, *28*, 978).

Gorjunova, A. 1954. [Mixed sowings of *Lens esculenta* with *Camelina* spp.]. *Zemledelie*, (*Agriculture*), *2* (7), 74–78. (*Field Crop Abstr.*, *8*, 142).

Hakam, M. M. 1974. Legume field crops improvement in the Arab Republic of Egypt. *Proc. 1st FAO/SIDA Semin., Improvement and production of field food crops for plant scientists from Africa and the Near East*. pp. 71–81. Rome: FAO, 688 pp.

Hakam, M. M. and Ibrahim, A. A. 1974. Cultural practices of grain legumes in the Arab Republic of Egypt. *Proc. 1st FAO/SIDA Semin., Improvement and production of field food crops for plant scientists from Africa and the Near East*. pp. 457–465. Rome: FAO, 688 pp.

Hamissa, M. R. 1974. Fertilizer requirements for broad beans and lentils. *Proc. 1st FAO/SIDA Semin., Improvement and production of field food crops for plant scientists from Africa and the Near East*. pp. 410–416. Rome: FAO, 688 pp.

Howard, I. K. and Sage, H. J. 1969. Isolation and characterization of a phytohemagglutinin from the lentil. *Biochemistry*, *8*, 2436–2441.

Janicek, G. and Hrdlicka, J. 1969. [Study of chemical composition of lentil (*Lens esculenta*)]. *Sbornik Vysoke Skoly Chemicko-Technologicke e Praze, E-Potraviny*, *23*, 55–61. (*Food Sci. Technol. Abstr.*, *3* (7), J910).

KAISER, W. J., DANESH, D., OKHOUVAT, M. AND MOSSAHEBI, G. 1972. [Virus diseases of lentil in Iran]. *Iran J. Plant Pathol.*, *8* (3), 75–84; (4), 32–33. (*Rev. Plant Pathol.*, *53*, 1654).

KANDÉ, J. 1967. Valeur nutritionelle de deux graines de légumineuses: Le pois chiche (*Cicer arietinum*) et la lentille (*Lens esculenta*). [Nutritional value of two grain legumes: chick pea (*Cicer arietinum*) and lentil (*Lens esculenta*)]. *Ann. Nutr. Aliment.*, *21* (2), 45–67.

KANNAIYAN, J. AND NENE, Y. L. 1973. Sclerotinia blight of lentil. *Curr., Sci.*, *42*, 32.

KANNAIYAN, J. AND NENE, Y. L. 1973. A new root rot disease of lentil. *Curr. Sci.*, *42*, 257.

KASASIAN, L. 1971. Vegetable crops. *Weed control in the tropics.* pp. 163–175. London: Leonard Hill Books, 307 pp.

KURIEN, P. P., DESIKACHAR, H. S. R. AND PARPIA, H. A. B. 1972. Processing and utilization of grain legumes in India. Symp. food legumes. pp. 225–236. *Tokyo, Jpn., Minist. Agric. & For., Trop Agric. Cent., Trop. Agric. Res. Ser.* 6, 253 pp.

MATHUR, H. G. 1958. Marketing of pulses in India. *Nagapur, India, Agric. Mark. Gov. India, Advis., Agric. Mark. Ser.* AMA 102, 182 pp.

MAUYRA, D. M. 1968. Lentil T–6 for double cropping in paddy areas. *Indian Farming*, *18* (3), 23–24.

MCFARLANE, J. A. 1970. Pulses. *Food storage manual.* pp. 335–346. Rome: FAO, World Food Programme, 799 pp., plus indexes, (loose leaf).

MEHTA, Y. R., GANGAWA, L. C. AND RAI, B. 1965. Lentil–6 can stand flood. *Indian Farming*, *15* (4), 7.

MUNJAL, R. L. AND VASUDEVA, R. S. 1962. Rabi crop diseases supplement. *Indian Farming*, *11* (11), 33–40.

NAQVI, R. H. 1963. Cultivation of lentils, called the poor man's meat in Pakistan. *Pak. Minist. Food & Agric., Food & Agric. Counc., Agric. Ser. Leafl.* 11(a), 5 pp.

NEZAMUDDIN, S. 1970. Masur *Lens culinaris* Medic. *Pulse crops of India.* (Kachroo, P. and Arif, M. eds.). pp. 306–313. New Delhi: Indian Counc. Agric. Res., 334 pp.

Lentil

PURSEGLOVE, J. W. 1968. *Lens esculenta* Moench. *Tropical crops: Dicotyledons.* Vol. 1. pp. 279–281. London: Longmans, Green and Co Ltd, 332 pp.

RAHMAN, Q. N., AKHTAR, N. AND CHOWDHURY, A. M. 1974. Proximate composition of foodstuffs in Bangladesh: (i) Cereals and pulses. *J. Sci. Ind. Res., (Bangladesh), 9,* 129–133.

ROQUIB, M. A. AND SEN, S. 1972. Study of improved methods of cultivation of pulses. Effect of sowing-date on grain yield of lentil (*Lens esculenta Moench.*). *Farm J.* (*India*), *14,* 17–19.

SAINT-CLAIR, P. M. 1972. Response of *Lens esculenta* Moench. to controlled environmental factors. *Wageningen, Meded. Landbouwhogesch.* 72–12, 84 pp.

SASTRI, B. N. 1962. *Lens* Mill. (Leguminosae). *The wealth of India: Raw materials.* Vol. 6 (L–M). pp. 60–66. New Delhi: Indian Counc. Sci. Ind. Res., 483 pp.

SEN GUPTA, K. 1962. Effect of spacing on the yield of grain in lentil (*Lens esculenta* Moench.). *Sci. Cult., 28,* 285–286.

SEN GUPTA, P. K. 1974. Diseases of major pulse crops in India. *PANS, 20,* 409–415.

SENEWIRATNE, S. T. AND APPADURAI, R. R. 1966. Lentils. *Field crops of Ceylon.* pp. 176–177. Colombo: Lake House Investments Ltd, 376 pp.

SHARAR, M. S., GILL, M. A. AND SHAFQAT, A. A. 1976. Lentil yield and quality as influenced by irrigation and fertilizer levels. *Pak. J. Agric. Sci., 13,* 231–234.

SHARMA, B. M. 1970. Effect of application of N, P and K on the grain yield of lentil *Lens culinaris* Medic. *Indian J. Agric. Sci., 40,* 512–515.

SHAW, F. J. F. AND BOSE, R. D. 1928. Studies in Indian pulses: Lentil (*Ervum lens* Linn.). *Mem. Dep. Agric. India,* (*Bot. Ser.*), *16,* 159–190.

SHUKLA, T. C. 1953. Root development in *Lens esculenta* Moench. *Curr. Sci., 22,* 17.

SINGH, D. 1957. Cultivation of lentil in Uttar Pradesh. *Gov. U.P. Kanpur, Tech. Bull.* 1 (NS), 12 pp.

SINGH, G. AND JOLLY, R. S. 1966. Lentil 9–12: A new masur for Punjab. *Indian Farming, 16* (2), 21.

SINGH, K. B. AND SINGH, S. 1969. Genetic variability and inter-relationship studies on yield and other quantitative characters in lentil *Lens culinaris* Medic. *Indian J. Agric. Sci.*, *39*, 737–741.

SMARTT, J. 1976. *Lens esculenta* Moench., *Lens culinaris* Medik., *Ervum lens* L., Lentil. *Tropical pulses*. p. 47. London: Longman Group Ltd, 348 pp.

SULSER, H. AND SAGER, F. 1976. Identification of uncommon amino acids in the lentil seed (*Lens culinaris* Med.). *Experientia*, *32*, 422–423.

SWAMINATHAN, M. S. AND JAIN, H. K. 1975. Food legumes in Indian agriculture. *Nutritional improvement of food legumes by breeding*. (Milner, M. ed.). pp. 69–82. New York/London: John Wiley and Sons, 399 pp.

TICHÁ, M., ENTLICHER, G., KOSTIR, J. V. AND KOCOUREK, J. 1970. Studies of phytohemagglutinins: (iv) Isolation and characterization of a hemagglutinin from the lentil, *Lens esculenta* Moench. *Biochem. Biophys. Acta*, *221*, 282–289.

TRIPATHI, B. D. AND DATE, W. B. 1975. Partial substitution of wheat flour by other flours for bread preparation: (ii) Use of pulse flour. *Indian Food Pack.*, *29* (3), 66–69.

VASENINA, G. G. 1972. [The effect of agrometeorological conditions on lentil plant development]. *Byulleten' Vsesoyuznogo Ordena Lenina Instituta Rastenievodstvaimeni NI Vavilova*, 2841–45. (*Field Crop Abstr.*, *26*, 5687).

VERTIGAN, W. A. 1966. Lentils in Tasmania. *Tasman. J. Agric.*, *37*, 56–58.

WESTPHAL, E. 1974. *Lens culinaris* Med. Pulses in Ethiopia, their taxonomy and agricultural significance. pp. 109–114. *Wageningen, PUDOC, Cent. Agric. Publ. Doc., Agric. Res. Rep.* 815, 278 pp.

WILSON, V. E. AND BRANDSBERG, J. 1965. Fungi isolated from diseased lentil seedlings in 1963–64. *Plant Dis. Rep.*, *49*, 660–662.

YEGNA NARAYAN AIYER, A. K. 1966. Lentils. *Field crops of India*. 6th ed. pp. 125–127. Bangalore City: Bangalore Printing & Publishing Co Ltd, 564 pp.

YOUNGMAN, V. E. 1968. Lentils a pulse crop of the Palouse. *Econ. Bot.*, *22*, 135–139.

ZOHARY, D. 1972. The wild progenitor and the place of origin of the cultivated lentil: *Lens culinaris*. *Econ. Bot.*, *26*, 326–332.

ZOHARY, D. 1976. Lentil. *Evolution of crop plants*. (Simmonds, N. W. ed.). pp. 163–164. London: Longman Group Ltd, 339 pp.

Common names — **LIMA BEAN, Burma bean, Butter bean[1], Madagascar (butter) bean[2], Rangoon bean, Sieva bean[3].**

Botanical name — *Phaseolus lunatus* L., syn. P. *limensis* Macf., *P. inamoenus* L.

Family — Leguminosae.

Other names — Abangbang (Ug.); Akpaka (pakera) (W. Afr.); Alotoko (Ug.); Apatram (Gh.); Avitas poroto (Peru); Awuje (Nig.); Caraota (Col.); Carolina bean, Civet bean (USA); Ckuku (Ug.); Curry bean (Afr.); Dafal (Ind.); Dau bien, D. ngu (Viet.); Double bean[4] (Ind.); Duffin bean; Fagiolo di Lima (It.); Feijão espadinho (Ang.); Fève creole (Fr.); Frijol comba (Mex.); Frijolito de Cuba (Sp.); Guaracaro (Venez.); Haba lima (Sp.); Habichuela[5] (Philipp.); Haricot de lima, H. du cap, H. du kissi (Fr.); Heerboontjie (S. Afr.); Java bohne (Ger.); Judia de lima, Judion (Sp.); Kabaro (Malag.); Kachang China, K. jawa, K. serendeng, K. serinding (Malays.); Kajang kaokara (Indon.); Kekara kratok (Malays.); Khasi kollu (Tam.); Kokondo (W. Afr.); Lobia[4] (Egy.); Lobiya (Ind.); Maharage[5] (E. Afr.); Mfini (Zan.); Mond-bohne (Ger.); Panguita (Col.); Patani[6] (Philipp.); Pithanga (Tam.); Pois amer (Fr.); P. chouche, P. doux (Ant.); P. du cap (Fr.); Pole bean[7] (USA); Poroto de lima, P. de manteca (Arg.); Pothudhambala (Sri La.); Raima-mama, Raimame (Jpn.);

---

[1] Used for large-seeded cultivars and also for the hyacinth bean, *Lablab purpureus*, and in Kenya for a white-seeded runner bean, *Phaseolus coccineus*.
[2] Used for large-seeded white cultivars.
[3] Used for small-seeded cultivars.
[4] Also sometimes used for the broad bean, *Vicia faba*.
[5] Also used for the haricot bean, *Phaseolus vulgaris*.
[6] Also used for the pea, *Pisum sativum*.
[7] More commonly used for climbing cultivars of the haricot bean, ***Phaseolus vulgaris***.

Roaj(galing) (Sud.); Sem[8] (Ind.); Sewee bean (USA); Sibatse simaron (Philipp.); Tagalo patani (Philipp.); Tua rachamat (Thai.); Yeguas (Mex.); Zabache (Philipp.).

Botany

A perennial, or annual herb, which shows considerable variation in the form of the vines, pods and seeds, as a result of field hybridization, or mutations, common to the species. The *pole* types are generally perennial, twining, usually 6–13 ft (1.8–4 m) tall, with an enlarged tap-root for starch storage. The *bush* types are normally annual, with a bushy dwarf habit, 1–3 ft (30–90 cm). The leaves are trifoliolate, the leaflets ovate 2–4.7–1.2–3.5 in. (5–12 x 3–9 cm), often hairy on the lower surface. The inflorescence is an axillary raceme, approximately 6 in. (15 cm) long, carrying numerous (up to 4 per node), white, or yellowish-white, flowers. The pods are oblong, generally curved, with a sharp beak, 2–4.7 in. (5–12 cm) long and 0.6–1 in. (1.5–2.5 cm) broad, somewhat pubescent, and containing 2–6 seeds. These are very variable in size, shape and colour, but are usually sub-divided into the small, *microspermus*—sieva, or baby limas, approximately 0.4 in. (1 cm) long, and the larger limas, *macrospermus*—lima, approximately 1 in. (2.5 cm) long, series of cultivars. They range from flat, to round potato types, and are often white or cream in colour, but red, purple, brown, black and mottled forms also exist.

Some authorities have divided the lima bean into two separate species, the large-seeded types, *Phaseolus limensis*, and the smaller, baby limas, or sieva beans, *Phaseolus lunatus*. There is, however, considerable evidence in favour of placing both types in one single species, *Phaseolus lunatus*. Van Eseltine (1931) suggested sub-dividing the species into the following five forms: (i) forma *macrocarpus*, the flat lima bean; (ii) forma *salcis*, the willow-leaf sieva lima bean; (iii) forma *lunonanus*, the bush type, sieve lima bean; (iv) forma *limenanus*, the bush type, large-seeded lima bean; (v) forma *solanoides*, the potato lima bean.

With the widespread cultivation of the lima bean many cultivars and selections have been developed. These include: Fordhook 242, a bush potato type, with

[8] Also used in parts of India for the hyacinth bean, *Lablab purpureus* and the runner bean, *Phaseolus coccineus*.

large, thick seeds; Henderson, a bush baby lima bean, with small, thin seeds, suitable for cultivation under more arid and hotter conditions, than most other types; the climbing or pole types, King of the Garden and Florida Speckled Butter, suitable for the production of green pods in the tropics; the climbing type Ventura with large, flat seeds, suitable for the production of dry beans; the semi-climbers Wilbar and Western, with small, thin seeds, and also suitable for dry bean production in the tropics.

Origin and distribution

According to Mackie (1943), the lima bean originated in Guatemala, but more recent evidence suggests that the small-seeded types originated in the Pacific foothills of Mexico and the larger, white-seeded types in Peru. Both types spread from these areas throughout the tropics and are now found throughout S. and C. America, the USA and southern Canada. In addition, they are one of the major food legume crops of the humid rain forests of Africa, and are of considerable importance in many parts of Asia, particularly Burma.

Cultivation conditions

Temperature—the lima bean is more exacting in its environmental requirements than most other cultivated species of *Phaseolus*. It requires a warm, equable climate, with monthly average temperatures preferably between 60° and 80°F (16–27°C). Temperatures below 55°F (13°C) retard growth, and high night temperatures, above approximately 70°F (21°C), result in quicker maturing plants, producing fewer, smaller seeds. At temperatures above 90° to 95°F (32–35°C) serious blossom-shedding and pod-drop can occur, particularly if the humidity is low. The lima bean is suscepible to frost damage and requires a frost-free period ranging from approximately 100 days for the early cultivars grown in the USA, to over 200 days for the large-seeded cultivars grown in Peru and Malagasy. For satisfactory seed germination the soil temperature should be at least 60° to 64°F (16–18°C); in fact with some cultivars germination may be very slow if the soil temperature is below 68°F (20°C), and it has been suggested that for maximum germination of large-seeded types the soil temperature should be between 70° and 80°F (21–27°C). The small-seeded sieva or baby limas are somewhat less critical in their temperature requirements and can be grown under hotter, more arid conditions, than the large-seeded types, which are often restricted to coastal areas, eg California and Peru, where the atmospheric humidity is high during the growing season. For example, for optimum results, the cultivar Fordhook 242 requires an average monthly temperature of between 60° and 70°F(16–21°C) with a maximum of 75°F (24°C).

Rainfall—the lima bean thrives in subhumid or humid areas of the tropics, with an annual rainfall of between 36 and 60 in. (900–1 500 mm), or even above 60 in. (1 500 mm). It is, however, reasonably tolerant of drought and can be grown successfully in areas with an annual rainfall as low as 20 to 25 in. (500–650 mm), provided that supplementary irrigation is available and the atmospheric humidity is reasonably high, preferably between 71 and 73 per cent, at flowering and pod-set. However, heavy rain at flowering can affect the pod-set adversely. When grown under semi-arid conditions the number of irrigations the crop receives varies considerably, according to the local climate, soil type, cultivar and plant stand. In California the crop does not usually receive any irrigation until at least 40 days after planting and then irrigations at two- or three-week intervals are normally sufficient. Furrow irrigation is most common; sprinklers can be used, but may result in a high incidence of bacterial leaf diseases and may affect pod-set adversely. In the production of dry lima beans, irrigation is usually stopped when about one-third of the pods have ripened, in order to induce the rest of the pods to mature.

Soil—the lima bean çan be grown on a wide range of soil types, provided that they are well drained, well aerated, and have a pH of 6.0 to 6.8, but for optimum results loam or clay loams drained to a depth of at least 6 ft (1.8 m) are required. The crop responds to the application of fertilizer and in California normally receives up to 250 lb/ac (280 kg/ha) of nitrogen depending upon the fertility of the soil. In Washington the application of 80–160 lb/ac (90–180 kg/ha) of nitrogen has been recommended on new lands, and on soils of low fertility, or after cropping for several years, the additional application of phosphoric acid 60–80 lb/ac (67–90 kg/ha) has been suggested. In Michigan growers are recommended to maintain a soil pH of 6.0–6.5 by the application of lime and to apply 20 lb/ac (22 kg/ha) of nitrogen, and depending upon the fertility of the soil, 25–200 lb/ac (28–225 kg/ha) of phosphoric acid and 25–250 lb/ac (28–280 kg/ha) of potash. In addition, if the soil pH is above 6.5 the addition of 5–8 lb/ac (5.6–9 kg/ha) of manganese may be required. In Africa, the application of farmyard manure 6 T/ac (15 t/ha), superphosphate 400 lb/ac (450 kg/ha), potassium sulphate 220 lb/ac (250 kg/ha) and calcium nitrate 89 lb/ac (100 kg/ha) has been suggested.

Lima bean

Lima beans are very sensitive to injury from mineral salts and for this reason the fertilizer should be applied in bands approximately 2 to 3 in. (5–7.5 cm) from the seed and at a depth of about 2 in. (5 cm).

Day-length—some cultivars long established in the tropics, and wild plants originating in the Caribbean area are short-day plants, but cultivars originating in more temperate zones, such as the USA, are day-neutral.

Altitude—low-lying areas are not usually suitable for the lima bean, which favours sloping, well-drained areas, between 3 000 and 6 000 ft (900–1 800 m) in the tropics, and between 3 000 and 4 000 ft (900–1 200 m) in more temperate regions (mean annual temperature 64° to 75°F (18–24°C)).

Planting procedure

Material—the use of certified seed meeting recognized standards of genetic purity, germination and cleanliness is recommended. Germination is epigeal and usually occurs in 5 to 6 days. Treatment with a combined insecticide/fungicide dressing such as lindane and Arasan, or dieldrin and Orthocide, is also recommended. In California the minimum germination rate for certified seed is 85 per cent, but in many other countries, particularly when untreated seed is used, the germination rate can fall below 50 per cent. Inoculation of the seed prior to planting does not normally give any significant response.

Method—the seeds can be planted manually or by machine. They are extremely susceptible to mechanical damage, so care must be taken with mechanical planting and in countries such as the USA indent cup-planters are often used. The seeds are planted in a moist, well-prepared seed-bed, at a depth of 1 to 2 in. (2.5–5 cm) in moist, heavy soils, and up 4 in. (10 cm) in lighter, drier soils. Trellis supports are provided for the climbing types. After planting, the crop usually receives at least two cultivations and efficient weed control is essential. Virtually all weed control studies have been carried out in the USA, where the pre-emergence use of trifluralin combined with diphenamid has been found to be very effective. Combinations of alachlor or pronamide combined with trifluralin are also reported to result in good broad spectrum weed control. The pre- or post-emergence use of dinoseb has also been suggested for the control of most annual grasses and broad-leaved weeds, but requires care, otherwise injury to the seedlings can result. Chloramben, chlorthal-dimethyl and nitralin are other herbicides which have been recommended for effective weed control.

Field-spacing—bush lima beans are usually planted in rows 24 to 30 in. (60–75 cm) apart, with 2 to 8 in. (5–20 cm) between the plants. Large-seeded cultivars, such as Fordhook 242, are normally planted 4 to 6 in. (10–15 cm) apart, and small-seeded ones such as Henderson, 3 to 5 in. (7.5–12.5 cm). Climbing types grown for their green immature pods are sometimes planted in hills 36 x 36 or 48 x 48 in. (90 x 90 or 120 x 120 cm) apart, with three or four seeds per hill, or alternatively they are planted in rows approximately 30 in. (75 cm) apart, with 6 to 12 in. (15–30 cm) between the seeds.

Seed-rate—this varies considerably and is very dependent upon the size of the seed, and the average rate of germination. Approximately 120–150 lb/ac (130–170 kg/ha) are required for large-seeded types and 50–70 lb/ac (56–78 kg/ha) for baby limas, although in the USA using high-germinating, treated seed, this rate may be reduced to 80 lb/ac (90 kg/ha) for large-seeded types and 40 lb/ac (45 kg/ha) for baby limas. In India a considerably lower seed-rate of approximately 7–11 lb/ac (7.8–12 kg/ha) is reported to be used for climbing types.

Pests and diseases

The lima bean is subject to many of the pests and diseases which affect the haricot bean, and other *Phaseolus* spp., although under humid, tropical conditions it is usually more resistant to attack than many other legume crops.

Pests—amongst the insect pests the lima bean pod-borer, *Etiella zinckenella*, is pantropic in distribution and is of considerable economic importance, the larvae boring into the pods and destroying the seeds. Effective control can be difficult sometimes. In Ghana regular spraying with Sevin has been found to give satisfactory results, in California spraying with carbaryl is recommended, in other areas DDT is sometimes used. Aphids, particularly the black bean aphid, *Aphis fabae*, can sometimes cause serious crop losses. Thrips, *Heliothrips fasciatus*, and lygus bugs, *Lygus* spp., can also be troublesome. Control of these pests is usually by spraying with DDT or organo-phosphorus compounds. In the USA the corn-seed maggot, *Hylemya platura*, is one of the most serious pests of the lima bean, particularly during the early part of the season when soil temperatures are low and germination liable to be slow. The best control is the use of seed treated with lindane or a thiram/lindane combination. The Mexican bean beetle, *Epilachna varivestis*, can cause crop losses in more arid areas, particularly to bush lima beans. The larvae and adult insects feed on the leaves resulting in skeletonized foliage. Spraying with rotenone is sometimes recommended for control, in addition to the organo-phosphorus insecticides.

The cowpea weevil, *Callosobruchus chinensis*, also attacks the lima bean both in the field and in storage, and can cause considerable crop losses. Like the haricot bean, the lima bean is also very susceptible to infestation from the bean bruchid, *Acanthoscelides obtectus*, particularly during storage.

Wireworms of the genus *Limonius* and several species of *Melanotus* are reported to cause considerable damage by boring into the germinating seeds, or the young shoots and stems. Pre-sowing insecticidal treatment of the seed is the best method of control. In California, springtails, *Onychiurus* spp., are reported to be troublesome at times, feeding on the roots, which results in stunting or delayed growth. Cutworms, such as the variegated cutworm, *Peridroma saucia*, and the grey cutworm, *Agrotis ipsilon*, can also be troublesome, but can be controlled effectively by spraying with DDT.

Root-knot nematodes, *Meloidoygne* spp., are a serious pest of the lima bean causing considerable reduction in yields, particularly when the crop is grown on sandy soils. Crop rotation with cereals can reduce drastically the nematode population in the soil, but often is not popular with growers in the USA. Soil fumigation with ethylene dibromide or pene-dichloropropane is sometimes practised, but is reported to be effective only for about two or three bean crops. The most satistactory control method appears to be to grow nematode-resistant cultivars such as the Western baby lima.

Diseases—the lima bean is subject to several diseases, amongst the most troublesome, particularly if wireworms are present in the soil, are the root rots caused by a complex of fungi, *Fusarium solani* f. *phaseoli*, *Phythium ultimum*, *Rhizoctonia solani*, *Thielaviopsis basicola* and *Macrophomina phaseoli*. Dressing the seed with fungicides such as Arasan, Phygon or Spergon, is reported to reduce the incidence of root rots significantly. Pod blight, *Diaporthe phaseolorum*, and downy mildew, *Phytophthora phaseoli*, are two destructive diseases which are of considerable economic importance. The former is very troublesome in the USA, and as it is seed-borne the use of certified disease-free seed is the most effective method of control. Downy mildew is widespread and can be very troublesome when the lima bean is grown in humid areas.

Bacterial blights can also cause severe crop losses, especially if there are outbreaks of heavy rain during the growing season. The three bacterial blights affecting the lima bean are: (i) the common bacterial blight caused by *Xanthomonas phaseoli*; (ii) halo blight, caused by *Pseudomonas phaseolicola*; (iii) bacterial spot caused by *Pseudomonas syringae*. The use of disease-free seed and crop rotation are the recommended control measures to use in those areas where blight epidemics are likely to occur. Rust caused by *Uromyces phaseoli* var. *typica* is troublesome in some tropical areas, although rare in the USA.

Seed pitting, or yeast spot, caused by *Nematospora phaseoli* can be very troublesome, as the seeds are damaged without any visible signs on the outside of the pods. The yeast spores enter through puncture holes made by insects such as lygus bugs, and require temperatures of between 70° and 80°F (25–27°C) for optimum growth. Anthracnose, *Colletotrichum lindemuthianum*, is another disease of economic importance, particularly in the USA. It is most effectively controlled by the use of disease-free seed and crop rotation. The lima bean is also susceptible to a number of virus diseases. In Puerto Rico bean golden yellow mosaic is of considerable importance. The vector is white fly, *Bemisia tabaci* race *sidae*. Plants affected by the virus fail to produce seed. This virus disease, or a closely related entity, appears to affect crops in certain Caribbean Islands and in S. America. In India the lima bean is reported to be affected by a very similar mosaic disease transmitted by thrips and aphids. In the USA small-seeded cultivars are reported to be susceptible to a strain of the cucumber mosaic virus.

When shelled immature lima beans are marketed, two defects, spotting and stickiness, can occur. Spotting is caused by the fungus *Cladosporium hebarium*, and stickiness by the bacteria *Pseudomonas ovalis*, *Achromobacter coadunatum* and *A. lipolyticum*. In the USA refrigerated storage, 32–40°F (0–4°C) alone, or combined with holding the beans in an atmosphere of 25 per cent, or more, of carbon dioxide, has been found to be a very effective method of controlling these diseases, but control may also be obtained on a commercial scale by washing the pods prior to shelling in a 4 per cent solution of calcium hypochlorite.

Growth period

Early-maturing baby lima beans produce mature seed about 90 to 110 days after sowing; standard lima beans, such as Ventura, produce a crop in 105 to 130 days; while the large-seeded white types of Peru and Malagasy may take from 200 to 270 days and produce a crop over several months.

Harvesting and handling

The method of harvesting depends upon the scale of production and the purpose for which the crop is grown.

Mature lima beans—are usually harvested when about three-quarters of the pods are yellow and dry and the rest have begun to turn green to yellow. In parts of Africa, Asia and S. America, with small-scale production, the pods are frequently picked by hand as they mature over a period of several months. The pods are spread out to dry and afterwards beaten to separate out the dry beans. With large-scale production, such as occurs in the USA, various types of mechanical harvesters are used. These consist essentially of steel knives, set so as to cut off the roots 2–3 in. (5–7.5 cm) below the soil surface, and rakes which place the cut vines on windrows to dry. Cutting and windrowing must be done when the pods are still moist with dew or high humidity, and tough, otherwise the pods shatter and the yield of seed is reduced considerably. The cut vines are left to dry from 7 to 21 days, normally about 10 days, after which they are threshed in specially designed bean threshers. Great care must be taken with the threshing operation as the seeds are very brittle and easily damaged. After threshing the seeds are carefully cleaned and sorted. Various types of equipment can be used for this operation, such as screen-air machines and gravity separators and many producers have now replaced the final hand-sorting operation by electronic eye sorters. In the USA about 5 per cent of the total weight of field-run lima beans is removed during the cleaning and sorting operations. The clean-sorted beans should have a moisture content of less than 15 per cent before they are packed in sacks, polythene bags, or bins, for warehouse storage. Precautions against insect infestation during storage are the same as those for haricot beans. In many tropical countries, where there is small-scale production for local use, the beans are sometimes stored in large earthenware jars or baskets, and covered with a layer of sand or ash to protect them against insect infestation.

Green lima beans for the fresh vegetable market—are often picked by hand at a stage when the seeds have become nearly fully developed, but while they are still sufficiently immature to be soft and tender, and before either the pods or the seeds have begun to lose their green colour. Usually a number of pickings are made over several weeks. The immature green seeds keep better in the pods, which are normally size-graded and packed in well-ventilated baskets or crates. They can be stored satisfactorily for about 7 days in well-ventilated conditions at 32° to 40°F (0–4°C) and a relative humidity of 90 per cent, but must be used promptly after removal from storage as the pods tend to discolour rapidly at ambient temperatures. Considerable quantities of shelled green lima beans are also marketed, particularly in the USA, and small machines have been designed which shell the beans in the field. However, mechanical shelling usually results in considerable bruising and splitting, so that the market value and storage life of the beans are much reduced. Green lima beans once removed from the pod are highly perishable and must be kept under refrigeration in baskets or polythene bags. Sound beans have a storage life of about 10 to 14 days at 32°F (0°C), approximately 8 days at 35°F (2°C), and 4–7 days at 40°F (4°C); the relative humidity should be 90 per cent. Hydrocooling has been reported to extend their storage-life. Green lima beans are very susceptible to the development of spotting and stickiness during storage, the causal organisms of which have been discussed in the section *Pests and diseases.*

Green lima beans for processing—in certain countries, notably the USA, considerable quantities of green lima beans are used by the vegetable processing industry and these are harvested mechanically. The plants are cut with specially designed lima bean knives mounted on the front bar of a tractor, and then hauled to a vining station, or the processing factory, where they pass through specially adapted pea viners. If the beans are shelled in the field they must be transported to the processing unit with the minimum of delay and are frequently packed with ice so as to reduce the rate of deterioration in quality to a minimum. In order to attain optimum quality and yield, a heavy concentration of pod-set is required, and special cultivars have been developed. Time of harvesting is also critical, usually within the limits of 1 or 2 days; after this there may be too high a percentage of over-mature beans, which have to be removed at the factory, often by expensive hand-sorting. In recent years many objective methods have been suggested by US processors as aids for selecting the correct stage for harvesting green lima beans.

Primary product

Seed—which is very variable as regards shape, size and colour. Baby lima beans normally have flat, thin seeds, approximately 0.4 in. (1 cm) long, and usually white in colour. Large-seeded lima beans are plumper, sometimes they are quite rounded as in the 'potato' types. They are approximately 1in. (2.5 cm) in length and can be white, cream, red, purple, brown or black, self-coloured or mottled. One hundred seeds can weigh from 1.6 to 7 oz (47–200 g). Both types of lima bean have a white hilum with translucent lines radiating from it to the edge of the seed-coat.

Yield

Under small-scale cultivation yields of dry seed usually vary from about 360–1 330 lb/ac (400–1 500 kg/ha), although in India lower average yields of 180–270 lb/ac (200–300 kg/ha) have been reported. In western Nigeria, under experimental conditions, yields of over 2 700 lb/ac (3 000 kg/ha) have been reported, indicating the potential of the crop in the humid tropics. In the USA yields of dry seed average between 3 000 and 4 000 lb/ac (3 400–4 500 kg/ha), while yields of green lima beans grown for processing average about 2 300 lb/ac (2 580 kg/ha).

Main use

The lima bean is a nutritious foodstuff. The mature dry beans are a valuable pulse crop, particularly in tropical Africa, where they are usually eaten boiled, fried in oil, or baked. Considerable quantities of lima beans, especially the large, white-types (Madagascar butter beans) are consumed in Europe and the USA. Recently the American foodstuffs industry has been carrying out considerable research and developmental work on the production of quick-cooking lima beans in an effort to increase their usage. Large quantities of green lima beans are consumed in the USA as a vegetable, similar to peas, either fresh, or in a variety of processed forms.

Subsidiary uses

Dry lima beans may be used for the production of a protein-rich bean flour. Recently the possibility of utilizing the product for the enrichment of bread has been investigated. In Japan they are one of the principal types of bean used for the production of bean paste. Lima beans are sometimes used for cattle food, but if they are fed in a raw state their use requires caution, otherwise poisoning may result. The seeds are also reported to be used occasionally in traditional medicine in Asia.

Secondary and waste products

The green immature pods are sometimes cooked and eaten as a vegetable in the tropics. The lima bean is reported to have been grown in Malaysia as a short duration cover or green manure crop. After vining the leaves and stems may be turned into hay or silage for animal feed. Lima bean silage is reported to have the following nutritive value: dry matter 27.3 per cent; protein 3.3 per cent; digestible protein 2.1 per cent; total digestible nutrients 14.2 per cent; nutritive ratio 5:8.

Special features

The fresh, green, immature beans consist of approximately 44 per cent edible portion. The proximate composition of the edible portion of Philippine lima beans has been given as follows: moisture 66.3 per cent; protein 8.3 per cent; fat 0.7 per cent; total carbohydrate 23.1 per cent; fibre 1.0 per cent; ash 1.6 per cent; calcium 28.0 mg/100 g; phosphorus 111.0 mg/100 g; iron 2.6 mg/100 g; sodium 2.0 mg/100 g; potassium 747 mg/100 g; vitamin A 65 iu/100 g; thiamine 0.15 mg/100 g; riboflavin 0.10 mg/100 g; niacin 1.20 mg/100 g; ascorbic acid 27.0 mg/100 g. An approximate analysis of mature lima beans has been given as: moisture 11.0 per cent; protein 19.7 per cent; fat 1.1 per cent; total carbohydrate 64.8 per cent; fibre 4.4 per cent; ash 3.4 per cent; calcium 84 mg/100 g; iron 5.2 mg/100 g; thiamine 0.46 mg/100 g; riboflavin 0.16 mg/100 g; niacin 1.8 mg/100 g. The seed-coat represents about 7 per cent of the whole bean. Total sugars amount to approximately 2.4 per cent of the edible portion and consist of the monosaccharides glucose, fructose and galactose, the disaccharide sucrose, the trisaccharide raffinose, and stachyose. The chief proteins present are α-globulin, β-globulin, plus a small quantity of albumin.

The average amino acid composition (mg/gN) has been reported as: isoleucine 310; leucine 509; lysine 465; methionine 78; cystine 63; phenylalanine 379; tyrosine 202; threonine 261; tryptophan 63; valine 322; arginine 371; histidine 197; alanine 291; aspartic acid 768; glutamic acid 818; glycine 262; proline 293; serine 409. With an increase in maturity the protein content is reported to increase and the sugar content to decrease. The protein content of the raw beans has a low digestibility which can be improved on cooking. Cooking also destroys the trypsin inhibitor and haemagglutinin present in the raw beans. The fatty oil present has the following characteristics: SG $^{25°C}$ 0.921; N_D $^{40°C}$ 1.4772; sap. val. 189.3; iod. val. 99.8; solidifying point 1°C; titre of mixed fatty acids 26°C; unsaponifiable matter approximately 1 per cent.

Lima bean

Lima beans also contain a cyanogenic glycoside, phaseolunatin and the enzyme limarase, which hydrolyses it in the presence of moisture to hydrocyanic acid (HCN), acetone and glucose. The concentration of the glycoside measured by the amount of HCN liberated on hydrolysis varies from 10 to more than 300 mg/100 g of bean. It is widely reported that coloured types of lima beans have a higher glycoside content, but data given by Montgomery (1964) show that there is no reliable correlation between cyanide and seed-coat colour. Many commercially grown lima beans yield about 1–8 ppm of hydrocyanic acid, well within the generally accepted safe limit of 10.20 mg/100 g. West Indian types and some wild forms, however, may yield as much as 300 mg/100 g, which can be fatal. Roasting, thorough boiling, or frying in oil, completely destroys the enzyme and renders the beans safe for human consumption.

In addition lima beans also contain a proteinase of the papain type, carotene oxidase, lecithin (0.62 per cent), cephalin (0.09 per cent), gum and tannin. Recent investigations have shown that the maturing seeds synthesize Se-methylseleno-cysteine, the selenium analogue of S-methylcysteine.

Processing

Fresh lima beans—may be canned, quick-frozen or dehydrated, and processed fresh limas are an important product of the US vegetable processing industry. For all these processed products the beans are first washed carefully and sorted; at this stage all debris, mis-shapen, damaged and over-mature beans are removed. In the USA specially designed separator/washers are used to carry out this operation, which reduces the hand-labour previously involved by over 60 per cent. The beans are next size-graded and then blanched in hot water at 190–200°F (88–93°C), the time depending upon the maturity of the beans and usually varying from 2 to 3 minutes for young, tender beans, to 7 or 8 minutes for more mature, tougher beans. For the canned product the beans are then filled hot into cans, 2 per cent hot brine is added and the cans are processed at 240–250°F (116–121°C). The suggested processing times for A2 (307 x 409) and smaller cans are 35 minutes at 240°F (116°C), if the initial temperature is 140°F (60°C) or more, and 40 minutes, if the fill temperature is between 70° and 140°F (21–60°C). For larger cans the processing time at 240°F (116°C) is 50–55 minutes. If the processing temperature is 250°F (121°C), then A2 and smaller cans are processed for 18 to 20 minutes and larger cans for 30 to 35 minutes. It is not necessary to exhaust the smaller size cans if the brine is added boiling, but many canners pass large cans through an exhaust box with the steam full on to prevent buckling. Immediately after processing the cans are water-cooled to between 95° and 105°F (35–41°C).

Quick-frozen lima beans are processed by packing the blanched beans in cartons and freezing them in plate-freezers. Dehydrated lima beans are often processed by treating the blanched beans with a 1.5 per cent sulphite solution, pH 7.2, to preserve their colour and then dehydrating them in a through-flow atmospheric drier, at a temperature of 120°F (49°C) for approximately 12 hours (until the moisture content is 5 per cent). The dehydrated beans are then packed under vacuum in plain cans.

Dry lima beans—are a popular pack in several countries, notably the USA and the UK. In the latter country butter beans (Madagascar or white Californian lima beans) are frequently processed as an out-of season product. In the USA lima beans are processed in brine, or in tomato sauce, similarly to haricot beans, by the vegetable canners. For a high quality product, which should be of a reasonably uniform size and colour, careful selection of the raw material is required; the dry beans should not be more than one year old, otherwise they do not absorb water and tend to give a hard pack. There is also a tendency for 'old' beans to shed their skins and to develop a pink spot, known as 'pink eye' when blanched. After preliminary sorting the beans are soaked in water until they have absorbed moisture to approximately 100 to 110 per cent of their dry weight. The soaking time varies from 10 to 12 hours in US canneries, to up to 16 hours or more in British canneries. After soaking, the beans are blanched for from 5 to 15 minutes at 190° to 212°F (88–100°C), the exact time varying with the nature of the bean. Over-blanching can result in splitting of the skins, a soft, mushy product, and matting of the beans in the cans. Immediately after blanching the beans are washed in cold water and inspected for visual defects such as skin splitting. The beans are then packed into cans covered with a brine containing 2 per cent salt; some canners add 2 to 3 per cent sugar, and some British canners also add a small quantity of caramel, or a permitted orange dye. Some US canners also add calcium chloride (100–200 ppm) to prevent excessive leaching of starch into the canning liquor. The processing time depends to some extent on the quality of the beans, but is normally between 40 and 60 minutes at 240°F (116°C). Immediately after processing the cans should be water-cooled until the average temperature of the contents is between 95° and 105°F (35–41°C). The use of cans with a sulphur absorbent lacquer throughout is recommended to help retard colour deterioration; it is essential if artificial colouring is used. If plain cans are used the product has a very limited shelf-life due to the greying of the beans and liquor.

Quick-cooking lima bean products—have been developed by the US food industry in recent years in an attempt to increase the usage of lima and other dry beans by housewives. Basically the process consists of: (i) intermittent vacuum treatment (Hydravac process) for 30 to 60 minutes in a solution of inorganic salts (generally 2.5 per cent sodium chloride; 1.0 per cent sodium tripolyphosphate; 0.75 per cent sodium bicarbonate and 0.25 per cent sodium carbonate); (ii) soaking for 6 hours in the same salt solution; (iii) rinsing; (iv) drying or quick-freezing. Beans processed in this way are reported to have an improved flavour and to require a considerably shorter cooking time.

Bean powder—for use in soups, soufflés, etc, is prepared from dry lima beans by first soaking them in water, as in the canning operation, then cooking and reducing them to a puree, before drum-drying to a moisture content of 10 per cent. The dried product is dried further to a moisture content of between 4 and 5 per cent in a vacuum-shelf drier and the resulting powder packed under nitrogen in sealed cans. The addition of 3 ppm of butylated-hydroxy-toluene has been found to help retain the flavour.

Lima bean protein isolate—may be prepared by extracting dry lima bean powder with 0.1 molar phosphate buffer solution at pH 7.2. The extract is centrifuged to remove starch and other insoluble materials, acidified to pH 5.0 with 0.2 molar phosphoric acid, and then heated at 212°F (100°C) for 10 minutes to coagulate the proteins and to inactivate the trypsin inhibitor. The pH of the protein curd is then adjusted to 6.4 with sodium hydroxide solution before it is freeze-dried to about 8.6 per cent moisture content. The resultant product has a protein content of 54.3 per cent.

Production and trade

Production—lima beans are a popular food legume in the USA, S. and C. America, in the humid lowlands of Africa, and in SE. Asia, particularly Burma. Most producing countries, however, do not show lima beans as a separate item in trade statistics, but include them under a general category 'dry beans', so that it is difficult to estimate global production. The major producers are the USA and the Malagasy Republic.

Lima beans: Production

Quantity tonnes

	Annual average 1965–69	Annual average 1970–74	1975
USA	50 439	43 323	37 369
of which:			
Large limas	33 656	23 848	18 503
Baby limas	16 783	19 475	18 866
Malagasy Republic	17 200	22 250	n/a

n/a Not available.

Trade—considerable quantities of lima beans enter international trade, especially the large white limas, often known as butter beans. Statistical data however are very fragmentary, the Malagasy Republic and the USA are the leading exporters, and the UK and Japan are the leading importers. The latter country does not show lima beans as a separate item in its trade returns, but imports in some recent years are estimated to have been of the order of 6 000 tonnes.

Lima beans: Major exporting countries

Quantity tonnes

	Annual average 1965–69	Annual average 1970–74	1975	1976
Malagasy Republic[1]	15 624	19 910	n/a	n/a
USA	5 771	9 832	5 063	8 183

[1] Mainly lima beans, but including a small proportion of other dry beans.
n/a Not available.

Lima beans: Importing countries

Quantity tonnes

	Annual average 1965–69	Annual average 1970–74	1975
UK	4 794	4 576	n/a
Canada	351	328	395

n/a Not available.

Lima bean

Prices—the average fob value for large white Madagascar butter beans exported from the Malagasy Republic was £81.75/t for the period 1970–73, compared with £73.40/t for 1965–69. The average value for large limas exported from the USA was £267.00/t in 1976 and £164.00/t in 1975, compared with the average value of £97.00/t for the period 1970–74 and £80.60/t for the years 1965/69. The value of baby limas was £170.00/t in 1976 and £182.00/t in 1975, compared with an average of £83.00/t for 1970–74 and £80.80/t for 1965–69.

Major influences

The lima bean is a food legume crop of considerable importance to the tropics, particularly to the low subhumid areas of Africa, where it is consumed as a vegetable and as a dry bean, despite the long time required for cooking. In the lowland tropics the lima bean is comparatively free from pests and diseases and can produce relatively high yields of dry seed (in excess of 2 700 lb/ac (3 000 kg/ha) with efficient cultivation. There is, however, need to develop improved high-yielding cultivars, with improved nutritional and culinary qualities which will mature more uniformly before the considerable potential of this bean is likely to be realized fully in the lowland tropics.

Bibliography

ALLARD, R. W. 1953. Production of dry edible lima beans in California. *Univ. Calif., Calif. Agric. Exp. Stn. Extn. Serv., Circ.* 423, 26 pp.

ANON. 1957. Frozen lima beans. *West Cann. Pack.*, *49* (5), 14–16.

ANON. 1958. New nematode-resistant lima bean. *Agric. Res.*, *6* (10), 5.

ANON. 1967. Increasing the production of protein foods: The lima bean (*Phaseolus*). *Ghana Farmer*, *11*, 152–153.

ARNON, I. 1972. Pulses or grain legumes. *Crop production in dry regions.* Vol. 2. pp. 217–260. London: Leonard Hill Books, 683 pp.

ARRAUDEAU, M. 1958. Note concernant le *Phaseolus lunatus.* [Note concerning *Phaseolus lunatus*]. *Bambey, Sénégal, Inst. Rech. Agron. Trop. Cult. Vivr., Ann. Cent. Rech. Agron., Bull. Agron.* 18, pp. 105–111.

AYKROYD, W. R. AND DOUGHTY, J. 1964. Legumes in human nutrition. *FAO Nutr. Stud.* 19, pp. 109; 116; 118. Rome: FAO, 138 pp.

BAUDET, J. C. 1977. The taxonomic status of the cultivated types of lima bean (*Phaseolus lunatus* L.). *Trop. Grain Legume Bull.*, (7), pp. 29–30.

BIRD, J., SANCHEZ, J., RODRIGUEZ, R. L. AND JULIA, F. J. 1975. Rugaceous (whitefly-transmitted) viruses in Puerto Rico. *Tropical diseases of legumes.* (Bird, J. and Maramosch, K. eds.). pp. 3–25. New York/London: Academic Press, 171 pp.

BROOKS, C. AND MCCOLLOCH, L. 1938. Stickiness and spotting of shelled green lima beans. *US Dep. Agric., Tech. Bull.* 625, 24 pp.

BURR, H. K., BOGGS, M. M., MORRIS, H. J. AND VENSTROM, D. W. 1969. Stability studies with cooked legume powders: (ii) Influence of various factors on flavor of lima bean powder. *Food Technol., 23*, 842–844.

CAPOOR, S. P. AND VARMA, P. M. 1948. Yellow mosaic of *Phaseolus lunatus* L. (Transmitted by whiteflies *Bemisia tabaci*). *Curr. Sci., 17*, 152–153.

CARANDANG, E. C. 1968. Patani. *Culture of vegetables.* pp. 78–80. Manila, Repub. Philipp., Dep. Agric. Nat. Resource, Bur. Plant Ind., 150 pp.

CHILDERS, N. F., WINTERS, H. F., ROBLES, P. S. AND PLANK, H. K. 1950. Vegetable gardening in the tropics. pp. 48-51. *US Dep. Agric., Fed. Exp. Stn. P.R. Circ.* 32, 144 pp.

CLORE, W. J. AND STANBERRY, C. O. 1951. Growing lima beans in irrigated central Washington. *Washington Agric. Exp. Stn., Bull.* 530, 19 pp.

CRUESS, W. V. AND LOW, D. 1956. New canned lima bean products. *Cann. Trade, 78*, (25), 44–48.

CULPEPPER, C. W. AND CALDWELL, J. S. 1945. The development of different parts of the lima bean pod in relation to tests for stage of development and eating quality of its seeds. *Fruit Prod. J., 24*, 331–336; 347; 368–372; 377; 379; 381.

CUNNINGHAM, H. S. 1947. Control of downy mildew of lima bean on Long Island. *New York State Agric. Exp. Stn. Geneva, Bull.* 723, 19 pp.

DEANON, J. B. (JR.) AND SORIANO, J. M. 1967. The legumes. *Vegetable production in Southeast Asia.* (Knott, J. E. and Deanon, J. R. eds.). pp. 66–96. Los Baños, Laguna: Univ. Philipp. Coll. Agric., 366 pp.

DUTCHER, A. W. 1957. Utilization of snap beans and lima beans. *Cann. Trade, 78* (48), 6–7; 15.

Lima bean

Eseltine, G. P. Van. 1931. Variation in the lima bean *Phaseolus lunatus* L., as illustrated by its synonym. *New York State Agric. Exp. Stn. Geneva, Tech. Bull.* 182, 24 pp.

Food and Agricultural Organization of the United Nations. 1970. Amino acid content of foods and biological data on proteins. *FAO Nutr. Stud.* 24, pp. 54–55. Rome: FAO, 286 pp.

Herbert, D. A. and Felmingham, J. D. 1969. Canning and freezing: Beans butter. *Food industries manual.* (Woollen, A. H. ed.). 20th ed. p. 16. London: Leonard Hill Books, 509 pp.

Holland, A. H., Kendrick, J. B. (Jr.), Lange, W. H. (Jr.) and MacGillivray, J. H. 1953. Production of green lima beans for freezing. *Calif. Agric. Exp. Stn., Ext. Serv. Circ.* 430, 22 pp.

Kee, W. E. (Jr.) and Fisher, V. J. 1976. Evaluation of quality measurement techniques for raw baby lima beans. *Hortscience*, *11*, 613–615.

Kramer, A. and Hart, W. J. (Jr.) 1954. Recommendations on procedures for determining grades of raw, canned and frozen lima beans. *Food Technol.*, *8*, 55–62.

Krishnamurthi, A. (ed.) 1969. *Phaseolus lunatus. The wealth of India: Raw materials.* Vol. 8 (Ph-Re). pp. 5–8. New Delhi: Publ. Inf. Dir. Indian Counc. Sci. Ind. Res., 394 pp.

Lange, W. H. (Jr.), Seyman, W. S. and Leach, L. D. 1956. Seed treatment of lima beans. *Calif. Agric.*, *10* (4), 3; 15.

Lopez, A. 1975. (i) Beans lima; (ii) Dried lima beans. *A complete course in canning.* (i) pp. 356–358; (ii) pp. 584–587. Baltimore, Maryland: The Canning Trade Inc, 755 pp.

Luh, B. S. and Maneepun, S. 1973. Utilization of dry lima beans for high protein bread. *Confructa*, *18*, 54–62.

Lutz, J. M. and Hardenburg, R. E. 1968. Lima beans. The commercial storage of fruits, vegetables and florist nursery stocks. p. 40. *US Dep. Agric., Agric. Handb.* 66, 94 pp.

Mackie, W. W. 1943. Origin, dispersal and variability of the lima bean *Phaseolus lunatus. Hilgardia*, *15*, 1–24.

Maneenpun, S., Luh, B. S. and Rucker, R. B. 1974. Amino acid composition and biological quality of lima bean protein. *J. Food Sci.*, *39*, 171–174.

MONTGOMERY, R. D. 1964. Observations on the cyanide content and toxicity of tropical pulses. *West Indian Med. J.*, *13*, 1–11.

MOODY, K. 1973. Weed control in tropical grain legumes. *Proc. 1st Int. Inst. Trop. Agric., Grain legume workshop.* pp. 162–183. Ibadan, Nigeria, Int. Inst. Trop. Agric., 325 pp.

NELSON, A. I., STEINBERG, M. P., NORTON, H. W., CLEVEN, C. C. AND FRITZSCHE, H. W. 1956. Studies on the dehydration of lima beans. *Food Technol.*, *10*, 91–95.

NIGAM, S. N. AND MCCONNELL, W. B. 1973. Biosynthesis of Se-methyl-selenocysteine in lima beans. *Phytochemistry*, *12*, 359–362.

PATEL, G. A. 1950. Double bean. *Poona Agric. Coll. Mag.*, *41*, 211–215.

POLLOCK, B. M. 1969. Imbibition temperature sensitivity of lima bean seeds controlled by initial seed moisture. *Plant Physiol.*, *44*, 907–911.

PORRITT, S. W. 1974. Beans lima. Commercial storage of fruits and vegetables. p. 27. *Can. Dep. Agric., Publ.* 1532, 56 pp.

PURSEGLOVE, J. W. 1968. *Phaseolus lunatus* L. *Tropical crops: Dicotyledons.* pp. 296–301. London: Longmans, Green and Co Ltd, 332 pp.

RACHIE, K. O. 1973. Relative agronomic merits of various food legumes for the lowland tropics. Potentials of field beans and other food legumes in Latin America. (Wall, D. ed.). pp. 123–129. *Cali, Colombia, Cent. Int. Agric., Trop. Ser. Semin.* 2 E, 388 pp.

RACHIE, K. O. AND ROBERTS, L. M. 1974. Grain legumes of the lowland tropics. *Adv. Agron.*, *26*, 1–132.

RACHIE, K. O. AND SILVESTRE, P. 1977. Grain legumes. *Food crops of the lowland tropics.* (Leakey, C. L. A. and Wills, J. B. eds.). pp. 41–74. Oxford: Oxford University Press, 345 pp.

RAPPAPORT, L. AND CAROLUS, R. L. 1956. Effects of night temperature at different stages of development on reproduction in the lima bean. *Proc. Am. Soc. Hortic. Sci.*, *67*, 421–428. (*Hortic. Abstr.*, *27*, 453).

REED, R. H. 1959. Processing limas for freezing: unit costs drop rapidly as grade-out drops and as total volume and effective utilization of plant increase. *Calif. Agric.*, *13* (3); 15.

REID, J. T. 1969. A mechanical harvester for southern peas, lima beans. *Agric. Eng.*, *50*, 412–413.

Lima bean

Rockland, L. B. 1962. Compositional studies on dry lima beans. *Rep. 5th dry bean res. conf.* (1961). p. 39. *US Dep. Agric., Agric. Res. Serv.*, 62 pp.

Rockland, L. B., Heinrich, J. D. and Dornback, K. J. 1971. Recent progress on the development of new and improved quick-cooking products from lima and other dry beans. *Rep. 10th dry bean res. conf.* (1970). pp. 121–131. *US Dep. Agric., Agric. Res. Serv.*, ARS 74–56, 146 pp.

Rockland, L. B. and Metzler, E. A. 1967. Quick-cooking lima and other dry beans. *Food Technol.*, *21*, 344–348.

Salunkhe, D. K. and Pollard, L. H. 1955. A rapid and simple method to determine the maturity and quality of lima beans. *Food Technol.*, *9*, 45–46.

Salunkhe, D. K., Pollard, L. H., Wilcox, E. B. and Burr, H. K. 1959. Evaluation of yield and quality in relation to harvest time of lima beans grown for processing in Utah. *Utah State Univ., Agric. Exp. Stn. Bull.* 407, 30 pp.

Sanchez, R. L. and Woskow, M. H. 1964. Sensory evaluation of canned lima bean varieties. *Rep. 7th dry bean res. conf.*, (1965). pp. 57–60. *US Dep. Agric., Agric. Res. Serv.*, ARS 74–32, 94 pp.

Singh, H. B., Joshi, B. S. and Thomas, T. A. 1970. The *Phaseolus* group: Lima bean (*Phaseolus lunatus* L.). *Pulse crops of India.* (Kachroo, P. and Arif. M. eds.). pp. 160–162. New Delhi: Indian Counc. Agric. Res., 334 pp.

Smartt, J. 1976. *Tropical pulses.* pp. 71; 90–95; 114–116. London: Longman Group Ltd, 348 pp.

Smith, M. A., McColloch, L. P. and Friedman, B. A. 1966. Market diseases of asparagus, onions, beans, peas, carrots, celery and related vegetables. pp. 16–20. *US Dep. Agric., Agric. Handb.* 303, 65 pp.

Sterling, C. 1955. Maturity indices in lima beans. *Food Technol.*, *9*, 395–398.

Sullivan, W., Cory, W. M. and Miller, M. D. 1941. Management practices with large limas in southern California. *Univ. Calif., Agric. Exp. Stn. Berkeley, Bull.* 657, 30 pp.

Thaung, M. M. and Walker, J. C. 1957. Studies on bacterial blight of lima bean. *Phytopathology*, *47*, 413–417.

Thompson, H. C. and Kelly, W. C. 1957. Lima bean. *Vegetable crops.* pp. 452–458. New York: McGraw–Hill Book Co Inc, 611 pp.

WESTPHAL, E. 1974. *Phaseolus lunatus* L. Pulses in Ethiopia, their taxonomy and agricultural significance. pp. 140–151. *Wageningen, PUDOC, Cent. Agric. Doc., Agric. Res. Rep.* 815, 278 pp.

WILEY, R. C. 1959. Slurry viscosity measurements as methods to determine maturity of lima beans and peas. *Food Technol., 13*, 694–698.

WILLIAMS, K. T. AND BENVENUE, A. 1958. Carbohydrate studies of the lima bean. *J. Ass. Of. Agric. Chem., 41*, 822–827.

YARNELL, S. H. 1965. Cytogenetics of the vegetable crops: (iv) Legumes. *Bot. Rev., 31*, 292–300.

ZAUMEYER, W. J. AND THOMAS, H. R. 1957. A monographic study of bean diseases and methods for their control. *US Dep. Agric., Tech. Bull.* 868, 255 pp.

ZAUMEYER, W. J. AND THOMAS, H. R. 1962. Bean diseases—how to control them. pp. 31–38. *US Dep. Agric., Agric. Res. Serv., Agric. Handb.* 225, 40 pp.

Common names	**LUPIN, Lupine.**
Botanical name	*Lupinus* L.
Family	Leguminosae.
Other names	Aci bakla (Turk.); Altramuz (Sp.); Chocho (S. Am.); Lupina (Czech.); Lupino (It., Sp.); Tremoçeiro, Tremoço (Port.); Turmas (Ind.); Turmus (Ar.); Vlči (Czech.); Wolfsbohne (Ger.); Wolfsboon (Dut.).

Botany

A large genus, with more than three hundred species of annual and perennial herbs, sometimes subshrubs, most numerous in the western parts of N. and S. America, with a second centre around the Mediterranean, thinly extending into the highlands of eastern Africa. Two species, *Lupinus albus L.* and *L. mutabilis* Sweet, have been cultivated as grain legumes for over 3 000 years, in the Mediterranean basin, and the highlands of S. America, respectively. Several other large-seeded species have come into cultivation more recently and are still in process of domestication.

The cultivated grain lupins are annual, erect, herbaceous plants, with a strong tap-root that penetrates deep into the soil and a well-developed root-system. The stem is coarse and thick and does not usually branch until the commencement of flowering. The leaves are palmate, normally with 6 to 8 leaflets. The papilionaceous flowers are borne in large, showy terminal racemes, or sometimes whorled, and can be white, pinkish, blue or purplish in colour. They are normally self-fertilized, although occasionally cross-fertilization occurs. The pods are flattened, mostly constricted or grooved crosswise between the seeds, often pubescent and usually containing 3 to 6 seeds.

The limits between the species of *Lupinus* are rather poorly marked, identification is sometimes difficult, and some botanical names have become misapplied, but the following are of importance as grain legumes:

(i) *Lupinus albus* L. (syn. *L. termis* Forssk.), the white or Egyptian lupin, is the oldest established cultivated species among the Mediterranean lupins. An erect annual 12–48 in. (30–120 cm) high, branching below the inflorescences at flowering. The flowers are white, tinged with blue or violet, not scented. The pods are 2.7–6.0 x 0.5–0.8 in. (70–150 x 12–20 mm), containing 3 to 6 seeds.

(ii) *Lupinus angustifolius* L., the narrow-leaved or blue lupin, an erect annual 8–60 in. (20–150 cm) high, with profuse lateral branching. The flowers are usually light to dark-blue tinged with purple, very occasionally pink, purple or white, in cultivated forms; not scented. The pods are 1.4–2.0 x 0.3–0.4 in. (3.5 –5.0 x 0.7–1.0 cm), with 4 to 7 seeds in the wild types, and up to 2.3 x 0.6 in. (6 x 1.5 cm), with 3 to 5 seeds in the cultivated forms.

(iii) *Lupinus cosentinii* Guss., the sand plain lupin, sometimes referred to as the W. Australian blue lupin. A robust, erect annual, 8–48 in. (20–120 cm) high with vigorous lateral branching. The stems are covered with very short hairs. The flowers are bright-blue, with a yellowish-white spot on the standard; sometimes very slightly scented (spicy), or unperfumed. The pods are 1.6–2.2 x 0.5–0.6 in. (4.0–5.5 x 1.25–1.5 cm) and may be densely covered with short fine hairs, or softly hirsute. They normally contain 3 to 5 seeds.

(iv) *Lupinus luteus* L., the yellow lupin, a herbaceous annual, 8–32 in. (20–80 cm) high; rosetted at first, but becoming erect with vigorous basal branching. The stems are hairy. The flowers are bright golden-yellow and sweet scented. The pods are 1.6–2.3 x 0.4–0.6 in. (4–6 x 1–1.5 cm), covered with short hairs and contain 4 to 6 seeds.

(v) *Lupinus mutabilis* Sweet, the pearl lupin, a robust showy annual 36–72 in. (90–180 cm) high, with both determinate and indeterminate forms. The flowers are usually blue, with white or sometimes yellow splotches on the standard petals. As the flowers mature, the petals turn first violet and then brown before dying, (the origin of the specific name *mutabilis*). The pods are hairy, normally 2.0–3.5 x 0.8 in. (5–9 x 2 cm) and contain 2 to 6 seeds.

In the past 40 years, or so, a great deal of systematic breeding work has been undertaken on the large-seeded grain lupins to develop alkaloid-free strains, the so-called 'sweet' lupins. A cultivar derived from the yellow lupin, *L. luteus*, called Weiko III, which is alkaloid-free, white-seeded and early-maturing, has become the dominant lupin grown in a number of European countries and a very similar strain Stellenbosch-Elsenburg Geel-1 is grown in S. Africa. A soft-seeded, alkaloid-free cultivar of the narrow-leaved lupin, *L. angustifolius*, originated in Sweden, but has since become cultivated widely in several countries, including the USA. It was introduced into Australia in 1959 but was found to have serious limitations. Since then an intensive breeding programme has been carried out in Australia. Recently developed cultivars of of *L. angustifolius* include the cultivar Uniwhite, released in 1967, which combines low-alkaloid content, soft-seedness and white seeds, with reduced pod shattering, Uniharvest released in 1971 and Unicrop released in 1973. These last cultivars are non-shattering and the latter is also early-maturing. In 1975 two more cultivars, Marri and Ultra, were released. Marri, which closely resembles Uniharvest, is resistant to grey leaf spot disease, *Stemphylium* spp. Ultra is the first commercial release of the Mediterranean white lupin, *L. albus*, in Australia. It is non-shattering and produces soft, white seeds.

Origin and distribution

The large-seeded grain lupins, with one exception, originated in the Mediterranean basin and N. and NE. Africa. Nowadays commercial lupin cultivation spans a wide range of latitudes, but only a restricted range of growing-season temperatures. The crop is produced as a summer annual in the cool temperate climates of N. Europe and New Zealand and as a winter annual in subtropical climates such as the south eastern USA. The distribution of the five principal species is as follows:

(i) *Lupinus albus*, cultivated throughout the Mediterranean region, the Upper Nile, Madeira and the Canary Islands. Occasionally cultivated in C. and SE. Europe, the USSR (Georgia), Ethiopia, S. Africa, Australia, the southeastern USA and S. America.

(ii) *Lupinus angustifolius*, originated in coastal areas bordering the Mediterranean, now found growing wild in France as far north as the Loire and has become naturalized to a small extent in S. Africa and SW. Australia. It is cultivated on a commercial scale in these areas, and in N. Europe, New Zealand, and the southeastern USA.

(iii) *Lupinus cosentinii*—found in Tunisia, Morocco, SW. Spain and S. Portugal, with isolated occurrences in certain Mediterranean islands such as Corsica, Sardinia and Sicily. It has become naturalized in SW. and S. Australia and New South Wales.

(iv) *Lupinus luteus*, probably originated in W. Spain and Portugal and has spread to coastal areas around the Mediterranean. It is also grown to a limited extent on sandy soils in N. Europe, in S. Africa and Australia.

(v) *Lupinus mutabilis* found mainly in the Andean regions of S. America, where it is cultivated on a small-scale.

Cultivation conditions
The climatic requirements of lupins are still only broadly known.

Temperature—for optimum results lupins require at least a five-month growing period during which the mean monthly temperatures are between 59° and 77°F (15–25°C), optimum 64° to 75°F (18–24°C). All Mediterranean species respond to vernalization. *L. angustifolius* is better adapted to relatively cool growing conditions and has a very strong tendency to flower abscission in areas with high average temperatures at flowering. This species is fairly tolerant of frost, surviving temperatures down to about 21°F (–6°C) in its vegetative growth stages. *L. albus* is also fairly tolerant of frost, but *L. luteus* will only tolerate light frost, while *L. cosentinii* is susceptible to frost damage and requires warmer growing conditions.

Rainfall—most lupin species require an evenly distributed rainfall of between 16 to 40 in./a (400–1 000 mm/a). They are usually very sensitive to drought conditions and require at least 5 months free from serious moisture stress. *L. cosentinii*, however, shows a degree of tolerance to moisture stress. Warm winds combined with low humidity at flowering can cause excessive flower abscission and poor seed-set. In Australia the optimum rainfall for the cultivar Uniwhite is generally considered to be between 18 and 24 in./a (450–600 mm/a), while Unicrop and Uniharvest require a minimum of 20 in./a (500 mm/a).

Soil—lupins characteristically grow on coarse-textured, well-drained, acid to neutral soils. However, there is some variation in the requirements of the various species. *L. luteus* is better adapted to soils of low fertility; above pH 6.5–7.0 this species often suffers from 'lime chlorosis', reported to be due to iron deficiency in Europe and manganese in Australia. *L. angustifolius* grows best on moderately acid to neutral soils and is less tolerant of low fertility, being susceptible to potassium and phosphorus deficiencies. In Australia it has also been reported to be highly susceptible to cobalt deficiency, but not to molybdenum, whereas *L. cosentinii* is reported to be more tolerant of soils of low phosphorus content, but to be susceptible to molybdenum deficiency. *L. albus* will tolerate mildly acid to slightly calcareous loamy soils and sandy loams of only moderate fertility. In the Sudan and

Egypt it is sometimes grown on saline soils. It is also susceptible to phosphorus deficiency and is reported to be the most sensitive of the grain lupins to waterlogging. *L. cosentinii* is also fairly tolerant of low fertility and mildly acid and neutral soils and to free lime. Compared with the other species it is generally more tolerant of mineral deficiencies, with the exception of molybdenum.

All lupin species have a relatively high requirement of phosphorus, and also need sulphur, for optimum results. The application of 150–200 lb/ac (170–225 kg/ha) of 22 per cent superphosphate has been recommended to Australian growers, while in S. Africa, on poor sandy soils, an additional dressing of 25–50 lb/ac (28–56 kg/ha) of potash has been found to be very beneficial.

Altitude—in Kenya sweet lupins are reported to grow successfully at an altitude of between 5 000 and 8 000 ft (1 500–2 400 m). *L. angustifolius* is found at altitudes up to about 5 000 ft (1 500 m) in the Iberian peninsula and N. Africa.

Day-length—all the Mediterranean grain lupin species are reported to be long-day plants, but the photoperiodic responses of certain cultivars requires further investigation, particularly the relationship between photoperiod and vernalization.

Planting procedure

Material—seed is used and germination is epigeal. Research workers in Australia recommend that the seed should be inoculated with the 'slow growing' root nodule bacteria *Phytomyxa*. Poor germination can be a problem, due to hardshell (impermeability of the seed-coat). Storage of seed with a moisture content of between 13 and 14 per cent in polythene bags, inside maize sacks, has been recommended as a method for increasing field germination. Recently pre-sowing treatment with sulphuric acid has been suggested for improving the germination rate of *L. cosentinii* seeds.

Method—the seeds are sown either manually, or by means of a seed-drill, at a depth of 1 to 2 in. (2.5–5 cm), in a clean seed-bed. A manual cereal drill or combine and a light covering harrow are frequently used in Australia. Date of sowing is critical and for optimum grain yields in Australia the seeds should be sown between mid-April and mid-May. Later sowing results in severely reduced yields. Weed control is essential, especially during the early stages of growth. The use of pre-emergence herbicides for weed control has been successful in Europe and Australia. In the latter country, simazine is recommended for the control of grassy and broad-leaved weeds. Other herbicides which have proved effective are di-allate (Avadex) and trifluralin (Treflan); 2, 4–D cannot be used, as lupins are highly susceptible to damage from this herbicide.

Field-spacing—a spacing of 6 in. (15 cm) between the plants and 7 in. (17.5 cm) between the rows is often used in Australia, but further study is required on the effect of spacing on grain yields.

Seed-rate—in Australia 60 to 80 lb/ac (67–90 kg/ha) has been recommended, recent research has shown that for higher yields the seed-rate should be at least 80 lb/ac (90 kg/ha). In S. Africa the seed-rate is usually between 60 and 70 lb/ac (67–78 kg/ha) for *L. angustifolius* and *L. luteus*, and 95 lb/ac (106 kg/ha) for *L. albus*. In the USA recommended seed-rates are: *L. albus* 160 lb/ac (179 kg/ha); *L. angustifolius*, 60–80 lb/ac (67–90 kg/ha); *L. luteus*, 40–60 lb/ac (45–67 kg/ha).

Pests and diseases

Pests—aphids, *Aphis* spp., are a widespread pest of lupins, especially of sweet cultivars of *L. luteus*, the bitter types are more or less immune. Infestation is most usual at the bud stage and can cause serious damage, resulting in reduced seed-setting. Aphids are of economic importance as the vector of several virus diseases infecting lupins, including bean yellow mosaic virus. Control by spraying is relatively easy although re-infestation can readily occur.

Budworms or earworms, *Heliothis* spp., are often a serious pest of winter-grown lupins in Portugal, S. Africa, Australia and SE. USA. In Western Australia seed losses of over 90 per cent have been reported in certain seasons. There, the incidence of this pest is very variable from year to year, possibly due to seasonal weather differences, and also due to natural control by a parasitic wasp. The larvae burrow into the green lupin pods and eat the developing seeds; once the seed-coat begins to toughen, attack is no longer possible. Losses can be reduced by early-sowing and the use of early-maturing cultivars, in addition to spraying with Dipterex or carbaryl compounds. There is a variation in the susceptibility of the species to attack by budworms. *L. luteus* is very highly susceptible, *L. albus* and *L. cosentinii* are highly susceptible and *L. angustifolius* moderately susceptible. Thrips, *Frankliniella* spp., are another fairly widespread pest of lupins. They are reported to reduce seed yields significantly in the USA, especially on the cultivar Borre.

In Australia red-legged earth mites, *Halotydeus destructor*, can be troublesome sometimes on sweet cultivars of *L. luteus*, and lucerne fleas, *Sminthurus viridis*, on sweet cultivars of *L. luteus*, *L. angustifolius* and *L. albus*; but spraying is usually only justified for exceptionally heavy infestations.
In the USA, the root weevil, *Sitona explicita*, is reported to be a serious pest of *L. angustifolius* and the lupin maggot, *Hylemya lupini*, is a troublesome pest of the species *L. angustifolius* and *L. albus*. Larvae of the white-fringed beetles, *Graphognathus* spp., attack the roots of lupins in the southeastern USA, where grasshoppers are also reported to cause occasional damage to the crop. In S. Africa snails and slugs often present a problem and unless they are effectively controlled with bait can cause the complete destruction of the crop.

Root-knot nematodes, *Meloidogyne* spp., are reported to affect lupin crops in the USA and parts of Australia, *L. angustifolius* appears to be the species most affected, *L. luteus*, *L. albus* and *L. cosentinii* being fairly resistant. Reports suggest, however, that nematodes are unlikely to become a serious field pest of lupins if proper rotation practices are carried out.

Diseases—lupins are subject to attack from a number of fungal diseases which could become one of the main factors limiting cultivation. Brown leaf spot, *Pleiochaeta setosa* (= *Ceratophorum setosum*), is one of the most widespread and destructive diseases of winter-grown lupins. The optimum temperature for growth is between 68° and 77°F (20–25°C), but it will grow at temperatures as low as 41°F (5°C), and is very prevalent under cool, humid conditions. It is carried over on infected seeds, and especially on crop residues, where it can survive for periods of over one year. Seed treatment and spraying are reported to give only moderate control. No genetic sources of resistance have been reported within susceptible species. *L. cosentinii* is more or less resistant to the disease. *L. albus* is very susceptible and this could be a major factor limiting its commercial cultivation. *L. angustifolius* and *L. mutabilis* are also fairly susceptible, and *L. luteus* moderately so.

Anthracnose, *Glomerella cingulata*, can be a problem on lupin crops grown in warm, humid conditions. It has been reported to be troublesome on *L. angustifolius* in the southern USA. The disease is seed-borne, but so far seed treatment has not been very effective. Mildew, *Erysiphe* spp. also infects lupins grown under warm humid conditions, particularly in the southern USA.

Grey leaf spot, *Stemphylium solani* and *S. botryosum*, have been reported to affect yields of *L. angustifolius* seriously in the USA, Australia and N. Europe. *L. albus* and *L. luteus*, however, appear to be immune. Several species of fungi can attack lupin roots at all stages of growth. Species of *Phythium* and *Rhizoctonia* produce a water-soaked decay of the roots and the stems at, or just below, the soil level. They are most active in warm weather. Between 6 and 20 per cent of the lupin crop in Egypt is reported to be affected by a root disease caused by *Rhizoctonia solani*. Phythium root rots infect mainly *L. albus* and *L. angustifolius*, the last species is also very susceptible to infection from *Rhizoctonia* spp. Another warm weather disease which causes root decay is southern blight, *Sclerotium rolfsii*. Once again *L. angustifolius* is the most susceptible species. In the USA ascochyta stem canker, *Ascochyta gossypii*, is sometimes troublesome on the narrow-leaved lupin, *L. angustifolius*, when it follows cotton in rotation. Other fungal diseases of lupins are, *Botrytis cinerea* and *Sclerotinia sclerotiorum*.

The fungal disease *Phomopsis rossiana*, which infects the stems of lupins, causes little economic damage to the crop, but is of considerable importance since it produces a toxin which can cause lupinosis in sheep, cattle and horses, when these animals are grazed on mature lupin stubble, particularly after the lupins have been thoroughly wet with heavy rain. The disease, which can result in the death of the animal, has been known for many years, but recently has been studied intensively in Western Australia, where *P. rossiana* is widespread. The upper and smaller stems of standing plants have been found to be most toxic, although leaf and pod material may develop a degree of toxicity. However, no toxic seeds have been found under natural conditions as yet, although seed can be infected by the fungus. Control of the fungus is difficult, since it can persist for at least two years in infected pieces of stalk and stubble fragments. Research indicates that spraying the green lupins with benomyl just before pod formation, reduces the amount of fungal infection. Another suggested technique is to make the lupin hay into fodder rolls. The recently released cultivar Ultra is reported to be resistant to *P. rossiana*.

Several virus diseases infect lupins, in particular bean yellow mosaic virus (BYMV) is of economic significance causing considerable crop losses. It is universal and in addition to lupins, affects a wide range of legume crops especially beans, *Phaseolus* spp., and peas *Pisum sativum*. The main symptom of BYMV in *L. luteus* and *L. cosentinii* is a mosaic mottling, with distortion of the leaflets and dwarfed bunchy growth. Infection before flowering seriously reduces seed-set and the plants often fail to mature properly. Infection at, or after flowering, does not affect seed-set, but small mis-shapen seeds are produced. The symptoms in *L. albus* are usually similar, but milder, and this species is less susceptible to infection. *L. angustifolius* is also less susceptible to infection, the symptoms in this species are the growing tip turns over to one side and blackens; if infection occurs before flowering the plant usually dies. With infection after flowering the pods blacken and fail to fill properly. Infection of *L. mutabilis* with BYMV is reported to be rare in be Australia. The disease is spread by aphids, and early sowing, combined with effective aphid control, is recommended as a means of reducing its incidence. The disease is also seed transmitted in *L. luteus*, so that when this species is grown it is essential to use only seed from disease-free crops. In Australia another virus disease, vascular wilt, has also been reported to be capable of causing severe reduction of yields in infected lupin plants.

Growth period

Lupins mature in approximately 105 to 180 days, depending upon the species, cultivar and local climatic conditions.

Harvesting and handling

Lupins are normally harvested when the pods have turned brown, preferably when the weather is moderately cool. A grain harvester with the drum speed suitably adjusted to minimize seed injury can be used satisfactorily. In the case of non-shattering types the drum speed can be set as that for wheat, ie approximately 900 r/m, provided that the concave is set slightly wider. For shattering types, such as Uniwhite, the drum speed should be reduced to 350–400 r/m and a wider adjustment of the drum is required.

After harvesting the seeds are cleaned to remove all plant trash, green seeds, etc, and are then dried either naturally, or in grain driers, to a moisture content of about 10 per cent to minimize the possibility of fungal spoilage. The dried seed may be stored in bags, or in bulk, in well-ventilated warehouses; its thick seed-coat makes it relatively resistant to attack from storage insects.

Primary product

Seed—which can vary considerably in size, shape and colour according to the species and cultivar. Bitter and sweet forms exist, and hard and soft-shelled seeds.

Seeds of the narrow-leaved lupin, *L. angustifolius*, are usually smooth, more or less spherical, approximately 0.24–0.3 in. (6–8 mm) long and 0.2–0.24 in. (5–6 mm) wide. The colour may be white, black, brown, or intermediate colours, sometimes grey with brown marbling and white spots. The cultivars Borre, Uniwhite, Unicrop and Uniharvest all have soft seeds and a low alkaloid content. Borre is dark-grey in colour, the other cultivars are white, although Uniwhite has very faint brown markings on the seed-coat. Seeds of the white lupin, *L. albus*, are usually soft, rectangular or square, with rounded corners, approximately 0.56–0.64 in. (14–16 mm) in length. Wrinkled and smooth forms exist and the colour is usually white or pinkish-white. The seeds are normally bitter, although sweet, low-alkaloid strains have been selected, but so far have achieved only limited cultivation. Seeds of the sand plain lupin, *L. cosentinii*, are hard and bitter, approximately 0.24–0.35 x 0.15–0.28 in. (6–9 x 4–7 mm), usually globular compressed, lightish-grey, or more commonly, brown with blackish markings, including a narrow arc around the hilum.

The seeds of the yellow lupin, *L. luteus*, are normally smooth, slightly flattened, approximately 0.24–0.3 in. (6–8 mm) long and 0.2–0.28 in. (5–7 mm) wide, mottled brown to black on a whitish background, often with a light-coloured arc around the hilum, or pure-white in some cultivars. The commercial cultivar Weiko III is soft, has a low alkaloid content, and is pure white in colour. The seeds of the pearl lupin, *L. mutabilis*, are bitter, smooth and shiny, ovoid to slightly compressed, usually 0.28–0.4 x 0.24–0.3 in. (7–10 x 6–8 mm), white or brownish-black in colour.

Yield

Yields of lupin seed can vary very considerably due to local climatic conditions, cultural practices and especially time of planting. In the USA yields normally fluctuate between 890 and 1 960 lb/ac (1 000–2 200 kg/ha) and in the Middle East between 1 070 and 1 335 lb/ac (1 200–1 500 kg/ha). In S. Africa yields ranging from 260 to 1 500 lb/ac (290–1 700 kg/ha) have been reported, depending upon the time of planting. In Australia, the cultivars Uniwhite, Uniharvest and Unicrop are considered to be capable of yielding at least 2 000 lb/ac (2 240 kg/ha) and with early planting, a high seeding-rate, (80 lb/ac (90 kg/ha)), and heavy application of superphosphate, are considered to have a potential of around 4 000 lb/ac (4 480 kg/ha). However, experience during the early 1970s showed that growers could sometimes have a complete, or partial crop failure, mainly due to excessive flower-drop, and average yields as low as 480 lb/ac (540 kg/ha) were reported in certain areas.

Main use

Ground lupin seed is utilized as a source of protein in concentrate mixtures for cattle, sheep, pigs and poultry. In Australia, sweet lupin seed is being used in increasing quantities as a substitute for groundnut cake, fish meal or soyabean meal in the production of high quality livestock rations. When bitter lupin seed is used it should not exceed 10 to 15 per cent of the concentrate mixture.

Subsidiary uses

Lupin seeds, particularly those of the white lupin, *L. albus,* are sometimes used in parts of southern Europe, the Middle East, Ethiopia and S. America as human food, after being soaked in water and boiled. In S. America the seeds of the pearl lupin, *L. mutabilis,* have been used traditionally as a source of protein by the rural population of the Andes and attention has been given to the possibility of utilizing this species as an oilseed. Lupin seed flour is sometimes used to fortify bread and recently research has been carried out into the possibilities of utilizing sweet lupin seed flour as a substitute for soyabean flour, as a binder or extender in various meat products, as an ingredient in noodles, pasta and various bakery products and for the preparation of a protein isolate. Some interest has also been shown in the use of the unmilled seed kernels as a replacement for split peas, lentils and other pulses. In Chile a bland, yellow flour, with a protein content ranging from 52 to 60 per cent has been produced experimentally, from *L. albus* and *L. luteus* seeds, with a view to using it as a protein supplement in biscuits, or 'ulpo', a gruel-like local foodstuff. In Australia lupin seed meal and flour are being used in the manufacture of dry dog biscuits and research has indicated that it could also be used as a protein supplement in some other types of pet foods.

Secondary and waste products

From time to time lupin seed has been used as a substitute for coffee. Lupin seed flour is reported to have been used as an egg substitute in Hungary. The seed of *L. angustifolius* and *L. albus* can be used as an excellent source of aspargine, used for the commercial production of tuberculin. The pearl lupin is sometimes grown as an ornamental plant in the USA and Australia.

Lupins are sometimes grown as a green manure crop, or for fodder or silage. The stubble left after the seed has been harvested is also used for grazing. The yield of plant material from a seed lupin crop is usually between 1.5 and 3 T/ac (3.75–7.5 t/ha). The approximate composition of the various parts of the plant has been given as: (i) *Stems:* crude protein 3.4–11.3 per cent; ether extract 0.2–1.5 per cent; crude fibre 26.1–54.6 per cent; N-free extract 35.2–47.8 per cent; ash 2.2–8.5 per cent; (ii) *Leaves:* crude protein 9.7–21.5 per cent; ether extract 1.1–2.3 per cent; crude fibre 21.3–29.9 per cent; N-free extract 45.0–52.2 per cent; ash 10.2–19.7 per cent; (iii) *Pods:* crude protein 3.5–17.3 per cent; ether extract 0.1–0.9 per cent; crude fibre 33.6–42.2 per cent; N-free extract 50.2–58.4 per cent; ash 2.1–4.0 per cent. The feeding of lupin hay or stubble to livestock, particularly sheep, requires caution, because of the possibility of the animals developing lupinosis, a degenerative liver condition of mycotoxal origin caused by the fungus *Phomopsis*, discussed in the previous section, *Pests and diseases*.

Special features

The seeds of all species of grain lupins have relatively high protein contents, usually between 34 and 42 per cent, and oil contents of 4 to 25 per cent. Bitter lupin seed normally has an alkaloid content of between 0.30 and 3.0 per cent, while the seed of the so-called 'sweet' lupins has an alkaloid content of 0.02 per cent, or less. An interesting feature is the high proportion of fibrous seed-coat which contains about 3 per cent protein, it ranges from about 24 per cent in the yellow lupin, *L. luteus*, to about 15 per cent in the larger-seeded white lupin, *L. albus*. The presence of β-sitosterol has been reported in the seeds of *L. albus* and *L. angustifolius*. The approximate composition of the whole seeds of various species of grain lupins, expressed as a percentage of dry matter, has been reported as follows:

(i) *L. angustifolius:* crude protein 34 per cent; fat 5–6 per cent; crude fibre 15 per cent; N-free extract 42–43 per cent; ash 3 per cent. The seeds contain two major proteins, α and β conglutin and a minor protein γ conglutin. The amino acid content, expressed as a percentage of the crude protein, is: lysine 5 per cent; methionine 0.8 per cent; cysteine 1.2 per cent; tryptophan 1.3 per cent. The seed yields a slow-drying fatty oil, with a bitter taste, which has the following characteristics: SG $^{15°C}$ 0.9463; N_D $^{19°C}$ 1.4790; iod. val. 104; sap. val. 183; acid val. 28; hydroxyl val. 9.9, unsaponifiable matter 8.0 per cent; solidification point −17° C. The component fatty acids are: saturated 10 per cent; oleic 47 per cent; linoleic 34 per cent; linolenic 2 per cent; erucic acid 7 per cent. The seeds of bitter cultivars of *L. angustifolius* contain up to 2.25 per cent of alkaloids, mainly D-Cupanine.

(ii) *L. albus:* crude protein 39–40 per cent; fat 9–11 per cent; crude fibre 9–12 per cent; N-free extract 36–37 per cent; ash 3 per cent. The amino acid content, expressed as a percentage of the crude protein, is: lysine 5.2 per cent; methionine 1.0 per cent; cystine 1.8 per cent; tryptophan 1.2 per cent. The seed yields an edible oil with the following characteristics. SG $^{15°C}$ 0.9229; N_D $^{19°C}$ 1.4758; iod. val. 107.6; sap. val. 182.9; acid val. 5.9; hydroxyl val. 6.3; unsaponifiable matter 3.9 per cent; solidification point −18°C. The component fatty acids are: saturated 10 per cent; oleic 61 per cent, linoleic 20 per cent, linolenic 2 per cent, erucic 7 per cent. The seeds of bitter cultivars are reported to contain the alkaloids lupine and hydroxylupanine. Samples of Egyptian seed have been found to be a good source of vitamins A, B and C.

(iii) *L. cosentinii:* crude protein 34 per cent; fat 4 per cent; crude fibre 18–20 per cent; N-free extract 39–41 per cent; ash 3 per cent.

(iv) *L. luteus:* crude protein 42 per cent; fat 5 per cent; crude fibre 16–17 per cent; N-free extract 32–33 per cent; ash 4 per cent. The amino acid content, expressed as a percentage of the crude protein, is: lysine 4.8 per cent; methionine 0.5 per cent; cystine 3.6 per cent; tryptophan 1.0 per cent. The seeds contain an edible oil with the following characteristics: SG $^{15°C}$ 0.9193; N_D $^{19°C}$ 1.4772; iod. val. 115.6; sap. val. 185.0; acid val. 20.0; hydroxyl val. 7.8; unsaponfiable matter 4.0 per cent; solidification point −16.5°C. The component fatty acids are: saturated 9 per cent; oleic 39 per cent; linoleic 45 per cent; linolenic 1 per cent; erucic 6 per cent.

(v) *L. mutabilis:* crude protein 43 per cent; fat 15–18 per cent; N-free extract 26–31 per cent; fibre 7–9 per cent; ash 4 per cent. Samples of seed from Peru and Chile, which were recently investigated with a view to determining the potential of this S. American lupin as an oilseed, were found to have average oil contents of 21.5 and 11.7 per cent, respectively. The component acids were, respectively: oleic 55.6–59.5 and 37.7 per cent; linoleic 21.5–25.1 and 40.1 per cent; palmitic 9.1–10.1 and 11.6 per cent; stearic 6.2–7.5 and 7.0 per cent; linolenic 1.7–2.2 and 1.9 per cent; arachidic 0.5–0.6 and 1.0 per cent; behenic 0.4–0.6 and 0.7 per cent.

Production and trade

Production—lupin seed constitutes less than 2 per cent of the world's reported production of grain legumes. Production for the period 1970–74 averaged 730 000 t/a, compared with 820 000 t/a for 1965–69, a decrease of approximately 10 per cent. In 1975 production fell to 577 000 t.

Lupins: Major producing countries

Quantity tonnes

	Annual average 1965–69	Annual average 1970–74	1975
USSR	550 000	524 000	350 000
Poland	157 000	105 000	94 000
Australia	400	36 000	87 000
S. Africa	44 000	21 000	13 000
Italy	21 000	12 000	8 000

Trade—lupins are of negligible importance in international trade and there are no statistical data. Shipments from Australia were reported to have exceeded 1 000 t in 1972 and to be increasing. Italy is reported to be the major market for these Australian exports, and Italian imports were estimated to be of the order of 2 000 t in 1973 and 3 500 t in 1974. In addition, in recent years the USA has imported small quantities of lupin seed—1973, 324 t; 1974, 265 t; 1975, 727 t; 1976 421 t.

Prices—in Australia in 1976 the average farm-gate price for lupin seed was £46.98/t and £50.33/t at port.

Major influences

Lupins hold considerable promise as a valuable crop for light-textured soils of the temperate and cooler subtropical regions. They are becoming of increasing importance in the USSR and Western Australia. The latter region is currently the world's leading centre for lupin research, particularly for the breeding of sweet, non-shattering, high-yielding cultivars. In 1974 there were reported to be some 28 000 ac (70 000 ha), with a potential production of 50 000 t, sown to lupin seed in Western Australia, and it has been estimated that if production continues to expand, by the early 1980s some 250 000 t/a of seed could be available for export.

Lupin

Because of their high seed protein content and high yield potential, sweet lupins could make a significant contribution to meet the world's growing protein requirements, particularly as a high grade animal feedingstuff. It has been predicted that the world market for animal feedingstuffs will continue to grow. In addition, the commercial development of inexpensive methionine supplementation has made it possible to utilize vegetable proteins, such as lupin seed, which hitherto were uneconomic because of their inadequate amino acid composition. Moreover, the oil content of lupin seeds considerably enhances their value for animal feed. Consideration has in fact been given recently to the possibility of developing the S. American species *L. mutabilis* as a vegetable oilseed. The expansion of lupin production, however, will depend upon the continued high prices of oilseed meals and the development of the full agricultural potential of the crop. There is need to increase the average seed yields of the cultivars already in commercial cultivation and to widen the range of adaption of the existing cultivated species. In practice seed yield in relation to total dry matter production is still poor compared with most established grain crops and there is also scope for considerable improvement in this area. In Western Australia, for example, the crop is barely profitable in many areas and generally returns are lower than for cereals and oilseeds.

Bibliograpy

ABDEL-FATTAH, A. F., ZAKI, D. A., EDREES, M., AND ABBASSY, M. M. 1974. Investigations on some constituents of *Lupinus termis* seeds. *Qual. Plant., Plant Food Hum. Nutr., 23*, 359–368.

ANON. 1953. Lupin seed flour as a substitute for eggs. *Food Trade Rev., 23* (5), 5.

ARNOLD, G. W., HILL, J. L., MALLER, R. A., WALLACE, S. R., CARBON, B. A., NAIRN, M., WOOD, P. MCR., AND WEELDENBURG, J. 1976. Comparison of lupin varieties for nutritive value as dry standing feed for weaner sheep and for incidence of lupinosis. *Aust. J. Agric. Res., 27*, 423–435.

ARNON, I. 1972. Pulses or grain legumes. *Crop production in dry regions.* Vol. 2. pp. 217–260. London: Leonard Hill Books, 683 pp.

AUSTRALIA, COMMONWEALTH SCIENTIFIC INDUSTRIAL RESEARCH ORGANIZATION. 1972. Lupins: sweet and sour. *Commonw. Sci. Ind. Res. Org., Rural Res.*, 77, 5–9.

AUSTRALIA, COMMONWEALTH SCIENTIFIC INDUSTRIAL RESEARCH ORGANIZATION. 1975/76. *Counc. Sci. Ind. Res., Division of protein, Annu. Rep.* 1975/76, p. 25. Melbourne, Australia, 30 pp.

BAILEY, R. W., MILLS, S. E. AND HOVE, E. L. 1974. Composition of sweet and bitter lupin seed hulls with observations on the apparent digestibility of sweet lupin seed hulls by young rats. *J. Sci. Food Agric.*, *25*, 955–961.

BRÜCHER, H. 1975. Improving the world protein supply by breeding new protein-rich plants. *Appl. Sci. Dev.*, *6*, 47–58.

CASTILLO, Y. R. 1965. Estudio sobre Lupinus (Chocho) en el Ecuador. [Study on lupins (chocho) in Ecuador]. *Arch. Venez. Nutr.*, *15*, 87–93. (*Nutr. Abstr. Rev.*, *37*, 299).

COMMONWEALTH BUREAU OF PASTURES AND FIELD CROPS. nd. Annotated bibliography 1358: (i) *Lupinus* spp. (general), 1948–72, 21 pp; (ii) *Lupinus albus*, 1948–72, 9 pp; (iii) *Lupinus angustifolius*, 1948–72, 7 pp; (iv) *Lupinus luteus*, 1948–72, 15 pp. *Commonw. Agric. Bur.*, *Bur. Pastures & Field Crops, Hurley, Maidenhead, Berks, England.*

DECKER, P., BOND, R. C. AND RITCHEY, G. E. 1948. Curing and storing lupine seed. *Univ. Florida, Agric. Exp. Stn. Press Bull.* 650, 4 pp.

ECKARDT, W. R. AND FELDHEIM, W. 1974. Lupinen, eine neue 'Olfrucht' fur Sudamerika? [Lupins, a new 'oilseed' for South America?]. *Z. Lebensmittel-unters. Forsch.*, *155*, 92–93. (*Nutr. Abstr. Rev.*, *45*, 4014).

EDRESS, M., ABBASSY, M. M. AND ABDEL-FATTAH, A. F. 1974. A study of sterols and vitamins of *Lupinus termis* seeds. *Qual. Plant.*, *Plant Food Hum. Nutr.*, *24*, 55–60.

EL-RAFEI, M. E. AND BADR, M. H. 1970. Studies on root rot disease of lupine. *Agric. Res. Rev.* (*U.A.R.*), *48*, (3), 100–105. (*Trop. Abstr.* *27* (4), v. 887e).

FAGAN, V. J. 1977. Sweet lupin seed meal in pig rations. *Tasman. J. Agric.*, *48*, 3–6.

FRANCIS, C. M., POOLE, M. L. AND CHOPPING, M. H. 1971. A new look at sweet lupins in Western Australia. *J. Dep. Agric. West. Aust.*, *12*, 85–90.

GARDINER, M. R. 1975. Lupinosis. *J. Dep. Agric. West. Aust.*, *16*, 26–30.

GARSIDE, A. L. 1975. Sweet lupins. *Tasman. J. Agric.*, *46*, 1–5.

Lupin

GARSIDE, A. L. 1977. Growing lupins for grain. *Tasman. J. Agric.*, *48*, 7–11.

GLADSTONES, J. S. 1958. The naturalised and cultivated species of *Lupinus* (Leguminosae) recorded for Western Australia. *J. Proc. R. Soc. West. Aust.*, *41*, 29–33.

GLADSTONES, J. S. 1960. Lupin cultivation and breeding. *J. Aust. Inst. Agric. Sci.*, *26*, 19–25.

GLADSTONES, J. S. 1967. 'Uniwhite'—a new lupin variety. *J. Dep. Agric. West. Aust.*, *8*, 190–196.

GLADSTONES, J. S. 1970. Lupins as crop plants. *Field Crop Abstr.*, *23*, 123–148.

GLADSTONES, J. S. 1972. Lupins in Western Australia. *West. Aust., Dep. Agric., Bull.* 3834, 37 pp.

GLADSTONES, J. S. 1974. Lupins of the Mediterranean region and Africa. *West. Aust., Dep. Agric., Tech. Bull. 26*, 48 pp.

GLADSTONES, J. S. 1975. Lupin breeding in Western Australia: The narrow-leaf lupin (*Lupinus angustifolius*). *J. Agric. West. Aust.*, *16*, 44–49.

GLADSTONES, J. S. 1976. The Mediterranean white lupin. *J. Agric. West. Aust.*, *17*, 70–74.

GORDON, W. C. AND HENDERSON, J. H. M. 1951. The alkaloid content of blue lupine (*Lupinus angustifolius* L.) and its toxicity on small laboratory animals. *J. Agric. Sci.*, *41*, 141–145.

GREENWOOD, E. A. N., FARRINGTON, P. AND BERESFORD, J. D. 1975. Characteristics of the canopy, root system and grain yield of a crop of *Lupinus angustifolius* cv. Unicrop. *Aust. J. Agric. Res.*, *26*, 497–510.

GUILLAUME, A. AND PROESCHEL, A. 1939. La graine de lupin (poudre entière et tourteau) et son emploi alimentaire. [Lupin seed (powder and press cake) and its use as a feed]. *Rev. Bot. Appl. Agric. Trop.*, *19*, 161–172.

HANSEN, R. P. 1976. Fatty acid composition of the total lipids from seeds of three cultivars of sweet lupin: *Lupinus albus* cv. '*Newland*', *L. albus* cv. WB2, and *L. luteus* cv. Weiko III. *N.Z. J. Agric. Res.*, *19*, 343–345.

HANSEN, R. P. AND CZOCHANSKA, Z. 1974. Composition of the lipids of lupin seed (*Lupinus angustifolius* L. var. 'Uniwhite'). *J. Sci. Food Agric.*, *25*, 409–415.

HENNING, P. D. 1949. Lupines in the winter-rainfall area. *Farming South Afr.*, *24*, 227–230.

HENSON, P. R. AND STEPHENS, J. L. 1958. Lupines: culture and use. *US Dep. Agric., Farmers' Bull.* 2114, 12 pp.

HILL, J. L. AND ARNOLD, G. W. 1975. The effect of lupinosis on the nutritional value of lupins to sheep. *Aust. J. Agric. Res., 26*, 923–935.

HORN, P. E. AND HILL, G. D. 1974. Chemical scarification of seeds of *Lupinus cosentinii* Guss. *J. Aust. Inst. Agric. Sci., 40*, 85–87.

HOVE, E. L. 1974. Composition and protein quality of sweet lupin seed. *J. Sci. Food Agric., 25*, 851–859.

HUDSON, B. J. F., FLEETWOOD, J. G. AND ZAND-MOGHADDAM, A. 1976. Lupin: An arable food crop for temperate climates. *Plant Food Man, 2*, 81–90.

HULME, S. A. 1957. Lupins. *Handbook for farmers in South Africa*. Vol. 2. *Agronomy and horticulture*. pp. 205–210. Pretoria, South Afr., Dep. Agric., 880 pp.

JUNGE, I., CERDA, P. AND SCHNEEBERGER, R. 1972. Sweet lupines: a new high protein crop for Chile. *League Int. Food Educ. Newsl.*, June, pp. 1-3.

JUNGE, I., SCHNEEBERGER, R., SANDOVAL, C., BAER, E. VON AND CERDA, P. 1973. *Lupine and quinoa: Research and development in Chile*. pp. 1–31. Concepción, Chile, Univ. Concepción Bio. Engin. Lab. Sch. Engin., 67 pp.

KLESSER, P. J. 1960. Virus diseases of lupins. *Bothalia, 7*, 207–231.

MANRIQUE, J. AND THOMAS, M. A. 1976. The effect of lupin protein isolation procedures on the emulsifying and water binding capacity of a meat protein system. *J. Food Technol., 11*, 409–422.

MASEFIELD, G. B. 1975. A preliminary trial of the pearl lupin in England. *Exp. Agric., 11*, 113–118.

MASEFIELD, G. B. 1976. Further trials of pearl lupins in England. *Exp. Agric., 12*, 97–102.

NEHRING, K. AND SCHIEMANN, R. 1951. Beiträge zur Kenntnis der Süsslupine: (ii) Mitteilung: Der Aufbau des Fettes der Süsslupine. [A contribution on the composition of the sweet lupin: (ii) Constituents of the fat of the sweet lupin]. *Fette Seifen, 53*, 10-15.

PALMER, J. 1976. Lupins for the north. *N.Z. J. Agric., 133* (3), 53.

PERRY, M. W. 1975. Field environment studies on lupins: (ii) The effects of time of planting on dry matter partition and yield components of *Lupinus angustifolius* L. *Aust. J. Agric. Res., 26*, 809–818.

Lupin

PERRY, M. W. AND GARTRELL, J. W. 1976. Lupin 'split seed': A disorder of seed production in sweet, narrow-leaved lupins. *J. Agric. West. Aust.*, *17*, 20–25.

PERRY, M. W. AND POOLE, M. L. 1975. Field environment studies on lupins: (i) Developmental patterns in *Lupinus angustifolius* L., the effects of cultivar, site and planting time. *Aust. J. Agric. Res.*, *26*, 81–91.

PLESSIS, S. J. DU AND TRUTER, J. A. 1953. Brown spot disease of lupins caused by *Pleiochaeta setosa* (Kirchn.) Hughes. *South Afr., Dep. Agric., Sci. Bull.* 347, 12 pp.

POMPEI, C. AND LUCISANO, M. 1976. Le lupin (*Lupinus albus* L.) comme source de protéines pour l'alimentation humaine: (i) Étude préliminaire. [The lupin (*Lupinus albus* L.) as a source of proteins for human nutrition: (i) Preliminary study]. *Lebensm.-Wiss u-Technol.*, *9*, 289–295.

POMPEI, C. AND LUCISANO, M. 1976. Le lupin (*Lupinus albus* L.) comme source de protéines pour l'alimentation humaine: (ii) Production d'isolats par coagulation acide et par ultrafiltration–diafiltration. [The lupin (*Lupinus albus* L.) as a source of proteins for human nutrition: (ii) Production of isolates by acid coagulation and ultrafiltration–diafiltration]. *Lebensm.-Wiss u-Technol.*, *9*, 338–344.

PRELLER, J. H. 1949. Lupines. *Farming South Afr.*, *24* (274), 25–29.

PRELLER, J. H. 1952. The sweet lupin for the highveld region. *Farming South Afr.*, *27* (316), 374.

PUCHER, G. W. 1948. Aspargine and glutamine. *Econ. Bot.*, *2*, 219.

QUINLIVAN, B. J. 1962. The effect of storage in polythene on the development of hard seed in West Australian blue lupin. *Aust. J. Exp. Agric. Anim. Husb.*, *2*, 209–212.

QUINLIVAN, B. J. 1971. Seed-coat impermeability in legumes. *J. Aust. Inst. Agric. Sci.*, *37*, 283–295.

RAHMAN, M. S., GLADSTONES, J. S. AND THURLING, N. 1974. Effects of soil temperature and phosphorus supply on growth and composition of *Lupinus angustifolius* L. and *L. cosentinii* Guss. *Aust. J. Agric. Res.*, *25*, 885–892.

REEVES, T. G. 1974. Lupins—a new grain legume crop for Victoria. *J. Agric. Vict. Dep. Agric.*, *72*, 285–289.

RUIZ, L. P. 1977. A rapid screening test for lupin alkaloids. *N.Z. J. Agric. Res., 20*, 51–52.

RUIZ, L. P. (JR.) AND HOVE, E. L. 1976. Conditions affecting production of a protein isolate from lupin seed kernels. *J. Sci. Food Agric., 27*, 667–674.

SMITH, P. M. 1976. Lupin. *Evolution of crop plants.* (Simmonds, N. W. ed.). pp. 312–313. London: Longman Group Ltd, 339 pp.

SOUTHWOOD, O. R. AND SCOTT, R. C. 1972. Sweet lupins: a potential high protein grain crop for N.S.W. *Agric. Gaz. N.S.W., 83*, 99–101.

SWART, L. G. 1955. Lupins in laying rations. *Farming South Afr., 30*, 404–406.

SWART, L. G. AND LIEBENBERG, C. R. 1954. Lupins in chick rations. *Farming South Afr., 29*, 227–229.

VOSLOO, W. A. 1960. Lupin seed meal can replace part of white fish meal. *Farming South Afr., 36* (1), 54–55.

VOSLOO, W. A. AND STEENKAMP, D. J. 1958. Lupin meal good pig feed. *Farming South Afr., 33* (10), 18–19.

VUUREN, P. J. VAN. 1962. How to identify lupins. *Farming South Afr., 38* (9), 61–64.

WALTON, G. H. AND FRANCIS, C. M. 1975. Genetic influences on the split seed disorder in *Lupinus angustifolius* L. *Aust. J. Agric. Res., 26*, 641–646.

WESTERN AUSTRALIA, DEPARTMENT OF AGRICULTURE. nd. *Lupins in Western Australia.* Perth, West. Aust., Dep. Agric. Leafl., 4 pp.

WESTERN AUSTRALIA, DEPARTMENT OF AGRICULTURE. 1975. Lupins. pp. 15–18. *Perth, West. Aust., Dep. Agric. Annu. Rep.* 1975, 41 pp.

WESTERN AUSTRALIA, DEPARTMENT OF AGRICULTURE. 1976. Lupins. pp. 16; 29–30. *Perth, West. Aust., Dep. Agric. Annu. Rep.* 1976, 53 pp.

WESTERN AUSTRALIA GRAIN POOL. 1976. *Lupin seed a new high protein seed.* Perth, West. Aust., The Grain Pool Leafl., 6 pp.

WESTPHAL, E. 1974. *Lupinus albus L.* Pulses in Ethiopia, their taxonomy and agricultural significance. pp. 114–121. *Wageningen, PUDOC, Cent. Agric. Publ. Doc., Agric. Res. Rep. 815*, 278 pp.

WOOD, P. MC. R. AND BROWN, A. G. P. 1975. Phomopsis the causal fungus of lupinosis. *J. Agric. West. Aust., 16*, 31–32.

ZYL, L. G. VAN. 1973. Lupins in the South-eastern Transvaal highveld. *Farming South Afr., 48* (12), 4–7; 10.

Common names MOTH BEAN, Mat (bean).

Botanical name *Vigna aconitifolia* (Jacq.) Maréchal, syn. *Phaseolus aconitifolius* Jacq.

Family Leguminosae.

Other names Aconite leaved (kidney) bean, Dew bean, or gram (Ind.); Haricot papillon (Fr.); Kallupayaru (Tam.); Kheri (Beng.); Kidney bean[1], Kumkumapesalu (Ind.); Math, Matki (Ind.); Matpe (Thai.); Meth-kalai, Minimulu (Ind.); Mittikelu (Malays.); Moot (Hind.); Naripayaru, Pani payeru (Tam.); Phillipesara, Tulkayrai (Ind.); Turkish gram (USA).

Botany

A perennial or annual, slender stemmed legume, with a creeping habit. The main stem is erect, about 6 to 12 in. (15–30 cm) in height, and from this, primary, secondary and tertiary branches arise, which trail along the ground. Normally up to 12 primary branches, ranging from 12 to 60 in. (30–150 cm) in length, are produced, and up to 25 or 30 secondary branches, and about 12 tertiary branches. The main stem and branches are angular and sparsely covered with stiff, more or less erect, hairs, which are white at first, but soon become rusty-brown. The leaves are alternate, trifoliolate, and may be deeply or slightly lobed, according to the cultivar. The leaflets are 2 to 5 in. (5–12 cm) long. In some forms the terminal leaflet is divided into 5 acuminate lobes and the lateral leaflets into 3 smaller lobes. The inflorescences are axillary and bear several small, bright-yellow flowers, approximately 0.3 x 0.2 in. (8 x 5 mm), which are normally self-fertilized. Each peduncle bears 1–6, usually 3, small pods, about 1 to 2 in. (2.5–5 cm) long, 0.2 in. (5 mm) broad and 0.1 in. (2.5 mm) thick, with a short, curved beak and covered with short, stiff hairs. They are usually buff or yellowish-brown in colour, although in certain forms they are marbled with black. Each pod normally contains 4–9 small seeds, which are yellow to brown, or mottled black, and are separated from each other by a whitish membrane.

---

[1] In countries other than India usually used for the haricot bean, *Phaseolus vulgaris*.

Origin and distribution

The moth bean is native to India, Pakistan and Burma, where it is found growing wild and under cultivation. It is a legume of considerable importance in the semi-arid areas adjoining tropical deserts in SE. Asia, and has also been introduced successfully into parts of the USA, notably California and Texas.

Cultivation conditions

Temperature—a high uniform temperature is required for optimum yields, and for quick, uniform germination of the seeds, a soil temperature of approximately 80°F (27°C). The moth bean is very susceptible to frost.

Rainfall—the moth bean is usually found in areas with an annual average rainfall of about 20–30 in. (500–750 mm). It cannot stand heavy rain and is not suited to the subhumid or humid areas of the tropics. Although established plants are tolerant of drought conditions, unless there is adequate soil moisture during the first 4 to 6 weeks after planting seed germination is very poor and there is a high mortality rate amongst the seedlings. It is frequently planted towards the end of the rainy season and grows mainly on stored soil moisture, although it can also be grown successfully as a hot season crop under irrigation. It has been suggested that the moth bean could be grown successfully under mixed cropping in W. Rajasthan, India, where the annual rainfall ranges between 4 and 10 in. (100–250 mm).

Soil—it can be grown on a wide range of soils and will thrive on light, sandy loams or heavy clay loams, although in India it is grown mainly on light sandy soils or red gravels. On light soils the application of phosphate, up to 40–50 lb/ac (45–56 kg/ha) is often recommended.

Altitude—in India the moth bean is grown successfully from sea level up to elevations of about 4 000 ft (1 200 m).

Day-length—although generally considered to be a short-day legume, Hartman (1969) recently investigated 19 lines under Hawaiian conditions and found them all to be day-neutral.

Planting procedure

Material—seed is used and germination is epigeal.

Method—in Asia the moth bean is frequently grown mixed with millets or cotton, when the seed is broadcast. The moth bean is sometimes grown as a pure crop in S. India, when it is sown at the end of June or the beginning of July with the onset of the monsoon. As the seed is small, a well-prepared seed-bed and a moist, friable soil about 7 in. (17.5 cm) deep is essential for high yields. In the USA, where the moth bean is grown occasionally as a forage crop, the seeds are normally planted 1.6 in. (4 cm) deep. Weeding, particularly during the early stages of growth, is essential and in arid conditions one or two light irrigations may be necessary to stimulate growth.

Field-spacing—in the USA the moth bean is usually drilled in rows 30–36 in. (75–90 cm) apart with 3 in. (7.5 cm) between the plants.

Seed-rate—in India the average seed-rate is 2–4 lb/ac (2.2–4.5 kg/ha) when grown mixed with another crop, such as millet, and 9–15 lb/ac (10–17 kg/ha) when grown as a pure crop. When grown for forage a seed-rate of 18 to 22 lb/ac (20–25 kg/ha) has been found to increase significantly the average (2 year) fresh and dry matter yields, compared with a seed-rate of 13 lb/ac (15 kg/ha). In the USA when grown for forage the average seed-rate is reported to be 6–8 lb/ac (6.7–9 kg/ha), although if seeded in very close drills up to 30 lb/ac (34 kg/ha) may be required.

Pests and diseases

The moth bean does not appear to suffer serious attack from insect pests in the field. In the USA it has been found to be susceptible to nematodes, although not to the extent of cowpeas. Like most beans, it is susceptible to attack from bruchids, *Callosobruchus* spp., during storage.

When grown in Asia it is reported to be subject to the following diseases, a leaf spot, *Cercospora cruenta*, a mildew, *Sphaerotheca humuli*, a blight which affects the leaves, stalks and pods, caused by *Phyllosticta phaseolina* or *Ascochyta phaseolorum*, and anthracnose, *Colletotrichum lindemuthianum*. The use of disease-free seed, and the destruction of all diseased plant debris have been found to be very effective in controlling blight, particularly if they are accompanied by mixed cropping with millets.

Growth period

Mature seed is produced approximately 90 days after planting.

Harvesting and handling

When grown as a grain legume the plants are usually cut with a sickle, dried for about one week, before being threshed and winnowed.

Primary product

Seed—which is small, approximately 0.2 in. (5 mm) long and 0.1 in. (2.5 mm) wide, somewhat reniform in shape, with rounded or truncate ends. The colour can range from a dull, whitish-green, to a brownish-buff, or mottled-black. The mottled-black seed is normally slightly larger. The hilum is central, approximately 0.1 in. (2.5 mm) long, white in colour, without any surrounding band. One hundred seeds weigh approximately 0.035 oz (1 g).

Yield

In India seed yields are reported to average 200–500 lb/ac (225–560 kg/ha), while in the USA average yields are reported to be 1 100–1 600 lb/ac (1 230–1 790 kg/ha). When grown as a forage crop, yields are reported to average 15–20 T/ac (37.5–50 t/ha) of green forage and 3–4 T/ac (7.5–10 t/ha) of hay.

Main use

The seeds are a nutritious foodstuff, particularly amongst the rural population of the more arid areas of Asia, where moth beans are eaten whole, usually after frying, or split and used for the preparation of dhal.

Subsidiary uses

Flour is sometimes prepared by grinding the seeds. It is usually used mixed with other flours in the preparation of unleavened bread. The seeds may be processed for starch, suitable for use in textile-sizing and calico-printing. The ripe seeds can also be utilized for livestock feeding. The moth bean is also grown as a forage crop. An approximate analysis of the green forage is: moisture 75.0 per cent; crude protein 3.0 per cent; fat 0.4 per cent; crude fibre 7.7 per cent; ash 3.9 per cent; N-free extract 10.0 per cent; calcium 0.90 per cent; phosphorus 0.16 per cent; potassium 0.95 per cent. The approximate composition of the hay is: moisture 10.0 per cent; crude protein 16.2 per cent; fat 2.4 per cent; crude fibre 16.0 per cent; ash 14.0 per cent; N-free extract 41.4 per cent. The moth bean is also occasionally grown as an anti-erosion measure.

Secondary and waste products

In Asia, the immature pods are sometimes boiled and eaten as a vegetable. In India the husks and pieces of seed left after the preparation of dhal are utilized for livestock feeding. The product obtained from the mills is reported to have a crude protein content of 20.5 per cent.

Special features
The approximate composition of the seed has been given as: moisture 8.6–10.8 per cent; protein 21.8–23.6 per cent; fat 0.8–1.2 per cent; carbohydrate 56.5–63.8 per cent; crude fibre 3.9–4.5 per cent; ash 3.1–4.2 per cent; calcium 0.22–0.32 per cent; phosphorus 0.10–0.70 per cent; potassium 1.0 per cent; magnesium 0.84 per cent; iron 0.009 per cent. The principal amino acids present are: lysine, leucine, isoleucine, phenylalanine and tyrosine. Like most grain legumes, the moth bean is deficient in cystine and methionine. The fat present in the seed is reported to have the following characteristics: melting point 10–15° C; refractive index (Zeiss) 63–65; SG $^{22-22°C}$ 0.8327; iod. val. 174.0; sap. val. 166.7; acid val. 1.0; thiocyanagen val. 158.2; unsaponifiable matter 1.1 per cent.

Processing
The moth bean may be processed into dhal similarly to the other Asian grain legumes.

Production and trade
India is the leading producer of moth beans. Production has averaged 302 700 t/a for the period 1967–68 to 1971–72, reaching a peak of 473 400 t in 1970–71. Thailand is also an important producer, but it is difficult to estimate current production, as moth beans are grouped with mung beans in the official statistics. Thailand is in fact the leading exporter of moth beans. Exports during the period 1970–74 averaged 42 000 t/a, compared with 19 300 t/a for 1965–69, an increase of 118 per cent. In 1975, however, exports amounted to 40 817 t and fell in 1976 to 30 824 t. About two-thirds of Thai exports are usually consigned to Japan, with the remaining one-third being consigned mainly to Malaysia, Hong Kong, Singapore and Sri Lanka. The fob value of Thai exports was £374.00/t in 1976 and £123.00/t in 1975, compared with an average value of £72.00/t for the five-year period 1970–74 and £49.60/t for 1965–69.

Major influences
Although the moth bean is considered to be one of the most drought tolerant of all the grain legumes it has never become a popular crop, even in semi-arid areas, because its low spreading habit makes harvesting very difficult and laborious.

Bibliography

BHOWN, A. S. AND GAUR, I. C. 1960. Physico-chemical study of *Phaseolus aconitifolius* (Moth). *Indian J. Appl. Chem.*, *23*, 157.

CHADA, Y. R. (ed.). 1976. *V. aconitifolia* (Jacq.) Maréchal. *The wealth of India: Raw materials*. Vol. 10 (Sp–W). pp. 471-475. New Delhi: Counc. Sci. Ind. Res. Publ. Inf. Dir., 591 pp.

CHOPRA, K. AND SWAMY, G. 1975. *Pulses: An analysis of demand and supply in India. Institute for social and economic change; Monograph 2*. New Delhi: Sterling Publishers, PVT, Ltd, 132 pp.

DESHPANDE, P. D. AND RADHAKRISHNA RAO, M. V. 1954. Nitrogen complex and amino acid composition of (i) amaranth (*Amaranthus gangeticus*) and aconite bean (*Phaseolus aconitifolius*). *Indian J. Med. Res.*, *42*, 77–83. (*Nutr. Abstr. Rev.*, *24*, 4276).

FARODA, A. S. 1972. Effect of seeding rates and row spacings on fodder production of moth bean (*Phaseolus aconitifolius*). *Ann. Arid Zone*, *11*, 183–186.

GHOSH, A. C. 1917. The fodder pulses meth, bhringi and mashyem kalai. *Agric. J. Bihar, Orissa*, *5* (1), 28–34.

HARTMAN, R. W. 1969. Photoperiod responses of *Phaseolus* plant introductions in Hawaii. *J. Am. Soc. Hortic. Sci.*, *94*, 437-440.

INDIAN COUNCIL OF AGRICULTURAL RESEARCH. 1969. Moth bean (*Phaseolus aconitifolius* Jacq.). *Handbook of agriculture*. 3rd ed. p. 194. New Delhi: Indian Counc. Agric. Res., 911 pp.

KENNEDY, P. B. AND MADSON, B. A. 1925. The mat bean (*Phaseolus aconitifolius*). *California, Agric. Exp. Stn. Bull.* 396, 33 pp.

MISRA, D. K. AND DAS, R. B. 1963. Legume-grass mixtures are good for pastures. *Indian Farming*, *13* (7), 9–10.

NARAYANAN, T. R. AND DABADGHAO, P. M. 1972. Dewgram. *Forage crops of India*. pp. 76–77. New Delhi: Indian Counc. Agric. Res., 373 pp.

PATEL, B. M., PATEL, R. B. AND SHUKLA, P. C. 1971. A note on nutritive value of kidney bean chuni (*Phaseolus aconitifolius* Jacq.). *Indian J. Anim. Sci.*, *41*, 925–926.

Moth bean

PIPER, C. V. 1914. Five oriental species of beans. pp. 28–30. *US Dep. Agric. Bull.* 119, 32 pp.

PURSEGLOVE, J. W. 1968. *Phaseolus aconitifolius* Jacq. *Tropical crops: Dicotyledons.* Vol. 1. pp. 286–287. London: Longmans, Green and Co Ltd, 332 pp.

RACHIE, K. O. AND ROBERTS, L. M. 1974. Grain legumes of the lowland tropics. *Adv. Agron., 26*, 1–132.

SAINI, R. AND MALIK, H. C. 1947. Studies in Punjab moth (*Phaseolus aconitifolius*). *Indian J. Agric. Sci., 17*, 281–288.

SATTAR, A. AND HAFIZ, A. 1952. Diseases of pulses (*Phaseolus* spp.). Researches on plant diseases of the Punjab. pp. 108–110. *Lahore, Pak. Assn. Adv. Sci., Sci. Monogr.*, 1, 158 pp.

SINGH, H. B., JOSHI, B. S. AND THOMAS, T. A. 1970. Moth bean *Phaseolus aconitifolius* Jacq. *Pulse crops of India.* (Kachroo, P. and Arif, M. eds.). pp. 156–158. New Delhi: Indian Counc. Agric. Res., 334 pp.

Common names	**MUNG BEAN, Green or Golden gram.**
Botanical name	*Vigna radiata* (L.) Wilczek, syn. *Phaseolus aureus* Roxb., *P. radiatus* L.
Family	Leguminosae.
Other names	Ambérique[1] (de Madagascar) (Fr.); Balatong,[1] Batong-hidjao[1] (Philipp.); Black gram[2]; Boubour (Fr.); Bundo (Jpn.); Cherupayaru (Ind., Malays.); Chickasano (Afr.); Chickasaw pea (USA); Chiroko[1] (Afr.); Dau xanh (Viet.); Dord (Hind.); Fagiolo mungo (It.); Goue mungboontjie (S. Afr.); Haricot doré, H. mungo[1] (Fr.); Jerusalem pea (Jam.); Judia de mungo (Sp.); Kachang hijan, K. hijau[1] (Malays.); Kanyensi (Zam.); Katjang djong, K. eedjo (Indon.); K. padi (Malays.); Kifudu[1] (Ug.); Lou teou (China); Lubia chiroko (E. Afr.); Mag, Mash[1] (Ind.); Mesh (Is.); Mongo (Philipp.); Moong (Ind.); Moyashi-mame (Jpn.); Mudgaha, Mug (Ind.); Muneta (Sri La.); Muñggo[2] (Philipp.); Mungobohne (Ger.); Nga-choi (China); Oregon pea (USA); Pachapayaru (Tam.); Pasalu (Malays.); Pasipayeru (Tam.); Passi payaru, Patcha-payru (Ind.); Pè-di-sein, Pè-di-wa, Pè-nauk (sein) (Burm.); Pesalu (Ind.); Sheuit tzehuba (Is.); Tientsin green bean (As.); Too-a kee-o, Tua kiew, T. kio, T. tawng, T. tong (Thai.); Yaenari, Yayenari (Jpn.).

Botany

A rapidly growing, erect or sub-erect, annual, usually 1–3 ft (30–90 cm) in height, and showing considerable variation in form and adaption; over 2 000 different types have been recognized. The root-system may be mesophytic or

[1] Also commonly used for urd, *Vigna mungo*.
[2] More correctly used to denote urd, *Vigna mungo*.

xerophytic. In the latter, the tap-root reaches to a great depth and the laterals arising from it extend obliquely downwards. The xerophytic form is frequently associated with an erect or semi-erect habit, and a longer growing season. The mesophytic form possesses numerous lateral roots near the soil surface and is normally associated with a more spreading habit; it is often found growing on alluvial soils in the SE. Asian sub-continent.

The mung bean is frequently much branched, a few cultivars have a spreading habit, although never as trailing as in urd, *Vigna mungo*, sometimes the tips of the branches are vining. The stems normally branch from the base and are angular and usually green, but occasionally splashed with purple. The leaves are trifoliolate with large, ovate, entire, or rarely lobed, membranous leaflets, 2–4 in. (5–10 cm) long, with scattered hairs on both sides and light to dark-green in colour, never yellowish. The yellow or greenish-yellow flowers are crowded in clusters, (10–20), on axillary or terminal racemes, usually the latter. The flowers are fully self-fertile and are normally self-pollinated. The seed-pods are sub-cylindrical 2–4 in. (5–10 cm) long and 0.15–0.24 in. (4–6 mm) wide, straight or slightly curved and normally containing 10 to 20 small seeds. When unripe the pods are various shades of green, but when fully mature may be grey, greyish-olive green, brown or buff. The seeds are globular or oblong, often green, but may be yellow, brown, or speckled with black. Two main types of mung bean are usually recognized: (i) *aureus*, yellow or golden gram, which has yellow seeds, is generally low in seed production, has a tendency for the pods to shatter, and is often grown for forage or green manure; (ii) *typica*, green gram, which has dark or bright-green seeds, is more prolific, ripens more uniformly, has less tendency to shatter, and is more popular as a grain legume. In addition, two other types are recognized in India, *grandis*, a black-seeded type and *bruneus*, a brown-seeded type.

The mung bean, *Vigna radiata*, is morphologically very similar to urd, *Vigna mungo*, and there has been considerable confusion in the nomenclature of the two species. Although Asian authors have always considered mung and urd as two distinct species, some European taxonomists have expressed doubt about this. However, despite being morphologically very similar there are marked incompatabilities to crossing and Otoul *et al.*, (1975) have recently found a chemotaxonomic argument giving support for a distinction at species level. Thus, it is increasingly clear that the mung bean and urd should be regarded as two distinct species.

Considerable work on the improvement of the mung bean has been undertaken in recent years, mainly in India, but also in Taiwan and tropical America. In India a number of improved high-yielding cultivars have been released including: Pusa Baisakhi, Krishna–11, Mung H 70–16, Kopargaon, all short-duration types suitable for multiple cropping programmes, and Jawahar–45, which is suitable for rainy season planting and for double cropping in rain-fed and limited water supply areas. Recently, a high-yielding, quick-maturing, non-shattering cultivar has been developed in the USA which produces semi-glossy, large seeds, particularly suitable for the production of bean sprouts.

Origin and distribution

The mung bean originated in SE. Asia and is an important pulse crop throughout southern Asia, especially in India, Burma, Thailand, Indonesia and the Philippines. It is also grown to a lesser extent in many parts of Africa and the Americas, particularly Oklahoma in the USA. In addition, it is now being grown on a small-scale in Australia.

Cultivation conditions

Temperature—the mung bean requires a warm equable temperature and is tolerant of high temperatures, thriving in areas with average temperatures of between 86° and 97°F (30–36°C). It is sensitive to frost.

Rainfall—for optimum results a well-distributed rainfall of 30 to 36 in./a (750–900 mm/a) is required, although reasonable yields can be obtained in areas with only 25 in./a (650 mm/a). The mung bean is fairly tolerant of drought. It is often grown in areas with limited rainfall, by utilizing the residual soil moisture after an irrigated crop such as rice, sometimes with supplementary irrigation 30–35 days after sowing. When grown in areas with a prolonged rainy season vegetative growth tends to be excessive, and rain at flowering is very detrimental to yield. In India, in areas with an annual rainfall of 40–50 in. (1 000–1 250 mm) difficulty is often experienced in picking the pods. However, moisture deficiency during flowering and seed ripening would appear to be conducive to the development of hard seeds.

Soil—the mung bean can be grown on a wide range of soil types provided that they are well drained, since it cannot stand waterlogging. For optimum results a deep loamy soil is required, but the mung bean is well adapted to clay soils and frequently is grown on black cotton soils in India. It is tolerant of alkaline and saline conditions. The fertilizer requirements of the mung bean have not been studied extensively. It responds to phosphorus and it has been suggested that in commercial production the most economic level of application is between 13 and 35 lb/ac (15–40 kg/ha) of phosphoric acid. However, on some lateritic soils with high phosphate fixation capacity, good response is obtained with applications of up to 89 lb/ac (100 kg/ha) of superphosphate. In India, the application of up to 356 lb/ac (400 kg/ha) together with ammonium sulphate 89 lb/ac (100 kg/ha) has been suggested. Banding the fertilizer has been found to be superior to broadcasting and experimentally foliar applications of phosphorus have been found to be effective in increasing yields.

Altitude—in India the mung bean is normally grown from sea level up to elevations of approximately 6 000 ft (1 800 m).

Day-length—although often classified as a short-day plant, day-neutral cultivars exist.

Planting procedure

Material—seed, which normally retains its viability for at least 2–3 years, is used. Germination is epigeal and usually between 85 and 95 per cent, although it can be affected by the hard seed-coat of certain strains. Inoculation with the appropriate rhizobium has been found to be beneficial where plants of the same host range have not been grown recently.

Method—a good tilth is required, and the seeds are normally planted about 1.6 in. (4 cm) deep, and may be broadcast or drilled in rows. In many parts of Asia, particularly India, the mung bean is frequently grown after rice or grown as a subordinate crop to cereals such as sorghum, maize or millet. In Taiwan the mung bean has proved to be highly profitable when grown with spring planted sugar cane. When grown as a subordinate crop the cultural practices for the main crop are followed and the seed is usually broadcast. When grown as a pure crop the mung bean is normally planted in rows, when grown during the wet season, planting on ridges is recommended to prevent waterlogging. The crop frequently receives one or two cultivations to suppress weed growth, etc, the first approximately 20 days after germination, the second after a further 14 days. Effective weed control, particularly during the early stages of growth, is very necessary. The pre-planting application of the herbicide trifluralin has been recommended in the USA and in the

Philippines. Chlorthaldimethyl and diphenamid are also reported to be effective. Post-emergence, the herbicides chlorthal, chloramben and linuron are said to give satisfactory results. In India, terbutryn, alachlor, dichlormate and nitrofen have shown promise in field trials.

Field-spacing—the distance between the rows can often vary from 10 to 35 in. (25–87.5 cm), although a spacing of 20 in. (50 cm) and 1.6–2 in. (4–5 cm) between the plants is sometimes recommended. In India, the rows are frequently 14 to 18 in. (35–45 cm) apart, with 9 in. (22.5 cm) between the plants. In the USA the rows are normally 18 to 42 in. (45–105 cm) apart.

Seed-rate—when sown broadcast as a subordinate crop the seed-rate is usually 2.7–3.6 lb/ac (3–4 kg/ha), when grown as a pure crop it can vary between 4.4 and 15 lb/ac (5–17 kg/ha), or even higher. In the USA for example, the seed-rate varies from 5–8 lb/ac (5.6–9 kg/ha) if the rows are 36–42 in. (90–105 cm) apart, to 15–20 lb/ac (17–22 kg/ha) if the rows are 14–18 in. (35–45 cm) apart. In India the seed-rate usually averages 8–10 lb/ac (9–11 kg/ha).

Pests and diseases

The mung bean is susceptible to many of the pests and diseases which attack other food legume crops in SE. Asia. In Africa and the USA it appears to be less susceptible to the pests and diseases which commonly affect the haricot bean and the cowpea.

Pests—in SE. Asia bean fly, *Ophiomyia* (*Melanagromyza*) *phaseoli*, is the most serious pest, sometimes resulting in the total loss of the crop. The larvae bore into the stem tissues and a rot at the collar often results. Good control is reported to be obtained by applying carbofuran in granular form to the soil at planting. Aphids, *Aphis craccivora*, can be troublesome sometimes in the Philippines, in addition to cutworms, *Mocis undata*, *Spodoptera litura* and *Chrysodeixis chalcites*, the pod borer, *Etiella zinckenella*, and the red spider mite, *Tetranychus cinnabarinus*. Satisfactory control of these pests is reported to be obtained by the use of the granular insecticides Cyolane, Cytrolane and Solvirex. The hairy caterpillar, *Diacrisia obliqua*, is also sometimes troublesome in SE. Asia. In India the larvae of *Amsacta moorei* are reported to cause serious damage to the crop in Madhya Pradesh. Another species *A. albistriga* is reported to be present in the south. Spraying with endrin is reported to give effective control. In addition, the beetle *Apion ampulum* is reported to be troublesome occasionally; its larvae feed on the developing seeds in the pods. Root-knot nematodes, *Meloidogyne* spp., can cause serious problems unless proper crop rotation is carried out. In the USA certain mung bean cultivars have been found to be hosts of the soyabean cyst nematode, *Heterodera glycines*.

Like most other grain legume crops the mung bean is susceptible to attack from a number of insect pests during storage. The most important is the cowpea weevil, *Callosobruchus chinensis*, which attacks the crop in the field and during storage. In India considerable losses are reported to occur due to attack from this and other bruchid beetles. The cowpea weevil has been found to affect adversely the yield and flavour of dahl made from mung beans. In NE. Thailand the storage weevil, *Acanthoscelides obtectus*, is reported to cause considerable damage to seed of certain cultivars, but there appears to be considerable variation in the degree of susceptibility of cultivars to attack.

Diseases—the mung bean is susceptible to a number of diseases the more important of which are a root rot caused by *Sclerotium rolfsii*, downy mildew, *Erysiphe polygoni*, leaf spots caused by *Cercospora* spp. and blight, *Macrophomina phaseoli*, halo blight, *Pseudomonas phaseolicola*, rust, *Uromyces appendiculatus*, bean blight, *Ascochyta phaseolorum*, anthracnose, *Colletotrichum lindemuthianum*, and several virus diseases including, leaf crinkle, yellow-mosaic, stunt and flower abortion. The cultivation of resistant cultivars, particularly if combined with good field sanitation, would seem to be the most effective method for controlling these diseases. For example, germplasm evaluation in Missouri, USA, identified the strains M–238 and M–330 as having good resistance to downy mildew and virus diseases.

Growth period

The mung bean normally takes from about 80 to 120 days to produce a seed crop, although some late-maturing types require up to 150 days. In India the newly evolved early-maturing cultivar Pusa Baisakhi produces a seed crop in 65 days, and the cultivar Jawahar–45 in 75 to 80 days. When grown as a vegetable, the green immature pods are usually picked about 50 to 70 days after sowing.

Harvesting and handling

When grown for mature seed, mung beans are harvested as the pods begin to darken. In India many cultivars do not ripen uniformly and the pods are normally picked by hand at five or six day intervals, until most of the pods are ripe, when the entire plant is uprooted. The cultivar Pusa Baisakhi has the advantage that 75 per cent of the crop can be harvested at the first picking and the remainder about 10 days later. Thus, the cost of harvesting is reduced considerably. After picking, the pods are left to dry in the sun prior to threshing by hand. The pods shatter easily so that careful handling is essential for the maximum recovery of seed. In the USA bean harvesters are frequently used and after the seed has been dried, it is threshed in a combine/thresher suitably adjusted to avoid splitting or cracking. During the harvesting operation it is essential to separate any green leaves or immature pods otherwise the seed can quickly become contaminated with moulds. Before being stored the seed should have a moisture content of less than 15 per cent, preferably about 10 per cent. Insect infestation is usually controlled by fumigation, often with an ethylene dichloride—carbon tetrachloride mixture. In rural areas mung beans are usually stored in hand-made receptacles made of mud, matting or wicker. In India, frequently dried neem leaves, *Azadirachta indica*, are burnt inside the covered receptacles as a precaution against insect infestation.

Primary product

Seed—the mung bean is small, approximately 0.1–0.2 x 0.12–0.15 in. (2.5–5 x 3–4 mm), broadly ellipsoid to globular, but flattened at the ends. It is commonly green in colour, but can be yellow (golden gram), brown, grey or a greenish-black, sometimes speckled with black patches. According to the cultivar the seeds may be bright and shiny, or dull. Usually the seed-coat is marked by innumerable fine wavy ridges or corrugations, which may be very prominent in some cultivars and barely discernible in others. The hilum is white, narrowly elliptical. Bright seeds are generally more popular and grey seeds the least popular. In India three grades are sometimes recognized: (i) *small*, 1 000 seeds weigh up to 0.8 oz (23 g); (ii) *medium*, 1 000 seeds weigh between 0.8 and 1.2 oz (23 g–35.5 g); (iii) *bold*, 1 000 seeds weigh more than 1.2 oz (35.5 g, or more).

Yield

Average yields of dry, mature seed at farmer level in most Asiatic countries, normally range between 180 and 620 lb/ac (200–700 kg/ha), although when grown as a subordinate crop in India yields may be as low as 100–200 lb/ac (112–224 kg/ha). In the USA yields are reported to average about 890 lb/ac (1 000 kg/ha). In India the improved cultivar Pusa Baisakhi is reported to yield on average 570 lb/ac (640 kg/ha) at farmer level, and to be capable of yielding over 1 000 lb/ac (1 120 kg/ha), when grown during the interval between the main rabi (dry season) and the kharif (wet season) crops, provided that it has been given an application of general fertilizer, and there has been efficient control of insect pests, particularly bean fly. A new cultivar G–65 developed from Pusa Baisakhi is reported to give an average yield of 725 lb/ac (810 kg/ha) at farmer level and 1 180 lb/ac (1 320 kg/ha) experimentally. In Taiwan a cultivar Tainan- 1 has produced 2 190 lb/ac (2 460 kg/ha) in trials, and several other cultivars have yielded about 1 655 lb/ac (1 860 kg/ha), or about triple that obtained by most farmers in SE. Asia, which indicates the potential productivity level of the mung bean.

Main use

The mung bean is a nutritious foodstuff, particularly in SE. Asia, where it is very popular because of its high digestibility and relative freedom from the flatulence effect commonly associated with many grain legumes. It is frequently used for feeding children, invalids or geriatrics. The seed may be eaten whole after boiling, or split and made into dhal. It is also used in various fried and spiced dishes such as noodles and 'balls'.

Subsidiary uses

Flour—after removing the seed-coat (approximately 12 per cent of the whole bean) and parching, mung beans may be ground into a nutritious flour, or used as a source of starch. The flour is widely used as an ingredient for various bakery products and snack foods. It is high in protein, particularly lysine, and is suitable for utilization as a source of protein. The approximate composition of mung bean flour has been given as: moisture 11.2 per cent; protein 23.6 per cent, (lysine approximately 6 per cent); carbohydrate 60.0 per cent; fat 1.3 per cent; crude fibre 0.9 per cent; ash 3.0 per cent.

Bean sprouts—mung beans are widely utilized, particularly in SE. Asia, and E. Africa, for the production of bean sprouts, a nutritious vegetable, prepared as follows. The beans are first soaked overnight in water, or for 4 hours in water at a temperature of 90°F (32°C). They are then placed in wooden or porcelain vats, covered to exclude the daylight, sprinkled with water at approximately four hourly intervals and kept at a temperature of 75°F (24°C) and a relative humidity of 60 to 70 per cent for 4 to 5 days, by which time the bean sprouts should be about 1.2 to 1.6 in. (3–4 cm) long. At this stage they are ready for eating, after washing and dehulling; they are often preserved by quick-freezing or canning.

Secondary and waste products

The dry seed is sometimes used for animal feeding, particularly poultry, when toasting or boiling is recommended to improve its nutritional value. The green immature seed-pods are occasionally picked and eaten as a vegetable. The mung bean makes quick, early growth and may be grown as a valuable green manure or cover crop. The flour is reported to be used occasionally in parts of SE. Asia as a soap substitute. It has also been suggested that sprouted mung beans could be utilized as a source of L–aspargine.

The leaves and stalks left after the beans have been harvested can be utilized for animal feeding. An approximate analysis of mung bean hay has been given as follows: moisture 9.7 per cent; crude protein 9.8 per cent; fat 2.2 per cent; crude fibre 24.0 per cent; ash 7.7 per cent; N-free extract 46.6 per cent; digestible crude protein 7.4 per cent; total digestible nutrients 49.3 per cent. The residue left after starch extraction is rich in protein and can also be utilized for animal feeding.

Special features

The mung bean has the following approximate composition: moisture 6.6–11.6 per cent; protein 19.7–24.2 per cent; total carbohydrate 60.3–67.5 per cent; crude fibre 4.2–4.4 per cent; ash 3.4–3.5 per cent; fat 1.0–1.3 per cent; calcium 118–145 mg/100 g; phosphorus 340–345 mg/100 g; iron 5.9–7.8 mg/100 g; potassium 1 028 mg/100 g. There is however considerable variation in the composition, according to the cultivar, climatic conditions and the standard of crop management, and protein contents as high as 31.2 per cent have been reported. High protein content tends to be associated with small seeds. The average amino acid content of the mung bean, (mg/gN) has been reported as: aspartic acid 716; cystine 44; threonine 209; serine 296; glutamic acid 865; proline 229; glycine 210; alanine 242; valine 259; methionine 33; isoleucine 223; leucine 441; tyrosine 156; phenylalanine 306; lysine 504; histidine 182; arginine 345.

A recent examination of W.I. mung beans showed that the total sugar content varied from 2.69 to 5.88 g per 100 g of seed; monosaccharides 0.38–1.00 g/100 g, sucrose 1.06–2.19 g/100 g, raffinose 0.38–0.69 g/100 g, stachyose 0.50–1.50 g/100 g. This considerable variation in the total sugar content and the individual sugars present is a possible explanation for the fact that some types of mung beans are non-flatulent, while others have a potential to be gas producers. Mung bean starch contains 28.8 per cent amylose and 71.2 per cent amylopectin.

A sample of oil from Pakistani mung beans was reported to have the following fatty acid composition: palmitic 28.1 per cent; stearic 7.8 per cent; arachidic 0.9 per cent; behenic 2.4 per cent; cerotic 6.3 per cent; oleic 6.4 per cent; linoleic 32.6 per cent; linolenic 14.4 per cent. The unsaponifiable matter contained 0.023 per cent stigmasterol on an air-dry basis.

Recently the presence of the peptide γ–glutamyl–S–methylcysteine and its sulphoxide has been reported. Numerous enzymes have been isolated from the mung bean, two trypsin inhibitors, in a crystalline form, and a saponin.

Processing

Dhal—in India dhal is prepared from mung beans by both dry and wet methods. In the dry method the seed is first partially bruised in 'chakkis,' or dhal rollers, treated with mustard, gingelly or groundnut oil, and water, and then left exposed to the sun for drying and softening of the seed-coat. The process is repeated two or three times until the seed-coat, or husk, can be removed easily by grinding. After grinding the split seeds are polished with magnesia powder and grit to remove the last traces of seed-coat and give the dhal an attractive appearance. In the wet method the seeds are soaked in water for 1 or 2 hours and then left to dry in the sun, after which they are passed through a grinder. The crushed seed is then treated with water and a vegetable oil, dried, then reground, and finally winnowed to remove all the pieces of broken seed-coat. These traditional methods of producing dhal are laborious and the yield and quality of the product sub-optimal. The Central Food Technological Research Institute, Mysore, (CFTRI), has developed improved methods and machinery for the economic production of dhal from different pulse crops, including mung beans, which have been described in the section dealing with the chick pea. The yield of dhal from mung beans by the CFTRI method is reported to average 82–83 per cent, compared with 62–65 per cent by the traditional methods.

Flour—mung bean flour may be produced by dry milling the seeds and sifting the resultant flour through a fine sieve. A superior product can be obtained by grinding the seeds in a pin mill with air separation of the seed-coat fraction.

Canned bean sprouts—the bean sprouts are blanched, packed into cans by hand, a brine solution is added, the cans are then exhausted at 180–190°F (82–88°C) closed and processed at 240°F (116°C) for 20 minutes, A2 (307 x 408) cans.

Production and trade

Production—mung beans are the principal beans grown in Asia, east of Pakistan. Unfortunately they are not shown as a separate item in the *FAO Production Yearbooks*, but are included in the item dry beans, which also includes the ubiquitous haricot bean, *Phaseolus vulgaris*, and several other beans. Moreover, when estimates are made of the production of mung beans, frequently related beans, especially moth beans, *Vigna aconitifolia*, are included.

Mung beans: estimated annual production (1973)

	Quantity tonnes
India	275 000
Thailand[1]	150 000
Indonesia	40 000
Pakistan	34 000
Philippines	15 000
Bangladesh	8 000
Taiwan	6 000

[1]Mung beans are one of three crops for promotion in Thailand's Third National Economic and Social Development Plan and it is anticipated that production could increase considerably by the mid-1970s.

Trade—although a high percentage of mung beans are utilized domestically considerable quantities enter international trade. Unfortunately statistical data are extremely fragmentary. Thailand and Burma are the leading exporting countries. Exports from Thailand reached 57 253 t in 1976 compared with 42 406 t in 1975. For the period 1970–74, exports averaged 48 000 t/a compared with 36 600 t/a for the period 1965–69, an increase of 31 per cent. Shipments from Burma have fluctuated considerably in recent years, being at the rather low level of between 6 000 and 8 000 t/a from 1970 to 1972, but reaching 22 000 t in 1973, then falling to 14 000 t in 1974 and dropping further in 1975 to 8 765 t.

Mung bean

Imports—Japan is the major import market for mung beans. Imports averaged 41 000 t/a for the period 1971–74, compared with 33 200 t/a for 1965–69, an increase of just over 23 per cent, and in 1975 amounted to 40 185 t. The United States also provides a significant market for mung beans. Imports for the period 1970–72 were estimated to be of the order of 5 400 t/a, and in 1975 amounted to 5 085 t, but fell to 2 351 t in 1976. Imports into the Philippines fluctuate considerably according to domestic production. In 1970 imports amounted to 207 t, in 1971 they rose to 1 400 t, in 1972 reached 3 300 t, but fell in 1974 to 481 t, and in 1975 were 310 t. Malaysia, Taiwan and Hong Kong also import significant quantities of mung beans from Thailand, but no reliable figures are available.

Prices—the wholesale price for average quality mung beans on the Calcutta market has shown an upward trend, averaging £102.60/t for the period 1970–74, compared with £84.76/t for 1965–69. The average fob value of exports from Thailand was £247.00/t in 1976 and £124.00/t in 1975 compared with £73.80/t for the years 1970–74, and £47.60/t for 1965–69. The average cif price of Japanese imports of mung beans, was £83.00/t for the years 1970–74, compared with £59.60/t for 1965–69. Import prices fluctuated during the late 1960s and early 1970s, but rose to £126.00/t in 1974 and reached £149.00/t in 1975.

Major influences

Although the mung bean is a well known food legume, particularly in SE. Asia, and its nutritional value has long been recognized, it has not been a popular crop with many farmers, because of the relatively low yields that have been obtained, and its susceptibility to attack from pests and diseases. Until comparatively recently it has received little attention from plant breeders, but the development by Indian workers of short-duration improved cultivars, with increased protein contents, indicates that the crop has considerable potential for improvement. There is need for plant breeders to intensify their efforts to develop improved cultivars, particularly disease-resistant strains capable of producing seeds with a high protein content. The work currently being undertaken at Missouri studying: (i) the range of adaptation of the mung bean, (ii) the range of adaptation of specific cultivars, and (iii) the characteristics of the mung bean plant influencing adaptation, could result in the development of much improved cultivars with relatively high productivity levels.

Not only is the mung bean a valuable grain legume, but it is also a valuable raw material for producing bean sprouts, used extensively in Chinese dishes. The demand for bean sprouts is world-wide and considerable. Japan imports over 40 000 t/a of mung beans, mainly for the production of bean sprouts. The mung bean would, therefore, appear to have possibilities in certain tropical countries as a cash crop for export.

Bibliography

ANON. 1960. Notes on the cultivation of mung and blackeye peas. *Farm J. Br. Guiana*, *21* (3), 17–18.

ANON. 1972. Pulse varieties developed at IARI. *Indian Farming*, *21* (10), 47.

ARAULLO, E. V. 1974. Processing and utilization of cowpea, chickpea, pigeon pea and mung bean. *Proc. symp. interaction of agriculture with food science, Singapore*. (MacIntyre, R. ed.). pp. 131–142. Ottawa, Int. Dev. Res. Cent., IDRC–033e, 166 pp.

ARWOOTH, N. L. 1972. Production and research on food legumes in Thailand. Symp. food legumes. pp. 93–100. *Tokyo, Jpn., Minist. Agric. & For., Trop. Res. Cent., Trop. Agric. Res. Ser.* 6, 253 pp.

ASIAN VEGETABLE RESEARCH AND DEVELOPMENT CENTRE. 1974. The research programme: The mung bean. *Taiwan, Asian Veg. Res. Dev. Cent., Annu. Rep.* 1972–1973, pp. 10–22.

ASIAN VEGETABLE RESEARCH AND DEVELOPMENT CENTRE. 1975. Mung bean. *Taiwan, Asian Veg. Res. Dev. Cent., Annu. Rep.* 1974, pp. 29–51.

AZIZ, M. A. and SHAH, S. S. 1966. Improvement of pulses in the former Punjab. *Agric. Pak.*, *17*, 267–287.

BENEMERITO, A. 1954/55. Bean sprout its nutritive values and uses. *Plant Ind. Dig.*, *12/13*, 10–13.

BHARGAVA, R. N. 1973. In Bihar: a new moong for relay cropping. *Indian Farming*, *23* (6), 19–20.

BHATNAGAR, P. S., SENGUPTA, P. K., SINGH, B. and GUPTA, R. N. 1964. New shining mung T–2. *Indian Farming*, *14* (9), 36.

BHULLAR, G. S. and SINGH, K. B. 1973. G–65 a new summer moong for Punjab. *Indian Farming*, *23* (4), 24; 39.

Mung bean

Bose, R. D. 1932. Studies in Indian pulses: Mung or green gram (*Phaseolus radiatus* Linn.). *Indian J. Agric. Sci.*, *2*, 607–624.

Bose, R. D. and Joglekar, R. G. 1933. Studies in Indian pulses: The root systems of green and black grams. *Indian J. Agric. Sci.*, *3*, 1045–1056.

Bott, W. and Kingston, R. W. 1976. Mung bean: An important new grain legume. *Qd. Agric. J.*, *102*, 438–442.

Carandang, E. C. 1968. Mungo. *Culture of vegetables*. pp. 67–69. Manila, Repub. Philipp., Dep. Agric. Nat. Resource, Bur. Plant Ind., 150 pp.

Castillo, M. B. 1975. Plant parasitic nematodes associated with mung bean, soybean and peanut in the Philippines. *Philipp. Agric.*, *59*, 91–99.

Chada, Y. R. (ed.). 1976. *V. radiata* (Linn.) Wilczek. *The wealth of India: Raw materials*. Vol. 10 (Sp–W). pp. 484–495. New Delhi: Counc. Sci. Ind. Res. Publ. Inf. Dir., 591 pp.

Chevalier, A. 1939. Une plante coloniale précieuse pour l'alimentation, le haricot doré ou boubour. [A precious colonial food plant: haricot doré or boubour]. *Rev. Bot. Appl. Agric. Trop.*, *19*, 313–322.

Chien-Pan, C. 1972. Current situation of food legume crops, production in Taiwan, the Republic of China. Symp. food legumes. pp. 11–22. *Tokyo, Jpn., Minist. Agric. & For., Trop. Agric. Res. Cent., Trop. Agric. Res. Ser.* 6, 253 pp.

Chopra, K. and Swamy, G. 1975. *Pulses: An analysis of demand and supply in India. Institute for social and economic change; Monograph 2*. New Delhi: Sterling Publishers, PVT, Ltd, 132 pp.

Choudhury, S. L. and Bhatia, P. C. 1971. Ridge-planted kharif pulses yield high despite waterlogging. *Indian Farming*, *21* (3), 8–9.

Commonwealth Bureau of Pastures and Field Crops. nd. *Phaseolus aureus*. Annotated bibliography 1210, 1950–70. *Commonw. Agric. Bur., Bur. Pastures & Field Crops, Hurley, Maidenhead, Berks, England*, 8 pp.

Commonwealth Bureau of Pastures and Field Crops. nd. *Phaseolus radiatus*. Annotated bibliography 1252, 1951–70. *Commonw. Agric. Bur., Bur. Pastures & Field Crops, Hurley, Maidenhead, Berks, England*, 4 pp.

Djang, S. S. T., Lillevik, H. A. and Ball, C. D. 1953. Factors affecting solubilization of the nitrogenous constituents of the mung bean, *Phaseolus aureus*. *Cereal Chem.*, *30*, 230–235.

EPPS, J. M. and CHAMBERS, A. Y. 1959. Mung bean (*Phaseolus aureus*), a host of the soybean cyst nematode (*Heterodera glycines*). *Plant Dis. Rep.*, *43*, 981–982.

GAPUZ, R. B., JESUS, F. J. DE and BLANCO, R. C. (JR.). 1954. A comparative study between soybean oil meal and mongo meal (raw and cooked) in chick starter mash. *Araneta J. Agric.*, *1* (2), 1–6.

GONZALES, O. N., BANZON, E. A., LIGGAYU, R. G. and QUINITIO, P. H. 1964. Isolation and chemical composition of mung bean (*Phaseolus aureus* Roxb.) protein. *Philipp. J. Sci.*, *93*, 47–56.

GUPTA, M. P. and SINGH, R. B. 1969. Variability and correlation studies in green gram *Phaseolus aureus* Roxb. *Indian J. Agric. Sci.*, *39*, 482–493.

HAWARE, M. P. and PAVGI, M. S. 1976. Field reaction of black gram and green gram to angular black-spot. *Indian J. Agric. Sci.*, *46*, 280–282.

HYMOWITZ, T., COLLINS, F. I. and POEHLMAN, J. M. 1975. Relationship between the content of oil, protein and sugar in mung bean seed. *Trop. Agric.*, (*Trinidad*), *52*, 47–51.

JAIN, H. K. 1972. Genetic improvement and production prospects of food legumes. Symp. food legumes. pp. 33–42. *Tokyo, Jpn., Minist. Agric. & For., Trop. Agric. Res. Cent., Trop. Agric. Res. Ser.* 6, 253 pp.

JESWANI, L. M. 1975. Varietal improvement of seed legumes in India. *Food protein sources*. (Pirie, N. W. and Swaminathan, M. S. eds.). pp. 9–18. London: Cambridge University Press, 260 pp.

KARIVARATHARAJU, T. V., RAMAKRISHNAN, V. and SUNDARARAJ, D. D. 1974. Effect of hard seed-coat on germination of green gram. *Indian J. Agric. Sci.*, *44*, 525–527.

KASASIAN, L. 1968. Chemical weed control in tropical root and vegetable crops. *Exp. Agric.*, *4*, 1–16.

KINGSTON, R. W. 1975. Berken a new mung bean variety. *Qd. Agric. J.*, *101*, 659–660.

KUHN, W. F. 1946. Growing mung bean sprouts for canning. *Food Ind.*, *18*, 1858–1860; 1996; 1998.

KURIEN, P. P., DESIKACHAR, H. S. R. and PARPIA, H. A. B. 1972. Processing and utilization of grain legumes in India. Symp. food legumes. pp. 225–236. *Tokyo, Jpn., Minist. Agric. & For., Trop. Agric. Res. Cent., Trop. Agric. Res. Ser.* 6, 253 pp.

Mung bean

LIGON, L. L. 1945. Mung beans: a legume for seed and forage production. *Okla. Agric. Exp. Stn. Bull.* B–284, 12 pp.

LITZENBERGER, S. C. 1973. The improvement of food legumes as a contribution to improved human nutrition. Potentials of field beans and other food legumes in Latin America. (Wall, D. ed.). pp. 3–16. *Cali, Colombia, Cent. Int. Agric. Trop., CIAT Ser. Semin.* 2E, 388 pp.

LUKOKI-LUYEYE. 1975. Distinction entre *Vigna radiata* et *Vigna mungo*. [Distinction between *Vigna radiata* and *Vigna mungo*]. *Bull. Rech. Agron., (Gembloux)*, *10*, 372–373.

MACKENZIE, D. R., HO, L., LIU, T. D., WU, H. B. F. and OYER, E. B. 1975. Photoperiodism of mung bean and four related species. *Hortscience*, *10*, 486–487.

MACKENZIE, D. R. and SHANMUGASUNDARAM, S. 1973. The AVRDC grain legume improvement programs. *Proc. 1st Int. Inst. Trop. Agric., Grain legume improvement workshop*. pp. 102–104. Ibadan, Nigeria, Int. Inst. Trop. Agric., 325 pp.

MADRID, M. T. (JR.) and VEGA, M. R. 1971. Duration of weed control and weed competition and the effect on yield: (i) Mung bean (*Phaseolus aureus* L.). *Philipp. Agric.*, *55*, 216–220.

MAMORIA, C. B. 1973. *Agricultural problems of India*. 7th ed. pp. 329–330. Allahabad: Kitab Mahal, 946 pp.

MATHUR, H. G. 1958. Marketing of pulses in India. *Nagapur, Agric. Marktg. Advis., Gov. India, Agric. Marktg. Ser.* AMA 102, 182 pp.

MATHUR, R. S., BANERJEE, A. K. and BAJPAI, G. K. 1965. The effect of vector control on yellow mosaic incidence on moong (mung bean) in India. *Plant Dis. Rep.*, *49*, 166–167.

MAZUNDAR, V. F., VASAVADA, C. R. and JOSHI, S. N. 1969. D45-6 a new protein rich moong variety. *Indian Farming*, *19* (4), 35–36.

MEKSONGSEE, L. A. and SWATDITAT, A. 1974. Development of a high-protein snack food (beanstalk). *Thai. J. Agric. Sci.*, *7*, 93–101.

MEW, I. C., WANG, T. C. and MEW, T. W. 1975. Inoculum production and evaluation of mung bean varieties for resistance to *Cercospora canescens*. *Plant Dis. Rep.*, *59*, 397–401.

MISRA, R. C. and SAHU, R. C. 1970. Embryology and seed structure in green gram (*Phaseolus aureus* Roxb.). *Indian J. Agric., Sci., 40*, 216–222.

MOODY, K. 1973. Weed control in tropical grain legumes. *Proc. 1st Int. Inst. Trop. Agric., Grain legume improvement workshop*. pp. 162–183. Ibadan, Nigeria, Int. Inst. Trop. Agric., 325 pp.

NARAYAN, T. R. and DABADGHAO, P. M. 1972. Greengram. *Forage crops of India*. pp. 75–76. New Delhi: Indian Counc. Agric. Res., 373 pp.

OTOUL, E., MARÉCHAL, R., DARDENNE, G. and DESMEDT, F. 1975. Des dipeptides soufres differencient nettement *Vigna radiata* de *Vigna mungo*. [Clearly differentiated sulphur-dipeptides of *Vigna radiata* and *Vigna mungo*]. *Phytochemistry, 14*, 173–179.

PABLO, S. J. and PANGGA, G. A. 1971. Granular systemic insecticides in the control of pests affecting mungo bean. *Philipp. J. Plant Ind., 36*, 21–28.

PADLAN, R. R. 1955. Germinated mungo meal as a protein supplement in a growing ration for chicks. *Philipp. Agric., 38*, 595–603.

PALO, A. V. 1972. Production of food legumes in the Philippines with special reference to leguminous vegetables. Symp. food legumes. pp. 189–195. *Tokyo, Jpn., Minist. Agric. & For., Trop. Agric. Res. Cent., Trop. Agric. Res. Ser.* 6, 253 pp.

PARPIA, H. A. B. 1975. Utilization problems in food legumes. *Nutritional improvement of food legumes by breeding*. (Milner, M. ed.). pp. 281–295. New York/London: John Wiley and Sons, 399 pp.

PINGALE, S. V., KADKOL, S. B. and SWAMINATHAN, M. 1956. Effect of insect infestation on stored Bengal gram (*Cicer arietinum* L.) and green gram (*Phaseolus radiatus* L.). *Food Sci., 5*, 211–213.

PIPER, C. V. and MORSE, W. J. 1914. Five oriental species of beans. pp. 16–25. *US. Dep. Agric. Bull.* 119, 32 pp.

POEHLMAN, J. M. (ed.). 1974. Mung beans. *Guide for field crops in the tropics and subtropics*. (Litzenberger, C. ed.). pp. 138–144. Washington, Off. Agric. Tech. Assist. Bur., Agency Int. Dev., 321 pp.

POEHLMAN, J. M., SECHLER, D. T., SWINDELL, R. E. and SITTIYOS, P. 1976. Performance of the fourth international mung bean nursery. *Univ. Missouri, Colombia, Agric. Exp. Stn. Spec. Rep.* 191, 36 pp.

POEHLMAN, J. M., SECHLER, D. J., WATT, E. E., SWINDELL, R. E. and AGGARWAL, V. D. 1975. Performance of the third international mungbean nursery. *Univ. Missouri, Colombia, Agric. Exp. Stn. Spec. Rep.* 180, 38 pp.

Mung bean

POEHLMAN, J. M., SECHLER, D. T., YOHE, J. M., WATT, E. E., SWINDELL, R. E. and BASHANDI MOHEB, M. H. 1973. Performance of the first international mungbean nursery. *Univ. Missouri, Colombia, Agric. Exp. Stn. Spec. Rep.* 158, 20 pp.

POEHLMAN, J. M., SECHLER, D. T., YOHE, J. M., WATT, E. E., SWINDELL, R. E. and BENHAM, E. 1974. Performance of the second international mungbean nursery. *Univ. Missouri, Colombia, Agric. Exp. Stn. Spec. Rep.* 171, 31 pp.

POEHLMAN, J. M. and YU-JEAN, F. F. 1972. Bibliography of mung bean research. *Univ. Missouri, Colombia, Agric. Exp. Stn. Spec. Rep.* 148, 15 pp.

RACHIE, K. O. 1973. Relative agronomic merits of various food legumes for the lowland tropics. Potentials of field beans and other food legumes in Latin America. (Wall, D. ed.). pp. 123–139. *Cali, Colombia, Cent. Int. Agric. Trop., CIAT Ser. Semin.* 2E, 388 pp.

RACHIE, K. O. and ROBERTS, L. M. 1974. Grain legumes of the lowland tropics. *Adv. Agron., 26*, 1–132.

RAGHAVENDRARAO, M. R. and SREENIVASAYA, M. 1946. Aspargine from Indian pulses. *Curr. Sci., 15*, 25–26.

RAYCHAUDHURI, S. P. 1968. Diseases of pulses. *Indian Farming, 17* (11), 39–43.

RETINAHM, A., RETHINAM, P., SANKARAN, S., RAJAN, A. V. and MORACHAN, Y. B. 1974. Chemical weed control in green gram (*Phaseolus aureus* Roxb.). *Madras Agric. J., 61*, 789–791.

SEN GUPTA, J. C. and MUKHERJI, D. K. 1949. Studies on the physiology of growth and development of mung (*Phaseolus aureus* Roxb.): (a) Effect of the time of sowing; (b) Vernalization and photoperiodism. *Indian J. Agric. Sci., 19*, 207–254.

SEN GUPTA, P. K. 1974. Diseases of major pulse crops in India. *PANS, 20*, 409–415.

SENEWIRATNE, S. T. and APPADURAI, R. R. 1966. Green gram. *Field crops of Ceylon*. pp. 166–169. Colombo: Lake House Investments Ltd, 376 pp.

SEVILLA EUSEBIO, J., GONZALES, R. R., EUSEBIO, J. A. and ALCANTARA, P. F. 1968. Studies on Philippine leguminous seeds as protein foods: (ii) Effect of heat on the biological value of munggo, paajab, tapilan and kadyos beans. *Philipp. Agric., 52*, 218–232.

SEVILLA EUSEBIO, J., MAMARIL, J. C., EUSEBIO, J. A. and GONZALES, R. R. 1968. Studies on Philippine leguminous seeds as protein foods: (i) Evaluation of protein quality in some local beans based on their amino acid patterns. *Philipp. Agric.*, *52*, 211–217.

SHIVASHANKAR, G., RAJENDRA, B. R., VIJAYAKUMAR, S. and SREEKANTARADHYA, R. 1974. Variability for cooking characteristics in a collection of green gram (*Phaseolus aureus* Roxb.). *J. Food Technol.*, *11*, 232–233.

SIEGEL, A. and FAWCETT, B. 1976. *Food legume processing and utilization* (*with special emphasis on application in developing countries*). Ottawa, Int. Dev. Res. Cent., IDRC–TS1, 88 pp.

SINGH, H. B., JOSHI, B. S. and THOMAS, T. A. 1970. The *Phaseolus* group, Green gram, *Phaseolus aureus* Roxb. *Pulse crops of India*. (Kachroo, P. and Arif, M. eds.). pp. 136–164. New Delhi: Indian Counc. Agric. Res., 334 pp.

SINGH, K. B. and MALHOTRA, R. S. 1970. Estimates of genetic and environmental variability in mung (*Phaseolus aureus* Roxb.). *Madras Agric. J.*, *57*, 155–159.

SINGH, L., SHARMA, D. and TOMAR, G. S. 1972. Jawahar–45 a versatile mung variety for kharif season. *Indian Farming*, *22* (1), 29.

SORIA, J. A. and QUEBRAL, F. C. 1973. Occurrence and development of powdery mildew on mongo. *Philipp. Agric.*, *57*, 153–177.

SWAMINATHAN, M. S. and JAIN, H. K. 1975. Food legumes in Indian agriculture. *Nutritional improvement of food legumes by breeding*. (Milner, M. ed.). pp. 69–82. New York/London: John Wiley and Sons, 399 pp.

SWATDITAT, A. and MEKSONGSEE, L. A. 1974. Development of high protein noodle products. *Thai. J. Agric. Sci.*, *7*, 175–187.

THANGARAJ, M. and SOUNDARAPANDIAN, G. 1974. Effect of pre-emergence herbicides on the control of weed in green gram (*Phaseolus aureus* Roxb.). *Madras Agric. J.*, *61*, 787–788.

THOMPSON, L. U. 1977. Preparation and evaluation of mung bean protein isolates. *J. Food Sci.*, *42*, 202–206.

TOMLINSON, J. and PLAXICO, J. S. 1962. An economic analysis of mung beans as a crop for sandy soils of central Oklahoma. *Okla. State Univ. Bull.* B–595, 31 pp.

Mung bean

Tourneur, M. 1958. L'ambérique et le mungo ne sont pas des *Phaseolus*. [Urid and mung are not *Phaseolus*]. *Riz Rizic.*, *4*, 131–148.

Venkataraman, L. V. and Jaya, T. V. 1976. Influence of germinated green gram and chick pea on growth of broilers. *J. Food Sci. Technol.*, *13*, 13–16.

Venugopal, K. and Morachan, Y. B. 1974. Response of green gram to seasons and graded doses of nitrogen and phosphorus fertilizers. *Madras Agric. J.*, *61*, 457–460.

Venugopal, K. and Morachan, Y. B. 1974. Studies on the uptake of nitrogen and phosphorus in two green gram varieties. *Madras Agric. J.*, *61*, 461–466.

Verdcourt, B. 1970. Studies in the *Leguminosae–Papilionoideae* for the 'Flora of tropical East Africa': IV. *Kew Bull.*, *24*, 507–569.

Wanbhen, A., Wisuthi, A. and Knapp, F. W. 1974. Stored grain insect studies: (i) Susceptibility of the bean and rice weevil to three insecticides; (ii) Resistance of mung bean and sorghum seed to laboratory infestations of bean and rice weevil. *Thai. J. Agric. Sci.*, *7*, 63–70.

Watt, E. E. and Maréchal, R. 1977. The difference between mung and urid. *Trop. Grain Legume Bull.*, (7), pp. 31–33.

Wen, D. Y. 1937. Studies on the production of soybean and mung bean sprouts. *Lingnan Sci. J.*, *16*, 627-629.

Westphal, E. 1974. *Phaseolus radiatus* L. Pulses in Ethiopia, their taxonomy and agricultural significance. pp. 151–159. *Wageningen*, *PUDOC*, *Cent. Agric. Publ. Doc.*, *Agric. Res. Rep.* 815, 278 pp.

Wrenshall, C. L., Meksongsee, L. A., Swatditat, A. and Udomsakdi, B. 1974. Mung bean flour preparation. *Thai. J. Agric. Sci.*, *7*, 37–48.

Yegna Narayan Aiyer, A. K. 1966. Greengram (*Phaseolus aureus* Roxb.). *Field crops of India.* 6th ed. pp. 120–122. Bangalore City: Bangalore Printing and Publishing Co Ltd, 564 pp.

Yohe, J. M. and Poehlman, J. M. 1972. Genetic variability in the mungbean L. (*Vigna radiata* (L.) Wilczek.). *Crop Sci.*, *12*, 461–464.

Common name	**PEA.**
Botanical name	*Pisum sativum* L.
Family	Leguminosae.
Other names	Afun tarbuti, Afuna (Is.); Ain-ater (Eth.); Akkerwt (Dut.); Alverja (Mex., Peru); Alverjón (Mex.); Amashaza (Ug.); Arveja (Sp.); Atari[1], Ater(o), Attur (Eth.); Basilla (Sud.); Batagadle, Batani (Ind.); Bazilla, Bazzelah[1] (Ar.); Bezelye[1] (Turk.); Bolakadala[1] (Sri La.); Borsó, (Hun.); Bsella (Egy.); Chicharo (Mex.); Citzaro (Philipp.); Danguleh (Eth.); Doperwt[1] (Dut.); Endō (Jpn.); Erbse (Ger.); Ertjie (S. Afr.); Ervilha (Port.); Erwt (en) (Dut.); Gishishato[1] (Eth.); Guisante (Sp.); Intongwe (Zam.); Kabuli mater[1] (Ind.); Kachang puteh[1] (Malays.); Kapri (Indon.); Kapucijners (Dut.); Katjang ertjis (Malays.); Mar, Martar[1] (Ind.); Mashâza (Rua.); Matar[1], Matar (mah), Matar-mar[1] (Ind.); Mauritius pea (Barb.); Njegele (Rua.); Obushaza (Ug); Pairu (Sri La.); Patani[1] (Tam.); Patanlu (Ind.); Pisello (It.); Pisson (Gr.); Pizelli (Cyp.); Pois (Fr.); Polong (Indon.); Saaterbse (Ger.); Sadaw-pè (Burm.); Sawawa (Malawi); Takarmany borsó (Hun.); Tukur-ater[1] (Eth.); Vilaiti matar[1], Watani[1] (Ind.).

Botany

An annual, climbing, herbaceous plant, showing very considerable variation in form and habit. Dwarf, medium and tall types exist, in which the stems are respectively, 6–36 in. (15–90 cm), 36–60 in. (90–150 cm), and 60–120 in. (150–300 cm) long. The tap-root is well developed and can grow to a depth of 40–48 in. (100–120 cm), it has many slender laterals, which form a circle of 20–30 in. (50–75 cm) diameter around the plant. Infrequently globular

[1] Usually used to denote garden peas.

nodules are present on the tap-roots of young plants. The stems are slender, angular, or round and hollow. Branching is very variable, some cultivars producing laterals freely, others rarely. Leaves are alternate, pinnate, with 1–3 pairs of ovate or elliptic leaflets ending in one or more tendrils. Normally these are approximately 0.6–2.2 x 0.5 in. (1.5–5.5 x 1.2 cm) in size; green and glaucous, but non-glaucous, yellow and variegated forms exist. At the base of the leaves are large leaf-like stipules, sometimes 4 in. (10 cm) in length and occasionally purple at the base. The infloresence is axillary, solitary, or in 2–3 flowering racemes. The flowers are large, butterfly-like, usually white, but may be pink or purple, and the flowers are normally self-pollinated. The pod is typical of the legumes, varying from 1.0–5.0 in. (2.5–12.5 cm) in length, and 0.5–1.0 in. (1.2–2.5 cm) wide, flat or cylindrical, short-stalked, straight or curved, and beaked. It can vary from yellowish-green to dark-green in colour, and when immature is fleshy and waxy. In most cultivars the pod is lined with a parchment-like membrane, the endocarp; this is absent in the edible-podded peas. The pod is usually dehiscent by both sutures and contains 2–10 seeds, which may be globose or globose-angular, smooth, or wrinkled, and of varying colours.

There are very many different types of peas and a formal classification is impracticable. A distinction may be made between the garden pea, var. *sativum* and the field pea, var. *arvense*.

The field pea, var. *arvense*, is the more hardy, and is frequently grown as a sprawling plant on a field-scale for its dried seeds. It usually has reddish-purple flowers and small pods and seeds. The garden pea, var. *sativum*, is less hardy and is grown for the tender immature green seeds, or for the green pods. The flowers are white and it has larger pods and seeds. Some authorities subdivide the garden pea into cv. *macrocarpon*, the edible-podded or sugar peas, which lack the parchment-like endocarp, and cv. *humile*, the ordinary garden pea. In addition, there is the Abyssinian pea, cv. *abyssinicum*, which is found in northern Ethiopia. It differs markedly from the field pea, in that it has leaves with one pair of leaflets, very small reddish-purple flowers and globose, glossy, sweeter seeds, with a black hilum.

There are many regional types and strains of peas and a great many cultivars have been produced in the temperate regions, particularly in Europe, N. America and the USSR. Not all of these have proved successful when grown in the tropics and subtropics and many experimental stations there have made selections from the more promising introductions and produced local strains adapted to indigenous methods of cultivation and utilization. In view of the very large number of cultivars and strains of peas grown it is not proposed to list or describe any of them in this digest.

Origin and distribution

Peas probably originated in south-western Asia and have been cultivated in Europe since the Bronze Age. They have spread to the temperate zones throughout the world and are grown as a cool-season crop in the subtropics, and at higher altitudes in the tropics.

Cultivation conditions

Temperature—a cool, but not excessively cold climate is required. Minimum temperature for seed germination is about 40°F (4°C), maximum 75°F (24°C). For optimum yields average temperatures should range from 55° to 66°F (13–18°C). When grown in the tropics peas do best in areas with a five-month cool-growing season. The plant can tolerate frost in the vegetative stage, but frost at flowering can cause heavy pod losses, and at pod-set is liable to produce deformed and discoloured seed. Temperatures above 80°F (27°C) shorten the growing period and adversely affect pollination.

Rainfall—an evenly distributed rainfall is required, preferably about 32–40 in./a (800–1 000 mm/a), although the crop is grown successfully in Australia in areas where the rainfall is as low as 16 in. (400 mm) a year, provided that the soils are deep and moisture retentive. In areas of low rainfall peas can be grown successfully under irrigation, requiring approximately 8–10 ac./in. (8.18–1028m^3) of water during their growth cycle. Under temperate conditions, maximum yields are reported to be obtained when the soil-moisture content is kept at 60 per cent of field-capacity during the period from emergence to just before flowering and at least 90 per cent during flowering.

Soil—peas can be cultivated over a wide range of soil types, provided that the drainage is good, as they cannot stand waterlogging. The crop does best on loams to clay loams, or sandy loams overlying clay. On light, sandy soils which do not hold moisture, yields tend to be reduced. The pH should range between 5.5 and 6.5, although some cultivars can tolerate pH 6.9–7.5. Peas are grown in the UK, for example, on chalky or limestone soils, although lime-induced chlorosis may occur through trace element deficiencies. Generally peas respond to fertilizers much less than most other legume crops.

Pea

Response to nitrogen is rare, where phosphorus and potassium reserves in the soil are adequate, the addition of nitrogen may in fact decrease yields. Generally the crop responds to applications of potash more frequently than to phosphate and in soils deficient in potassium the application of 220 lb/ac (250 kg/ha) of a 0:1:2 NPK fertilizer has been recommended. For optimum results the fertilizer should be applied about 2 in. (5 cm) to the side and 1 in. (2.5 cm) below the seed. Molybdenum is essential for nodulation and can affect the total solids content of the seed. In cases of deficiency 2 oz/ac (141 g/ha) of ammonium molybdate usually corrects the deficiency. Peas grown on alkaline soils may suffer from manganese deficiency, which shows itself as a dark discolouration in the centre of the seed, (marsh spot). The application of 14 lb/ac (16 kg/ha) of manganese sulphate has been suggested to correct this deficiency.

Altitude—in the tropics peas seldom yield well below about 4 000 ft (1 200 m). In Uganda the crop thrives best at an altitude of 6 000 ft (1 800 m) or more, and in Kenya, optimum yields are obtained at 7 000–9 000 ft (2 100–2 700 m).

Day-length—reference has been made in the literature to the cultivation of short-day adapted cultivars in Africa.

Planting procedure

Material—the use of disease-free, fungicide treated seed is recommended. In certain tropical areas the inoculation of field pea seed has been found to be beneficial, particularly where the peas are being grown on land which has not been previously sown to a leguminous crop, or where the soil is of very low fertility. Germination is hypogeal and usually between 80 and 95 per cent. Under normal storage conditions pea seed will retain its viability for 1–2 years.

Method—peas should not be grown on the same land more frequently than every 3 to 5 years because of problems of soil-borne diseases. Because of their soil-improving properties, peas frequently precede a cereal crop such as wheat or barley, and in the tropics are often the first crop planted after a fallow period. They are sensitive to poor soil conditions and for optimum yields require a fine soil tilth over a firm weed-free seed-bed. The seeds are usually drilled 2–3 in. deep (5–7.5 cm) in rows, although in certain areas, eg India, field peas are sometimes grown mixed with cereals, when the seeds are broadcast and covered with soil by harrowing. Weed competition may seriously reduce yields and frequent hoeing is often necessary, particularly during the early growth stages. With large-scale production, herbicides are used. Dinoseb ammonium is widely used when peas are grown as a dry grain legume crop. Dinoseb amine was used formerly for garden peas, but in recent years has been replaced by the substituted triazine herbicides.

Field-spacing—field peas are frequently sown in rows 12–24 in. (30–60 cm) apart with 6–7 in. (15–17.5 cm) between the plants. Garden peas are often sown about 2–3 in. (5–7.5 cm) apart. In New Zealand garden peas are usually drilled in rows 7 in. (17.5 cm) apart, with 15–20 in. (37.5–50 cm) between the rows when the crop is harvested mechanically, and 21–30 in. (52.5–75 cm) for hand-picking. In India, in spacing trials with field peas, the maximum yield was obtained when 3 x 3 in. (7.5 x 7.5 cm) square grid was used. In trials in the USA, optimum spacing for garden peas for processing was found to be 2–4 in. (5–10 cm) apart, with 4–8 in. (10–20 cm) between the rows.

Seed-rate—this is very dependent upon the size of the seed and the level of crop management. In many countries the average seed-rate for field peas ranges between 60–90 lb/ac (67–100 kg/ha) for small-seeded cultivars and from 120 to 150 lb/ac (130–170 kg/ha) for large-seeded types. The average seed-rate for garden peas for processing usually ranges between 200 and 250 lb/ac (225–280 kg/ha).

Pests and diseases

Pests—peas are subject to several insect pests in the field and during storage. The pea aphid, *Acyrthosiphum pisum*, (= *Macrosiphum pisum*) is widespread throughout N. America, Europe, Australia, and parts of Africa and Asia. It is sporadically a serious pest and is of considerable economic importance, not only because it can cause stunting of growth, but also because it is the vector of more than 25 virus diseases. Certain cultivars show a degree of resistance to aphid attack but effective control is obtained usually by spraying with organo-phosphorus insecticides, nicotine sulphate or rotenone. The pea moth, *Laspeyresia nigricana*, attacks peas grown for processing in Europe and N. America, and is one of the most serious pests of field and garden peas in the UK. The larvae feed on the developing seeds and infection with fungal diseases often follows, frequently spoiling seeds that have been left undamaged. Early planting, rotation and deep ploughing are of value in controlling this pest. In cases of infestation the following insecticides are reported to give good control: azinphos-methyl, carbaryl, fenitrothion or tetrachlorvinphos.

Pea

In the UK, pea and bean weevils, *Sitona* spp., and the pea midge, *Contarinia pisi*, can also be troublesome. The pea cyst eelworm, *Heterodera goettingiana*, is also often a serious pest and difficult to control. Infested plants make poor growth and turn yellow prematurely, sometimes dying before the pods are filled. Care should be taken to avoid overcropping with leguminous crops since once a field is infested it may be 10–12 years before a good crop of peas can be grown on the land.

One of the most widespread and injurious pests of peas is the pea weevil, *Bruchus pisorum*. It is nearly world-wide in distribution and has been reported as doing serious damage to field and culinary peas in Canada, USA, S. Africa, USSR, India, Japan and Australia. The larvae feed on the seed and infestations can average from 30 to 70 per cent if control measures are not applied. Dusting with rotenone during the early bloom period is probably the safest and most effective control measure, although DDT and parathion are frequently used. In Australia the red-legged earthmite, *Halotydeus destructor*, can be troublesome on field peas in some areas, particularly during seedling emergence. Spraying with phosmet, or treatment of the seed with a systemic insecticide, eg dimethoate, are recommended as control measures. Climbing cutworms, *Heliothis punctigera*, can also cause severe damage to field pea crops in Australia. The main damage is caused by the caterpillar entering the pods and eating the seed. Spraying with DDT or endosulphan is reported to give satisfactory control. In India the leaf-eating caterpillar, *Spodoptera exigua* (= *Laphygma exigua*), is of economic importance, sometimes defoliating plants. The larvae of the pod borer, *Heliothis armigera*, also can be troublesome, eating both leaves and seeds. Dusting with BHC or DDT is reported to give reasonable control of both these pests. In addition, peas are susceptible to attack during storage from a number of insects, including bruchids, *Callosobruchus* spp.

Diseases—peas are subject to a number of diseases, which because of their serious effect on yield and market value, are a major factor of crop production.

Ascochyta blight, leaf or pod spot, caused by the *Ascochyta* complex, is of major economic importance, occurring in most areas where peas are grown. *Ascochyta pinodella* causes purplish-black lesions on the stem. *A. pisi* produces large, sunken, brownish lesions on the leaves and pods, sometimes penetrating the pod-wall and causing brown stains on the peas. *Mycosphaerella pinodes* produces numerous, tiny, irregular shaped purplish-black spots on all parts of the plant. All the organisms of the *Ascochyta* complex are seed-borne and can survive in the soil from season to season. Reasonable control of ascochyta blight can be achieved by the use of disease-free seed, efficient sanitation in the field and at least a three-year crop rotation. Recently control has been achieved by treating the seed with benomyl, but the cost of the treatment makes it unlikely that it could be used widely.

Root rot, *Aphanomyces euteiches*, a soil-borne fungal disease, affects crops in the UK, northern France, Scandanavia, parts of the USA, the nonchernozem zone of the USSR and Tasmania. It is of major economic importance in the USA, particularly in Wisconsin, Minnesota and New York State. Incidence can be very severe in wet seasons at soil temperatures of 71° to 82°F (22–28°C) and in soils with a high water-retaining capacity. As yet no resistance has been incorporated into commercial pea cultivars and there are no effective fungicides to control the disease economically in the field. The only control is the avoidance of land known to be infested with the pathogen, since crop rotation of even 10 years is not always an effective control of this disease. Other root rots of peas, of somewhat lesser importance, are *Fusarium solani*, f.sp. *pisi*, *Phythium ultimum*, and *Rhizoctonia solani*. All are more severe during abnormally wet seasons and on poorly drained soils. Peas grown on impoverished soils are also more subject to attack. Crop rotation is usually recommended to control root rots.

Pea wilt, *Fusarium oxysporum* f. *pisi* race 1, is another widespread soil-borne disease which can be troublesome. It causes yellowing of the leaves and stunted growth, and may be controlled by growing resistant cultivars; nowadays a large number of resistant strains, particularly of processing peas, are available commercially. In the USA in the past 30 years a disease known as near-wilt caused by the fungus, *Fusarium oxysporum* f. *pisi* race 2, has been observed. It is closely related to pea wilt and also causes yellowing of the leaves and stunted growth, but usually attacks the plant at a later stage of growth so that crop losses are not normally as great.

Downy mildew caused by the fungus *Peronospora pisi* is widely distributed and can be troublesome in certain seasons in the UK, New Zealand, Australia and parts of the western USA. It is favoured by relatively high humidity and temperatures of between 50° and 70°F (10–21°C), but is checked by temperatures of 80°F (27°C) and dry weather. The use of resistant cultivars is the most effective means of controlling this disease, which is characterized by a white downy, or cottony growth, on all parts of the plant, including the inside of the pods. Although not usually serious in the UK, powdery mildew, *Erysiphe polygoni*, sometimes causes considerable damage to pea crops in India, Australia and the USA. It is characterized by the formation of a white dust, or powder, on the leaves and less frequently on the stems and pods. The leaves become yellow, dwarfed and sometimes malformed, and in severe cases of infection there are small brown spots or streaks on the pods. Dusting the plants with sulphur at regular intervals is reported to give reasonable control, especially if it is combined with crop rotation and good field sanitation.

Bacterial blight, *Pseudomonas pisi*, is a seed-borne disease which can cause serious losses in certain seasons in N. America, E. and S. Africa, Australia, New Zealand, Argentina, Uruguay, Colombia, Nepal, Japan, S. and C. Europe and the USSR. It affects all parts of the plant above the ground causing water-soaked lesions, and if infection starts from the seed, or if the plants are not more than 3 in. (7.5 cm) high, they can die without producing a crop. Later infection reduces the yield of pods considerably. It is most severe when humidity is high, and on occasions can affect 25 to 30 per cent of the crop. Control is difficult and there is need to develop resistant cultivars.

Peas are subject to a number of virus diseases, the most important are common pea mosaic, pea enation mosaic, pea stunt, yellow bean mosaic and pea streak. Each is caused by a distinct virus and crop losses are serious in many areas, particularly where there are large aphid populations. In general field and processing peas are more resistant to virus diseases than horticultural crops. The cultivation of resistant cultivars is the most satisfactory means of controlling virus diseases.

Growth period
The time for a crop to reach maturity depends upon the cultivar, climatic conditions and time of sowing, and on average can vary from 90 to 160 days for the production of field peas, and from 56 to 84 days for the production of green peas or the pods. If grown for fodder the plants can be cut just as they are forming pods, usually about 75 days from sowing.

Harvesting and handling
Dry peas—are harvested when the leaves begin to yellow and the lower pods begin to wrinkle and the peas harden. If harvesting is delayed once the crop has reached maturity the peas lose colour and become brittle, which reduces

their market value. Excessive losses, due to shelling, often can be avoided by cutting the crop in the early morning when the pods are more turgid. For large-scale production specially designed pea-cutters are used which lift the haulms off the ground and place them in windrows, or on tripods or racks, to dry. When the crop has dried sufficiently, usually in 7 to 10 days, combine harvesters are brought into the field and the haulms are hand-forked into them for threshing. Threshing requires considerable care otherwise splitting losses can be appreciable. In order to minimize these the cylinder speed should be reduced to between 500 and 600 r/m, or even less. Other adjustments, such as fitting a breast plate, or the removal of alternate beater bars from the cylinder, may also be necessary. In the UK treatment of the dry pea crop immediately before harvesting with desiccants, such as sulphuric acid, or Diquat, followed by immediate threshing has recently become popular. When dried peas are grown as a grain legume crop in the tropics the crop is frequently cut by hand and threshed by hand flails.

After threshing the peas are sorted and cleaned of extraneous matter, usually by passing them through a series of screens, or by the use of electronic sorters. Dried peas can be stored successfully in bags or in bulk, provided that they have been adequately dried and cooled. For prolonged storage the moisture content should be between 10 and 16 per cent. Artificial drying is sometimes necessary and should be carried out in two stages, with an interval of at least 48 hours between each, if the initial moisture content is above 22 per cent. If the peas are to be used for seed the drying temperature should not exceed 100–110°F (38–43°C) if the moisture content is below 24 per cent, or 95–100°F (35–38°C) if above 24 per cent. Peas for edible purposes can be dried at slightly higher temperatures, preferably between 110° and 120°F (43–49°C); at temperatures above 135°F (57°C) the texture toughens and splitting is liable to occur. Peas are susceptible to insect infestation, particularly bruchids, and the treatment, before filling, of all storage structures and containers, with an insecticide such as malathion, or fumigation with methyl bromide is normally recommended.

Green peas—are harvested at the immature stage when the pods are well filled, but the peas are sweet and soft. For marketing in the pod the crop is picked by hand, either by going over the vines once and stripping off all the pods into sacks or nets, or by selective picking, which involves going over the vines several times; sometimes as many as 7–8 pickings, spread over 5–7 weeks are made. In most agriculturally advanced countries peas for processing are grown under contract to the processing companies, who have very stringent requirements as regards harvesting dates and maturity. The prime object in harvesting processing peas is to obtain the maximum yield of peas at the optimum stage of maturity for processing and to deliver them to the processing plant with the minimum degree of deterioration, thus, speed, cleanliness, and good organization are essential. Instruments, such as the tenderometer, maturometer and texturometer, are used to determine the optimum stage at which the crop should be harvested. Some processors use a system of accumulated heat units to determine the harvesting date of pea crops in an attempt to regulate supplies to the factory. The number of degree hours above 40°F (4°C) required to bring a specific cultivar to maturity is calculated, (usually between 30 000 and 35 000 hour/degree units). The method is open to criticism since temperature is not the only factor affecting growth, and photoperiod should also be taken into consideration. Processing peas are either harvested with a pea-cutter and then fed into a stationary pea-viner, or harvested and shelled in one operation in mobile pea-viners. In the viners the pods are broken by a set of beaters and the green peas fall through the screens and sieves and are collected on shallow trays for transporting to the processing unit.

Once harvested green peas quickly toughen and deteriorate in flavour, because of sugar loss. They keep better in the pod and once shelled prompt cooling to about 32°F (0°C) is advisable. This can be achieved by the use of iced water or crushed ice, or a combination of both, or refrigerated air. Peas should be stored at this temperature and a relative humidity of 90–95 per cent, preferably 95 per cent, as they deteriorate rapidly. Under these conditions they can be stored in a reasonably sound condition for 7 to 14 days. In an atmosphere of 5–7 per cent carbon dioxide they remain in a reasonable condition for about 20 days at 32°F (0°C) and 13–15 days at 40°F (4°C).

Primary product

Seed—which shows very wide variation in shape, size, colour and composition. Pea seed may be green, yellowish-green, blue-green, whitish-grey, brown, or mottled, rounded, or flattened in shape, and may vary in size from approximately 0.14 to 0.2 in. (3.5–5.0 mm). The hilum is very distinct and can be colourless, brown or yellow. The surface of the pea may be smooth, wrinkled, indented or dimpled. One hundred seeds weigh approximately 0.5 to 0.8 oz. (15–25 g).

The main types of peas grown commercially are: the Dun and Maple peas, grown for animal feed and the Blue and Marrowfat peas grown principally for human consumption. Dun peas are round and a greyish-green colour, while Maple peas are round and usually a speckled yellowish-brown. Blue peas are smooth, round, hard and dark-green in colour; Alaska, which was formerly the most widely grown canning pea in the USA, is a typical example. Marrowfat peas are usually square in shape, wrinkled or dimpled and pale-green in colour; Harrison's Glory, formerly widely grown in the UK, is a typical example.

Yield

There is considerable variation in the yield of peas, according to environmental conditions, cultivar and the standard of crop husbandry.

Dried peas—a yield of about 1 780 lb/ac (2 000 kg/ha) is generally regarded as satisfactory. In 1973 the world average yield was 1 030 lb/ac (1 160 kg/ha) and the following average yields were recorded in various countries:

Uganda 360 lb/ac (400 kg/ha), Ethiopia 840 lb/ac (940 kg/ha), Canada 1 440 lb/ac (1 620 kg/ha), Mexico 510 lb/ac (570 kg/ha), USA 1 500 lb/ac (1 680 kg/ha), Argentina 860 lb/ac (960 kg/ha), Uruguay 1 870 lb/ac (2 110 kg/ha), India 530 lb/ac (600 kg/ha), UK 2 910 lb/ac (3 270 kg/ha), USSR 1 360 lb/ac (1 530 kg/ha), Australia 1 040 lb/ac (1 170 kg/ha), New Zealand 2 200 lb/ac (2 470 kg/ha).

Green peas—yields of green peas are also very variable and in addition to the factors listed above, the yield is also dependent upon the utilization of the crop. The yield of green peas grown for freezing or dehydration is less than that for canning. A good average yield of green peas in the pod is about 6 000 lb/ac (6 700 kg/ha) and that of shelled green peas 4 000 lb/ac (4 500 kg/ha), although in countries such as the UK and Belgium yields of green peas have averaged about 9 000 lb/ac (10 000 kg/ha) in recent years, compared with average yields of 2 540 lb/ac (2 850 kg/ha) in India, 2 460 lb/ac (2 760 kg/ha) in Canada, and 610 lb/ac (680 kg/ha) in Bolivia.

Main use

Peas are a nutritious foodstuff. When fully mature and dry they are a valuable food legume, often being ground into a flour and used extensively in the manufacture of soups. In India they are consumed split as dhal, or roasted, or parched. Large quantities of dry Marrowfat and Alaska type peas are canned and eaten as a vegetable. Fresh green peas are almost universally accepted as a nutritious vegetable. Green peas are in fact the leading processed vegetable; very large quantities being grown in countries such as the USA and the UK, for canning, freezing or dehydrating.

Subsidiary uses

Dry, fully mature field peas are used extensively, especially in the USSR and parts of Europe, for animal feed. Cracked, shrivelled and peas unsuitable for human consumption are used for this purpose in the USA. Peas are highly palatable to livestock and when their price is similar to that of cereals they can be used as an economic grain substitute. They can be included in rations for poultry as the only source of protein and have been used to replace 10 per cent of the protein components in pig growing rations.

Attempts have recently been made to produce a bland protein-rich concentrate from peas suitable for the enrichment of biscuits, beverages and meat analogs.

Secondary and waste products

Certain cultivars, usually known as mange-tout, or sugar peas, are grown on a market garden scale for their green immature pods, which are eaten as a vegetable, similar to French beans. In parts of Asia, the young foliage of garden peas is also sometimes eaten as a vegetable.

In some countries, eg India, tall-growing, round white, or grey-seeded, field peas are grown for fodder, or for making silage. For every tonne of shelled peas produced for processing, approximately 4–5 tonnes of pods and haulms are obtained, which are usually dried or ensilaged and used as a livestock food, or as a soil improver. Pea waste obtained from the vining machines contains approximately 12 per cent digestible crude protein.

Special features

There is considerable variation in the composition of field peas depending upon the cultivar, the stage of maturity, the climatic conditions and the standard of crop husbandry. An approximate composition has been given as: moisture 11.0 per cent; protein 22.5 per cent; fat 1.8 per cent; total carbohydrate 62.1 per cent; fibre 5.5 per cent; ash 2.6 per cent; calcium 64 mg/100 g of edible portion; iron 4.8 mg/100 g; vitamin A 100 iu/100 g, thiamine 0.72 mg/100 g; riboflavin 0.15 mg/100 g; niacin 2.4 mg/100 g, ascorbic acid 4 mg/100 g. The average amino acid composition, (expressed as mg/gN), has been reported as: isoleucine 267; leucine 425; lysine 470; methionine 57; cystine 70; phenylalanine 287; tyrosine 171; threonine 254; valine 294; arginine 595; histidine 143; alanine 253; aspartic acid 685; glutamic acid 1 009; glycine 253; proline 244; serine 271. The protein content of peas, however, can range from about 19.7 to as high as 28 per cent, and it has been shown that when peas were fertilized with sulphur their methionine content increased from 1.29 to 2.18 g per 100 g of protein.

Garden peas are also a nutritious foodstuff, their approximate composition is as follows: moisture 72.1 per cent; fat 0.1 per cent; protein 7.2 per cent; total carbohydrate 19.8 per cent; ash 0.8 per cent; calcium 20.0 mg/100 g of edible portion; phosphorus 80.0 mg/100 g; iron 1.5 mg/100 g; carotene 0.30 mg/100 g; thiamine 0.32 mg/100 g; riboflavin 0.15 mg/100 g; nicotinic acid 2.5 mg/100 g; ascorbic acid 25 mg/100 g.

The major proportion of the carbohydrate content of peas is starch, which contributes to the alcohol-insoluble solids content (AIS). Sucrose is the most important sugar present, since it determines the sweetness of the peas. Other sugars whose presence has been identified are stachyose, glucose, fructose and galactose. As peas mature there is a steady increase in starch content and a decrease in sucrose. An analysis of fully developed green peas gave a sugar content of 5.9 per cent (dry basis) and a starch content of 32.9 per cent, while fully mature ripe peas, had a sugar content of 4.1 per cent and a starch content of 43.4 per cent. For processing as a vegetable, good quality peas should have a high sugar content and low starch content. In general wrinkled types of peas have high sugar contents with a predominantly amylose type of starch content. Smooth-seeded types generally have relatively lower sugar contents and the starch is predominantly amylopectin. Pea starch has a low viscosity and gel strength. The grains are kidney-shaped, or irregularly oval, up to 40 microns in diameter, and are extremely resistant to swelling in hot water. Peas also contain pectin, 2.5 per cent as calcium pectate.

The proteins of both field and garden peas consist principally of a globulin, legumin, with smaller amounts of another globulin, vicilin, and an albumin named legumelin. Methionine and cystine are the main limiting amino acids. The effect of processing, eg autoclaving or canning, on the nutritive value of pea protein is somewhat controversial and there are numerous published papers, with some rather conflicting results, but it is generally accepted that the nutritive value is impaired by heat treatment.

Pea

Peas contain a number of enzymes, several of which cause deterioration during processing and subsequent storage, due either to their effect on the nutritive value, or their role in the development of off-flavours and odours. The following enzymes have been reported to be present in peas and to be responsible for deterioration of their quality: catalase, peroxidase, ascorbic acid oxidase, chlorophyllase, lipase, lipoxidase, α-hydroxy-acid dehydrogenase and pyruvic decarboxylase.

Dry peas contain 1–2 per cent of a pale, golden-yellow oil which has the following constants: SG $^{15·5°C}$ 0.919; N_D $^{35°C}$ 1.4766; solidifying point −12°C; sap. val. 184–185; iod. val. 106; unsaponifable matter 1.0–1.5 per cent. The oil is reported to contain the glycerides of palmitic, oleic and possibly arachidic acids. It also contains meta-xylohydroquinone, which has shown some promise as a contraceptive.

Peas also contain appreciable quantities of saponin. Three odorous pyrazine derivatives, thought to be of major significance in the flavour of peas, have been isolated. They are 3-isopropyl-2-methoxypyrazine I, 3-S-butyl-2-methoxypyrazine II and 3-isobutyl-2-methoxypyrazine III. Field peas are reported to contain 80 units of haemagglutinizing activity and an antitrypsin activity of 8.4 units. Autoclaving for 5 minutes at 250°F (121°C) is claimed to reduce the antitrypsin activity significantly.

Processing

Peas may be canned, dehydrated, frozen or made into a bland protein-rich flour.

Canning—canned peas, both fresh (green) and dried (processed), are very popular processed vegetables and are often an important source of lysine in the diet. In countries, such as the UK and the USA, with large canning industries, the production of peas for canning is a highly organized industry and very sophisticated specialized machinery has been developed so that the process is almost completely automatic from cutting the vines in the field to the stacking of the final product in the cannery warehouse. In many other countries, eg India, peas are canned on a much smaller scale and most of the processing operations are carried out manually.

For the production of good quality canned green peas, the peas should be freshly harvested at the tender sweet stage, when their AIS content is between 11 and 16 per cent, preferably around 12.2 per cent. (ie at a maturometer reading of 250). The peas are first cleaned to remove pieces of pod, etc, washed in water, graded for size and inspected. They are then blanched in hot water which is kept above 170°F (77°C) at all times. The average time for blanching is 2 to 3 minutes at 210°F (99°C), or 3 to 4 minutes at 180°F (82°C), depending upon the maturity of the peas. The maximum time is 5 minutes, peas requiring a longer period are too mature for canning. The use of alkalizing agents, such as calcium hydroxide or calcium glutamate, to control the pH of the blanching water has been suggested as a means of retaining the bright-green colour of fresh peas. After blanching the peas are thoroughly washed in cold water to reduce their temperature to below 90°F (32°C), as quickly as possible. They are then quality graded by being passed through a brine bath in which any over-mature peas sink to the bottom of the bath. After a further final inspection the peas are filled into cans and covered with a hot solution of canning brine, which usually contains approximately 2.5 per cent of high-grade vacuum salt, practically free from traces of magnesium and calcium salts. The brine also contains a varying quantity of sugar, depending upon the type of pea canned, its maturity, and the preference of consumers. In the UK most canners add 2 to 4 per cent of sugar, in the USA 3.1 to 8.7 per cent, with higher levels, 7.5 to 8.7 per cent, preferred by many consumers. Flavouring, such as mint, or spearmint, may be added at this stage, and also green colouring matter, if this is permitted under local food legislation. Peas are extremely susceptible to hardening when canned and the water used for the brine solution should have a hardness of 8° or less. The cans are usually exhausted to give a can-closing temperature of not less than 170°F (77°C) and then processed in retorts, or automatic pressure cookers at 240° to 260°F (116–127°C). The time of processing depends upon the can size and the type of cooker employed, but as a guide A2 (307 x 408) cans require 40 minutes at 240°F (116°C) and A2½ (401 x 411) cans 45 minutes, when automatic cookers are used the temperature can be reduced by 5 or 10 minutes, and for each 5 degree (F) rise in temperature, the time may be reduced by about 8 minutes. Fully lacquered vegetable cans are usually used. After processing, the cans are thoroughly cooled before being stored. If processed inadequately, canned peas are very susceptible to 'flat souring'. When this occurs the cans remain normal, but the contents develop a sour taste and are unfit to eat.

For the canning of dried or processed peas, small Blue or Marrowfat peas are used. This pack has the advantage that it can be produced out of season. The dried peas, which have been previously cleaned and graded, are first soaked for 16 to 18 hours in three times their own quantity of soft water, (about 8° hardness), maintained at a uniform temperature. After this soaking, all hard peas are removed and the softened ones thoroughly washed before they are blanched in hot water at a temperature of from 200° to 208°F (93–98°C) for approximately 5 minutes. After blanching the peas are washed to remove starch, and then inspected, when any stained or split peas are removed. The peas are next filled into cans and brine containing approximately 2 per cent salt and 4 per cent sugar is added. Mint or spearmint may also be added at this stage and colouring. The pack is normally exhausted at 170–180°F (77–82°C) for 8 to 10 minutes, the cans sealed, and then processed at 240°F (116°C) 10 lb/in.^{2} (60.7 kg/cm^2) pressure for 20 to 40 minutes, depending upon the can size.

Dehydration—green peas are very heat sensitive and although dehydration is widely practised the process requires great care in order to preserve the culinary quality and colour. For optimum results the peas should be harvested at tenderometer readings of 100 to 105, although the dehydrator's scale normally ranges from 85 to 120. Very immature peas, however, wrinkle badly and do not rehydrate satisfactorily, while peas over about a 115 tenderometer reading tend to be too starchy. The peas should be cooled to about 40°F (4°C) immediately after harvesting and vining, preferably with refrigerated air since iced water tends to toughen the skins.

At the dehydration plant the initial stages of cleaning, washing and grading are similar to those used for the canning of green peas. Various dehydration techniques are employed, but the peas are usually pricked, or scarified, to assist evaporation of the water in the drier, before being blanched for 1.5 to 3 minutes in steam, or hot water at 210°F (99°C). Various additives such as sodium sulphite, sodium carbonate, sodium chloride and sucrose, are often added to the blanching water. If a steam blanch is used this is followed by a sulphite spray. The use of sulphite helps to enhance the colour of the dehydrated pea and also prolongs its shelf-life. In the UK concentrations up to 1 250 ppm in the end-product are usual. After blanching the peas are dehydrated immediately, if this operation is delayed, as when tunnel-driers are used, they must be cooled either by cold air, or a water spray, otherwise there is a significant increase in the bacterial count. Through conveyor-band, belt-trough and tunnel-driers are all used to dehydrate peas. If a through conveyor-band drier is used the inlet temperature in the first zone does not normally exceed 214°F (101°C), and reduces to 195°F (91°C) in the second zone and to 185°F (85°C) in the third zone. The peas are usually removed with approximately 15 per cent moisture content, to finish drying in bins. At this stage the product is extremely heat sensitive and scorching and colour regression can occur easily unless precautions are taken to control carefully temperature and airflow. Usually the peas remain in the drying bins for about 4 hours to reduce their moisture content to approximately 5 per cent. Belt-trough driers normally operate at 235°F (113°C) inlet temperature initially, but this is reduced considerably in the later stages. Tunnel-driers with a parallel airflow usually operate at a maximum temperature of 210°F (99°C) at the 'wet' inlet and 160°F (71°C) at the 'dry' tunnel inlet. The outlet temperature is usually between 140° and 145°F (60–63°C) and the drying cycle is from 6 to 7 hours. After drying the peas are sorted and graded according to size and colour. Very small and split peas are normally removed and ground to produce a high grade pea powder suitable for soup manufacture. Dehydrated peas are hygroscopic and very light sensitive and for this reason are often packed in laminated foil pouches for retail sale, sometimes with the addition of a desiccant such as calcium oxide. For bulk packs, strong multi-ply paper sacks incorporating not less than 5 plies, with a 300 gauge black pigmented polythene liner on the inside of the sack, are frequently used.

Quick-freezing—quick-frozen green peas are a popular vegetable and the mainstay of the large frozen vegetable industries of the USA and the UK. As with the dehydrated product, only tender, immature peas with a sweet flavour and an AIS content of approximately 9–12 per cent can be processed satisfactorily, and cooling immediately after harvesting is essential. Although iced water has been used for this operation many processors in the UK use refrigerated air. On arrival at the factory the peas are carefully cleaned, washed and graded, before being blanched. Some packers favour blanching for 50–60 seconds in water at 210°F (99°C), others prefer 90 seconds at 200°F (93°C). After blanching the peas are cooled immediately and are then quality graded by passing through brine density separators. After a further washing the graded peas are frozen. In the UK, flo-freezers, which work on the principal of fluidization are commonly used. The peas are picked up in a stream of air which is refrigerated down to −35°F (−37°C) and carried along in a stream in constant motion. Thirteen minutes after entering the freezer the peas fall out, individually frozen, and are then packed and stored at −20°F (−29°C).

Dehydrofreezing—a process has been developed by which green peas are partially dried to 50 per cent of their blanched weight and then frozen.

Uniformity of size is an important factor in the drying operation and blanching has been found to be critical for the success of the process.

Pea flour—dry peas can be used to produce a nutritious flour, but require debittering in order to produce a palatable product. Autoclaving the peas for a short period, approximately 10 minutes followed by immediate cooling, gives a debittered pea which can be readily ground into a flour suitable for soup manufacture.

Dried split pea products—recently attempts have been made to develop a protein-rich appetizing snack food from split peas. Basically the process consists of pre-conditioning the split peas by soaking them in water, followed by submerging them in hot cooking oil for a short period (2–3 min). A crisp, crunchy product which can be used as a substitute for salted nuts, potato crisps, etc, is produced.

Production and trade

Production—in most years dry peas constitute about 25 per cent of the total estimated world output of grain legumes. World production of dry peas for the period 1970–74 averaged 11 326 000 t/a, compared with 10 149 000 t/a in 1965–69, an increase of almost 12 per cent. In 1975 production was 10 221 000 t, and in 1976 rose to 13 427 000 t.

Peas (dry): Major producing countries

Quantity tonnes	Annual average 1965–69	Annual average 1970–74	1975	1976
People's Repub. China	3 140 000	3 848 000	4 700 000[1]	4 817 000[1]
India	860 000	611 000	564 000[1]	564 000[1]
Zaire	80 000	143 000	219 000[1]	n/a
USA	183 000	167 000	166 000	132 000[1]
Hungary	107 000	100 000	116 000	110 000[1]
Morocco	37 000	61 000	99 000	112 000
Ethiopia	122 000	131 000	52 000	52 000[1]

[1] Estimate.
n/a Not available.

In addition, there is an appreciable production of green immature peas as a vegetable crop. Statistical data are more fragmentary and a considerable proportion of the crop is produced by smallholders so that the total world output is probably considerably underestimated. World production for the period 1970–74 was estimated to have averaged 4 257 000 t/a, compared with 3 878 000 t/a, in 1965–69, an increase of about 10 per cent. Production in 1975 amounted to some 4 873 000 t and in 1976 to 4 786 000 t.

Peas (green): Major producing countries

Quantity tonnes

	Annual average 1965–69	Annual average 1970–74	1975	1976
USA	1 277 000	1 170 000	1 276 000	1 116 000
UK	494 000	619 000	692 000	715 000[1]
France	361 000	453 000	572 000	584 000[1]
Italy	247 000	260 000	264 000	253 000
USSR	n/a	142 000	175 000[1]	200 000[1]
Hungary	101 000	186 000	184 000	192 000[1]
Australia	115 000	109 000	126 000[1]	131 000[1]

[1] FAO estimate.
n/a Not available.

Trade—food quality dried peas are a world-wide trade item, while feed peas are an important animal feedingstuff, particularly in Europe. The total quantity of dried peas entering international trade has fluctuated considerably during the past decade and in recent years the trend has tended to be downwards. World trade for the period 1970–74 is estimated to have averaged some 357 800 t/a, compared with some 483 400 t/a for 1965–69, a fall of about 26 per cent.

Peas (dry): Major exporting countries

Quantity tonnes

	Annual average 1965–69	Annual average 1970–74	1975	1976
USA	101 922	103 617	74 762	91 997
Netherlands	40 405	31 996	23 003	27 583
Morocco	12 774[1]	25 835	28 530	n/a
New Zealand	10 475	18 850	n/a	n/a
Canada	12 832	18 102	14 204	n/a
Australia	3 955	15 098	3 826	n/a
Belgium/Luxembourg	20 988	12 668	9 802	11 131

[1] Two year average.
n/a Not available.

Pea

Peas (dry): Major importing countries

Quantity tonnes

	Annual average 1965–69	Annual average 1970–74	1975	1976
Netherlands	237 790	119 136	68 952	66 999
UK	68 553	40 595	23 519	25 041
German Fed. Repub.	52 546	42 375	31 870	39 526
France	17 595[1]	35 813	13 557	n/a
Japan	20 776	30 037	31 960	n/a
Belgium/Luxembourg	22 838	29 080	21 266	25 952

1 Three year average.
n/a Not available.

In spite of their highly perishable nature, limited quantities of green peas enter international trade, but statistical data are very fragmentary, and it is difficult to estimate the total tonnage exported.

Peas (green): Major exporting countries

Quantity tonnes

	Annual average 1965–69	Annual average 1970–74	1975	1976
Netherlands	4 400	3 600	5 300	n/a
Spain	2 700	4 400	4 400	n/a
Mexico	2 800	4 200	n/a	n/a
Italy	1 800	2 000	2 000	2 768
Morocco	600	900	900	n/a

n/a Not available.

Peas (green): Major importing countries

Quantity tonnes

	Annual average 1965–69	Annual average 1970–74	1975	1976
France	4 600	4 800	5 300	n/a
Belgium/Luxembourg	4 000	4 000	7 000	7 628
German Fed. Repub.	1 700	2 000	1 900	2 316
Netherlands	200	700	1 000	n/a
United Kingdom	n/a	800	4 200	335

n/a Not available.

Prices—in the USA producer prices for dried peas remained relatively stable for the period 1965 to 1972, the annual average 1965–69 was £38.41/t and in 1972 it was £48.11/t. In 1973, however, prices rose to the exceptionally high level of £133.06/t due to a world shortage of animal feedingstuffs. In Uganda the minimum farm price fixed by the government averaged £33.70/t for the period 1970–74, compared with £35.00/t for 1968–69.

The average cif price of imports of dried peas into the UK was £225.00/t in 1976 and £187.00/t in 1975, compared with £85.00/t for the period 1970–74 and £71.00/t for 1965–69. Japanese imports are generally of low-quality, including culls and splits, and their average price was £134.00/t in 1975 compared with £77.40/t for the period 1970–74 and £47.80/t for 1965–69. The average cif price of imports of green peas into the German Federal Republic has shown a steady upward trend and was £245.13/t in 1976 and £220.68/t in 1975, compared with £110.35/t for the years 1970–74 and £67.42/t for 1965–69.

Major influences

Dry peas are one of the most important of the food legumes. Although they have been grown traditionally in the more temperate regions they are also suitable for cultivation at intermediate and higher elevations in the tropics and as a winter crop in Mediterranean type ecologies. Their susceptibility to various diseases, however, could handicap their further development as a food legume, unless high-yielding disease resistant cultivars can be selected.

Although the protein content can vary considerably according to environmental conditions, it is generally regarded as satisfactory, moreover, recent research has indicated that it should be possible to increase the protein content by about 2 or 3 per cent by selection. Except for a low methionine content, peas have a satisfactory essential amino acid profile and are a good source of lysine. Recently consideration has been given in Canada to the possibility of developing the production of field peas as a protein crop for the formulated feed market. However, there is need to improve average yields and so far it has not been established whether protein levels can be increased, and maintained at high levels, if yields are increased. Unless this is possible then the value of peas as a protein crop, particularly for the formulated feed market, is questionable.

Bibliography

AGARWAL, P.C., RODRIGUEZ, R. AND SAHA, N. K. 1969. Studies on some important varieties of green peas of northern India: (i) Physicochemical characteristics; (ii) Canning trials. *Indian Food Pack.*, *23*, 12–16; 17–21.

ALI-KHAN, S. T. AND ZIMMER, R. C. 1972. Growing field peas. *Can. Dep. Agric. Publ.*, 1493, 9 pp.

ALLEN, A. G. AND FRAPPELL, B. D. 1970. Green peas for processing. *Tasman. J. Agric. 41*, 140–144.

ANON. 1952. Pea vine residue makes excellent cattle fodder. *Agric. Gaz. N.S.W.*, *63*, 220.

ANON. 1955. When to harvest canning peas. *Aust. Food Manuf.*, *24* (*12*), 22; 24; 26; 28.

ANON. 1960. Successful dehydro-freezing of peas. *Cann. Pack.*, *129* (*12*), 36- 37.

ANON. 1963. New quick-freezing method for peas. *Food Process. Packag.*, *32*, 371–375.

ANON. 1972/73. Field peas—Cinderella crop of the 70's. *Can.*, *Natl. Res. Counc.*, *Sci. Dimension*, pp. 10–15.

ANON. 1975. US dry peas and lentils—a thriving industry. *Int. Fruit World*, *34* (*1*), 71–78.

ARNON, I. 1972. Pulses or grain legumes. *Crop production in dry regions.* Vol. 2. pp. 217–260. London: Leonard Hill Books, 683 pp.

ARTHEY, V. D. 1964. Quick-frozen peas: some factors affecting their quality. *Agriculture* (*UK*), *71*, 252–255.

AUSTIN, G. T. AND AUSTIN, H. F. 1963. Debittering split peas for bland, high-protein foodstuffs. *Food Technol.*, *17*, 1319–1320.

BAJAJ, S., MICKELSEN, O., BAKER, L. R. AND MARKARIAN, D. 1971. The quality of protein in various lines of peas. *Br. J. Nutr.*, *25*, 207–212.

BAKKER-ARKEMA, F. W., PATTERSON, R. J. AND BEDFORD, C. L. 1969. The manufacturing, utilization and marketing of instant legume powders. *Rep. 9th dry bean res. conf.* (1968). pp. 35–45. *US Dep. Agric.*, *Agric. Res. Serv.*, ARS 74–50, 94 pp.

BEN-SINAI, I. M., BEN-SINAI, M., AHMED, E. M. AND KRAMER, A. 1965. The food and fodder value of pea plant parts (*Pisum sativum* L.) as related to harvest time and variety. *Food Technol.*, *19*, 856–859.

BHARDWAJ, S. N., SHARMA, D. N. AND NATH, V. 1971. Varietal differences in drought tolerance of field pea (*Pisum sativum* L. var. *arvense* Poir.). *Indian J. Agric. Sci.*, *41*, 894–900.

BHATIA, B. S. 1962. Studies on dehydration of green peas. *Food Sci.*, *11*, 286–287.

BHATIA, B. S., KUPPUSWAMY, S., GURURAJA RAO, R. AND BHATIA, D. S. 1963. Dehydration of green peas. *Indian J. Technol.*, *1*, 250–254.

BLAND, B. F. 1971. Peas. *Crop production: Cereals and legumes.* pp. 326–382. London/New York: Academic Press, 466 pp.

BONGIRWAR, D. R. AND SREENIVASAN, A. 1977. Development of quick cooking peas. *J. Food Sci. Technol.*, *14*, 17–23.

BRIEN, R. M., CHAMBERLAIN, E. E., COTTIER, W., CRUICKSHANK, I. A. M., DYE, D. W., JACKS, H. AND REID, W. D. 1955. Diseases and pests of peas and beans in New Zealand and their control. pp. 8–33. *N.Z. Dep. Sci. Ind. Res. Bull.* 114, 91 pp.

BRINDLEY, T. A., CHAMBERLIN, J. C. AND SCHOPP, R. 1958. The pea weevil and methods for its control. *US Dep. Agric., Farmers' Bull.* 1971, 24 pp.

BUBENZER, G. D. AND WEIS, G. G. 1974. Effect of wind erosion on production of snap beans and peas. *J. Am. Soc. Hortic. Sci.*, *99*, 527–529.

BURROWS, V. D., GREENE, A. H. M., KOROL, M. A., MELNYCHYN, P., PEARSON, G. G. AND SIBBALD, I. R. 1972. Field peas. *Food proteins for grains and oilseeds. A development study projected to 1980.* pp. 64–67. *Report of a study group appointed by the Hon. Otto E. Lang, Minister responsible for the Canadian Wheat Board.* Ottawa, House of Commons, 214 pp.

CHOW, C. AND BELL, J. M. 1976. Effects of various heat and pH treatments on digestibility of protein in pea protein concentrate (*Pisum sativum*). *Can. J. Anim. Sci.*, *56*, 559–566.

COMMONWEALTH BUREAU OF PASTURES AND FIELD CROPS. nd. *Pisum sativum:* seed composition. Annotated bibliography 1370, 1953–72. *Commonw. Agric. Bur., Bur. Pastures & Field Crops, Hurley, Maidenhead, Berks., England*, 8 pp.

Pea

Cousin, R. 1974. *Le pois.* Paris, Inst. Natl. Rech. Agron., 251 pp.

Crampton, M. J. and Goulden, D. S. 1974. New peas resistant to wilt. *N.Z. J. Agric., 128*, 12–13.

Das, R. C. and Hota, B. N. 1967. Growth regulator effects on growth, yield and quality of *Pisum sativum* (Asch. & Graebn.). *Proc. int. symp. subtrop. trop. hortic.* pp. 379–382. Bangalore, Hortic. Soc. India, 907 pp.

Davies, D. R. 1976. Peas. *Evolution of crop plants.* (Simmonds, N. W. ed.). pp. 172–174. London: Longman Group Ltd, 339 pp.

Dudley, J. E. (Jr.) and Bronson, T. E. 1956. The pea aphid on peas and methods for its control. *US Dep. Agric., Farmers' Bull.*, 1945, 13 pp.

Ewing, E. 1957. Factors for resistance to pre-emergence decay of peas (*Pisum sativum*). *Proc. Caribb. Reg. Am. soc. hortic. sci., 5th annu. meet.* pp. 52–54. Mexico, D. F., 75 pp.

Feinberg, B. 1973. Vegetables. *Food dehydration.* (Arsdel, W. B. Van, Copley, M. J. and Morgan, A. I. eds.). 2nd ed. Vol. 2. pp. 1–82. Westport, Connecticut: Avi Publishing Co Inc, 529 pp.

Frings, J. F. J. 1976. The rhizobium—pea symbiosis as affected by high temperatures. *Wageningen, Meded. Landbouwhogesch.*, 76–77, 76 pp.

Gane, A. J. 1970. Dried peas. *Agriculture (UK), 77*, 398–401.

Gane, A. J. 1972. *Vining peas in England.* Peterborough, England, Processors and Growers Res. Organ., 53 pp.

Gautham, R. C. and Singh, R. K. 1971. A note on effect of chemical and mechanical weeding on the quality of pea (*Pisum sativum*). *Mysore J. Agric. Sci., 5*, 475–477.

Greensmith, M. 1971. Peas. *Practical dehydration.* pp. 89–99. London: Food Trade Press Ltd, 174 pp.

Harter, L. L., Zaumeyer, W. J. and Wade, B. L. 1957. Pea diseases and their control. *US Dep. Agric., Farmers' Bull.* 1735, 28 pp.

Jadhav, P. S., Jain, T. C. and Prasannalakshmi, S. 1975. *Sorghum, millets, peas: A bibliography of the Indian literature, 1969–73.* pp. 56–59. Hyderabad, India, Int. Crops Res. Inst. Semi-arid Trop., 116 pp.

Jain, N. L. and Choudhury, B. 1963. Some studies in the dehydration of peas. *Indian J. Hortic., 20*, 129–134.

JAISWAL, S. P., KAUR, G., KUMAR, J. C., NANDPURI, K. S. AND THAKUR, J. C. 1975. Chemical constituents of green pea and their relationships with some plant characters. *Indian J. Agric. Sci.*, *45*, 47–52.

KAUR, G., SHUKLA, F. C. AND SINGH, D. 1976. Studies on the varietal differences in physico-chemical characteristics of some varieties of peas (*Pisum sativum*). *Indian Food Pack.*, *30* (*4*), 5–8.

KHAN, S. A. AND KHAN, S. 1967. Drying of peas in Peshawar region. *Agric. Pak.*, *18*, 31–38.

KOTASTHANE, S. R. 1975. Fungicidal control of powdery mildew of pea. *Sci. Cult.*, *41*, 450–452.

KRISHNAMURTHI, A. (ed.). 1969. *Pisum* Linn. *The wealth of India: Raw materials*. Vol. 8 (Ph-Re). pp. 124–139. New Delhi: Counc. Sci. Ind. Res., Publ. Inf. Dir., 394 pp.

LENKA, D. AND GAUTAM, O. P. 1971. Method of planting for increased grain production of pea (*Pisum sativum* L. var. *arvense* Poir.). *Indian J. Agric. Sci.*, *41*, 280–283.

LOCK, A. 1969. Canning of vegetables—peas. *Practical canning*. pp. 288–340. London: Food Trade Press Ltd, 415 pp.

LONGDEN, P. C. AND AUSTIN, R. B. 1970. Harvest methods for seed crops of vining peas. *Exp. Hortic.*, (21), pp. 42–48.

LOPEZ, A. 1975. Peas. *A complete course in canning*. 10th ed. pp. 395–403. Baltimore, Maryland: The Canning Trade Inc, 755 pp.

LYNCH, L. J. AND MITCHELL, R. S. 1953. The definition and prediction of the optimal harvest time of pea canning crops. *Melbourne, Aust., Counc. Sci. Ind. Res., Bull.* 273, 43 pp.

LYNCH, L. J., MITCHELL, R. S. AND CASIMIR, D. J. 1959. The chemistry and technology of the preservation of green peas. *Adv. Food Res.*, *9*, 61–151.

MANER, J. H. 1973. Investigations of plants not currently used as major protein sources. *Proc. symp., alternative sources of protein for animal production* (1972). pp. 87–118. Washington, Natl. Acad. Sci., 183 pp.

MCAULIFFE, J. D. 1962. Field peas: diseases. *J. Agric. South Aust.*, *65*, 330–335.

Pea

McAuliffe, J. D. and Webber, G. D. 1961. Field peas. *J. Agric. South Aust.*, *65*, 168–171; 222–226.

McFarlane, J. A. 1970. Pulses. *Food storage manual*. pp. 335–346. Rome: FAO, World food programme, 799., pp. plus indexes, (loose leaf).

Ministry of Agriculture, Fisheries and Food, 1965. Weed control in peas. *Lond., Minist. Agric., Fish. Food, Advis. Leafl.* 376 (Revised), 7 pp.

Ministry of Agriculture, Fisheries and Food. 1971. Pea cyst eelworm. *Lond., Minist. Agric., Fish. Food, Advis. Leafl.* 462 (Revised), 4 pp.

Ministry of Agriculture, Fisheries and Food. 1972. Thrips on peas. *Lond., Minist. Agric., Fish. Food, Advis. Leafl.* 170 (Revised), 4 pp.

Ministry of Agriculture, Fisheries and Food. 1973. Pea moth. *Lond., Minist. Agric., Fish. Food, Advis. Leafl.* 334, (Revised), 4 pp.

Ministry of Agriculture, Fisheries and Food. 1974. Pea midge. *Lond., Minist. Agric., Fish. Food, Advis. Leafl.* 594, 4 pp.

Muehlbauer, F. J. and Dudley, R. F. 1974. Seeding rates and phosphorus placement for Alaska peas in the Palouse. *Washington State Univ. Coll. Agric. Res. Cent., Bull.* 794, 4 pp.

Muehlbauer, F. J. and Schalk, J. M. (eds.). 1974. Field peas (including garden peas and edible podded types), (*Pisum sativum*). *Guide for field crops in the tropics and the subtropics*. (Litzenberger, S. C. ed.). pp. 154–161. Washington, Off. Agric. Tech. Assist. Bur. Agency Int. Dev., 321 pp.

Murphy, R. F. 1973. Screening of pea cultivars for dehydration. *Ir. J. Agric. Res.*, *12*, 293–325.

Murphy, R. F. 1974. Evaluation of dried pea cultivars. *Ir. J. Agric. Res.*, *13*, 301–306.

Murray, K. E., Shipton, J. and Whitefield, F. B. 1970. 2-methoxy-pyrazines and the flavour of green peas (*Pisum sativum*). *Chem. Ind.*, (27), pp. 897–898.

Nourse, H. C. 1973. Field peas in South Australia. *South Aust., Dep. Agric. Ext. Bull.* 3, 19 pp.

Pal, A. B. 1976. Note on the differential response of some pea cultivars to heat units. *Indian J. Agric. Sci.*, *46*, 104–105.

PAPAVIZAS, G. C. AND AYERS, W. A. 1974. Aphanomyces species and their root diseases in pea and sugarbeat. *US Dep. Agric. Res. Serv., Tech. Bull.* 1485, 158 pp.

PARKIN, E. A. AND BILLS, G. T. 1955. Insecticidal dusts for the protection of stored peas and beans against bruchid infestation. *Bull. Entomol. Res., 46*, 625–641.

PURSEGLOVE, J. W. 1968. *Pisum sativum* L. *Tropical crops: Dicotyledons.* Vol. 1. pp. 311–315. London: Longmans, Green and Co Ltd, 332 pp.

RAWSTHORNE, J. H. 1974. Calibration of the pea tenderometer, and a comparison of tenderometer units from different countries. *Madrid, 4th Int. Congr. Food Sci. & Technol., Work Doc. 3, Sensory properties of foods: Evaluation methods, Pap.* 21, pp. 54–56.

REYNOLDS, J. D. 1957. Some developments in threshed-pea growing. *Agric. Rev., 3 (8)*, 22–26.

ROBINSON, R. R., HALFHILL, E., LANDIS, B., RETAN, A. AND PORTMAN, R. 1974. Insects of peas. *USA, Pacific Northwest Coop Ext. Publ.*, PNW 150, 19 pp.

ROGERS, J. L. AND BINSTED, R. 1972. The processing and freezing of vegetables: Peas. *Quick-frozen foods.* 2nd ed. pp. 228–232. London: Food Trade Press Ltd, 499 pp.

RYALL, A. L. AND LIPTON, W. J. 1972. Commodity requirements of unripe fruits and miscellaneous structures. *Handling, transportation and storage of fruits and vegetables.* Vol. 1. pp. 130–131. Westport, Connecticut: Avi Publishing Co Inc, 473 pp.

SAXENA, J. K., TRIPATHI, R. M. AND SRIVASTAVA, R. C. 1975. Powdery mildew resistance in pea (*Pisum sativum* L.). *Curr. Sci., 44*, 746.

SCHELTEMA, J. H., SYKES, S. M. AND LAST, J. H. 1961. Acceptability of frozen peas in relation to maturity and other factors. *Melbourne, Aust., Counc. Sci. Ind. Res. Organ., Div. Food Preserv. Tech., Pap. 26*, 22 pp.

SCHROEDER, W. T. AND PROVVIDENTI, R. 1964. Evaluating *Pisum sativum* for resistance to pea mosaic. *New York State Agric. Exp. Stn. Geneva, Bull.* 806, 10 pp.

SEN, B. AND KAPOOR, I. J. 1975. Systemic fungicides for the control of wilt of peas. *Veg. Sci., 2*, 76–78.

SEN GUPTA, P. K. 1974. Diseases of major pulse crops in India. *PANS*, *20*, 409–415.

SETTY, L. AND SIDDAPPA, G. S. 1961. Composition and properties of dried peas in relation to their suitability for canning. *J. Sci. Food Agric.*, *12*, 537–541.

SHAH, W. H. AND TREMAZI, S. A. 1968. Production of canned peas. *Pak., Counc. Sci. Ind. Res., Sci. Ind.*, *6*, 111–116.

SHARGA, A. N. AND JAUHARI, O. S. 1970. Studies on the effects of foliar application of molybdenum on nodulation and quality of garden pea (*Pisum sativum* L.). *Madras Agric. J.*, *57*, 216–221.

SHERWOOD, R. T. AND HAGEDORN, D. J. 1958. Determining the common root rot potential of pea fields. *Wisconsin Univ., Agric. Exp. Stn. Bull.* 531, 12 pp.

SHIPTON, J., WHITEFIELD, F. B. AND LAST, J. H. 1969. Extraction of volatile compounds from green peas (*Pisum sativum*). *J. Agric. Food Chem.*, *17*, 1113–1118.

SIEGEL, A. AND FAWCETT, B. 1976. *Food legume processing and utilization* (*with special emphasis on application in developing countries*). Ottawa, Int. Dev. Res. Cent., IDRC–TS1, 88 pp.

SINGH, H. B. AND JOSHI, B. S. 1970. Peas *Pisum sativum* L. *Pulse crops of India.* (Kachroo, P. and Arif, M. eds.). pp. 256–305. New Delhi: Indian Counc. Agric. Res., 334 pp.

SINGH, H. B., JOSHI, B. S., PANT, K. C. AND CHANDEL, K. P. S. 1973. An exotic pea shows promise. *Indian Farming*, *23* (6), 25.

SINGH, H. B., PANT, K. C., JOSHI, B. S. AND CHANDEL, K. P. S. 1976. Evaluating a pisum collection. *Rome, FAO, Plant Genet, Resour.*, (32), pp. 17–35.

SMARTT, J. 1976. *Tropical pulses.* pp. 47–49; 112; 117. London: Longman Group Ltd, 348 pp.

SMITH, M. A., MCCOLLOCH, L. P. AND FRIEDMAN, B. A. 1966. Market diseases of asparagus, onions, beans, peas, carrots, celery and related vegetables. pp. 31–35. *US Dep. Agric., Agric. Handb.* 303, 85 pp.

TAYLOR, J. D. AND DYE, D. W. 1976. Evaluation of streptomycin seed treatments for the control of bacterial blight of peas (*Pseudomonas pisi* Sackett 1916). *N.Z. J. Agric. Res.*, *19*, 91–95.

TILT, J. AND TAYLOR, B. L. 1954. Field peas. *Tasman. J. Agric., 25*, 230–239.

TORREY, M. 1974. Vegetables: Peas. *Food technology review 13: Dehydration of fruits and vegetables*. pp. 171–176. New Jersey/London: Noyes Data Corp, 286 pp.

VINCENT, C. L. 1958. Pea plant population and spacing. *Washington Agric. Exp. Stn. Bull.*, 594, 9 pp.

WAGER, H. G. 1964. Physiological studies of the storage of green peas. *J. Sci. Food Agric., 15*, 245–252.

WECKEL, K. G., MATHIAS, W. D., GARNATZ, G. F. AND LYLE, M. 1961. Effect of added sugar on consumer acceptance of canned peas. *Food Technol., 15*, 241–242.

WESTPHAL, E. 1974. *Pisum sativum* L. Pulses in Ethiopia, their taxonomy and agricultural significance. pp. 176–193. *Wageningen, PUDOC, Cent. Agric. Publ. Docum., Agric. Res. Rep.* 815, 278 pp.

YARNELL, S. H. 1962. Cytogenetics of the vegetable crops: (iii) Legumes, garden peas *Pisum sativum* L. *Bot. Rev., 28*, 465–537.

YOUNG, J. M. AND DYE, D. W. 1970. Bacterial blight of peas caused by *Pseudomonas pisi* Sackett 1916, in New Zealand. *N.Z. J. Agric. Res., 13*, 315–324.

YOUNGS, C. G. 1970. Peas for food and feed. *The feed protein market: a background study*. (Hudson, S. C. ed.). Appendix III, 7 pp. Ottawa, Canadian Wheat Board, 92 pp.

YOUNGS, C. G. AND CRAIG, B. M. 1974. Production and utilization of field peas (*Pisum sativum*). *Madrid, 4th Int. Congr. Food Sci. & Technol., Work Doc. 8a, New food sources of key nutrients, Pap.* 24, pp. 67–69.

ZARKADAS, C. G., HENNEBERRY, G. D. AND BAKER, B. E. 1965. The constitution of leguminous seeds: (v) Field peas (*Pisum sativum* L.). *J. Sci. Food Agric., 16*, 734–738.

ZAUMEYER, W. J. 1962. Pea diseases. *US Dep. Agric., Agric. Res. Serv., Agric. Handb.* 228, 30 pp.

ZIMMERMANN, G. AND LEVY, C. 1962. Correlation between alcohol-insoluble and lysine availability in canned peas. *J. Agric. Food Chem., 10*, 51–53.

ZOHARY, D. AND HOPF, M. 1973. Domestication of pulses in the Old World. *Science, 182*, 887–894.

Common names **PIGEON PEA, Angola pea, Congo bean or pea, Dhal**[1]**, No or Non-eye pea, Red gram, Yellow dhal.**

Botanical name *Cajanus cajan* (L.) Millsp., syn. *C. indicus* Spreng.

Family Leguminosae.

Other names Adhaki (Ind.); Ads sudani (Sud.); Alberga, Alverja (Salv.); Ambrevade (Maur., Malag.); Angolische erbse (Ger.); Apena (Ug.); Arahar, Arhar (Ind.); Burusa/u (Ug.); Cachito (Gu.); Cadios (Philipp.); Chicharo de arbol (Mex.); C. de paloma (Col.); Chinchoncho (Venez.); Cytise des Indes (Fr.); Dau chieu, D. trieu, D. xay (Viet.); Embrevade (Fr.); Ervilha de Congo (Ang.); Frijol chino[2] (Gu.); F. de palo (Salv.); F. gandul (Cuba); F. japones (Gu.); Gandal (P.R.); Gandul (C. Am.); Gandures (P.R.); Garbanzo falso (Nic.); Goodé (Indon.); Goongo pea (Cuba, P.R.); Green pea (W.I.); Guando/u (Lat. Am.); Guandu de fava larga (Braz.); Guisante de paloma, G. enano (Sp.); Gungo pea (Jam.); Ihora (Ind.); I. parippu (Sri La.); Imposo (Zam.); Kachang dal, K. hiris, K. kayu (Malays.); Kadios, Kadyos (Philipp.); Kakunda bakishi (Zar.); Kandalu, Kandi, Kandulu (Ind.); Katjang goode, K. (h)eeris, K. kadjoo (Indon.); Katschang (Malays.); Ki-mame (Jpn.); Lenteja françesa (Guam); Lopena (Ug.); Lubia adassi (Sud.); Mbani (Tanz.); Mbazi (Zan.); Nkol (Zar.); Ohota-farengota (Eth.); Oror (Ind.); Otile (Nig.); Pai-si-gong (Burm.); Paripu (Sri La.); Pay-in-chong, Pèsigon, Pè-sin-gôn (Burm.); Pois cajan,

---

[1] Also frequently used for lentils, *Lens esculenta*.
[2] Sometimes used in Venezuela for the cowpea, *Vigna unguiculata*.

P. d'Angola, P. de Congo, P. nain, P. pigeon[1] (Fr.); Puerto Rican pea (Haw.); Puso-poroto (Peru); Quimbolillo (C. Rica); Quinchoncho (Venez.); Rahar (Ind.); Salbocoghed (Som.); Strauch erbse (Ger.); Thogari (Ind.); Thovaroy (Sri La.); Thuvara(i) (Tam.); Thuvaran (Ind.); Timbolillo (C. Rica); Togare/i, Tovarai, Tua-re (Thai.); Tur(a), Tuvar(a)(i), Tuvaray, Tuvarika, Tuver(a) (Ind.); Vio vio, Waken kurawa (Nig.); Yewof-ater (Eth.).

Botany

An erect, woody, short-lived, perennial shrub, often grown as an annual, which shows considerable variation in form under different environmental and cultural conditions. It can vary in height from 2 to 12 ft (0.6–3.6 m) and the spread of the branches normally ranges from less than one-quarter to more than one-half of the height of the plant. The root-system varies according to the type: tall, upright cultivars produce long, vertical, deeply penetrating tap-roots, while spreading, bushy cultivars produce shallower more spreading roots. Under favourable conditions both have large clusters of nodules. The stems are angular, hairy and branched. The point on the main stem where branching begins, and the number of secondary branches, varies, but in most types branching begins between the 6th and 10th node, ie 6–10 in. (15–25 cm) above the ground. The leaves are pinnately trifoliolate, with lanceolate or elliptic leaflets. They show considerable variation in size and shape, normally ranging from 4 to 7 in. (10–17.5 cm) in length and are of various shades of green, pubescent or glabrous. The flowers are borne on short axillary or terminal racemes and are usually about 0.8 in. (2 cm) in length. They vary in colour from pale-yellow to orange, often with the standard striped or splotched with dark-red or purple. They may be self-pollinated, but 5–40 per cent (av 20 per cent) cross-pollination can occur. The pods are somewhat flattened, dehiscent, 1.6 to 4 in. (4–10 cm) in length and 0.24 to 0.6 in. (6–15 mm) in breadth, green, purple or maroon, or green, splotched with purple or maroon. They are more or less downy, obliquely constricted between the seeds and terminate in a slender beak. In some types of pigeon peas they are pendant, in others they are quite erect. They contain 2 to 8 (commonly 4) seeds which vary in size, shape and colour.

[1] Also sometimes used in the Antilles for the rice bean, *Vigna umbellata*.

Pigeon pea

Pigeon peas show great diversity in their habit, growth period, and the colour, shape and size of the pods and seeds. Some authorities recognize two separate varieties: (i) *Cajanus cajan* var. *flavus*; (ii) *C. cajan* var. *bicolor*. The former is earlier maturing, semi-dwarf in habit and has yellow flowers and green pods, usually with 2–3 seeds. The tur cultivars of India are typical. The latter, includes most of the truly perennial types, the plants are later maturing and tend to be larger, more bushy, with purple-speckled yellow flowers and green pods, often blotched with dark-red and containing 4 or 5 distinctly speckled seeds. The arhar cultivars, extensively cultivated in N. India, are typical of the variety *bicolor*. Innumerable hybrids have arisen by cross-pollination of the original varieties with the result that every possible gradation now exists. More recently it has been suggested that pigeon peas could be classified into the following five groups: (i) tall compact; (ii) tall open; (iii) medium-height compact; (iv) medium-height open; (v) dwarf bushy. Late-maturing cultivars are usually type (i) and early or intermediate-maturing cultivars, type (iv). Another classification of the species has been suggested by Akinola and Whiteman (1972); they suggest three major groups: *Group A*, early-maturing types in which pod ripening is both extensive and intensive; *Group B*, early-maturing types, in which pod ripening is extensive; *Group C*, late-maturing types, in which pod ripening is intensive.

Numerous improved cultivars of pigeon peas have been developed, many of which have been selected to suit local environmental and climatic conditions. In India for example, a number of high-yielding, short-duration cultivars have been developed including: Kanke–9, Kanke–3, Co–1, UPAS–120, Pusa ageta, Sharda, Makta and T–21. At the International Institute of Tropical Agriculture (IITA) the seed of three elite, semi-dwarf, short-season cultivars, CITA–1, 2 and 4, has recently become available in small quantities and the high-yielding lines 4738, 4761 and 4F–12061 are currently being tested and evaluated. In Puerto Rico a very popular commercial cultivar is Kaki, or Caqui, named because of the khaki colour of the pod, and in Hawaii the cultivar New Era is extensively cultivated. Another cultivar Amarillo is day-neutral in Florida and has the advantage that it can be sown at different dates and harvested throughout the year.

Origin and distribution

The centre of origin of the pigeon pea has been the subject of much discussion. Some authorities consider that it is of Indian origin. There is evidence, however, that it was cultivated in Egypt before 2 000 BC and many authorities consider that it may have originated in the region between Egypt and East Africa and to have become widespread in Africa and the SE. Asian sub-continent in prehistoric times. The pigeon pea was introduced into the New World in the 16th century and in the Pacific during the 18th century. It is now widely cultivated between 30°N and 30°S, although about 92 per cent of the total world production is in India.

Cultivation conditions

Temperature—the pigeon pea is susceptible to frost damage at all stages of growth. The most favourable temperature range is from 64° to 85°F (18–29°C). Under dry conditions temperatures as low as 50°F (10°C) may be tolerated by some cultivars, although the seed-set of others may be reduced considerably. Provided that there is adequate moisture and the soil is reasonably fertile the pigeon pea may be grown in areas with average temperatures as high as 95°F (35°C). For optimum seed yields bright sunshine is essential, excessive cloud or shade produces spindly growth and poor seed-set.

Rainfall—an average, annual rainfall of between 24 and 40 in. (600–1 000 mm) is most suitable. High yields are obtained when there is good rainfall during the first 2 months of growth followed by a dry period during flowering and harvesting. The pigeon pea is drought resistant and economic yields are obtained in areas with an average rainfall of about 15 in./a (380 mm/a), provided there is adequate soil moisture. In India, it is sometimes grown under irrigation when it may receive 25 ac/in. (2 570 m^3) in 7–8 applications. Although drought resistant and grown in arid areas the pigeon pea may be grown successfully in humid areas with an annual rainfall in excess of 100 in. (2 500 mm), provided that there is no standing water, even for a few days. In such areas the plant tends to make vegetative growth at the expense of seed production.

Soil—the pigeon pea can be grown on a wide range of soil types provided that they are not deficient in lime and are well drained, as it cannot stand waterlogging. For optimum results deep loam almost neutral soils are required, excessive acidity inhibits nodulation and the plants may become chlorotic or suffer dieback when grown on soils outside the pH 5–7 range, or when there is a deficiency of phosphorus or manganese. Many cultivars are tolerant of saline soils, from 6 to 12 mmhos/cm. Little is known of the precise manurial requirements of the pigeon pea. In most countries responses to the application of nitrogen and potassium have been very erratic, applications of nitrogen in some instances, eg mixed cropping with millet in India, have been found to depress yields. Most studies indicate the need for phosphorus and under tropical conditions the application of between 18 and 89 lb/ac (20–100 kg/ha) of phosphoric acid is generally recommended. Although experiments in Venezuela, on poor soils, indicate that applications as high as 270 lb/ac (300 kg/ha) would be required. Experiments have shown that the application of sulphur alone, or in combination with phosphorus, can increase significantly seed yield and the nitrogen content of the root nodules. Indian workers report that with the application of up to 100 ppm sulphur the methionine content of the seed increases, but that the effect decreases at higher levels of sulphur. Pigeon peas are susceptible to zinc deficiency, which is often controlled successfully in India by the soil application of 2–4 ppm zinc, or by foliar spraying with 0.5 per cent zinc sulphate with 0.25 per cent lime.

Day-length—short-day, day-neutral and intermediate forms have been identified. Most established cultivars are short-day plants and sowing date is critical affecting their growth period and height. Many of the day-neutral cultivars give lower average yields and poorer quality seeds.

Altitude—in India the pigeon pea is grown at elevations up to 6 000 ft (1 800 m), in Venezuela it is reported to be grown up to an elevation of 10 000 ft (3 000 m), in Indonesia up to 6 500 ft (2 000 m), in Jamaica up to 3 500 ft (1 100 m). In Ethiopia best results are obtained between 5 400 ft and 6 000 ft (1 700–1 800 m), in Uganda between 2 000 ft and 4 000 ft (600–1 200 m) and in Hawaii, between 100 ft and 2 500 ft (30–750 m). Plants tested at elevations between 3 500 ft and 5 000 ft (1 100–1 500 m) in Hawaii failed to set seed.

Wind—the pigeon pea is fairly resistant to wind and is sometimes planted in double rows as a windbreak. It is however, rather sensitive to sea spray and does not thrive if planted near the seashore.

Planting procedure

Material—sound, healthy seed should be used. Inoculation with suitable rhizobium strains increases the yield significantly. Germination is hypogeal and is usually good, 85–95 per cent with fresh seed. In areas of high humidity the seed is reported to begin to lose its viability four months after harvest.

Method—pigeon peas are used both in short- and long-term cropping systems, and when grown as a food crop in India and Africa are often intermixed with other crops such as maize, sorghum, or millet, when they are left to mature on residual soil moisture after the cereal has been harvested. In this case the seed is dropped into plough furrows either by hand or through a one-furrow seed-drill—one row of pigeon peas to every three to five rows of cereal. The pigeon peas then receive the same weeding and other cultural operations as the principal crop. When grown as a pure crop, the seed may be broadcast, or planted in rows, in moist, well-prepared soil at a depth of 1–2 in. (2.5–5 cm). Ridge planting has been recommended in soils liable to become waterlogged. In the first 60 days of growth effective weed control is essential, particularly in areas of limited rainfall. Chloramben applied pre-emergence has been suggested for effective weed control, although its use is reported to cause slight injury to the crop. In the W.I. satisfactory weed control has been obtained by the pre-emergence use of prometryne, particularly if combined with paraquat applied as a post-emergence spray. In the Dominican Republic, a mixture of alachlor plus linuron has proved to be an economic treatment, and terbutryne is reported to give 95 per cent weed control for up to 9 weeks after application. At IITA promising results have been obtained with chlorthal-dimethyl applied pre-emergence and Basalin applied pre-plant, followed by diphenenamid applied pre-emergence. When grown as a forage crop, defoliating and cutting back at three to five-month intervals is widely practised. The use of growth regulators is being investigated in Jamaica, but so far their use has proved to be uneconomic.

Field-spacing—varies according to the size of the mature plant and the purpose for which it is required. It has been shown that the growth habit readily adapts itself to the space available, but close spacing tends to increase plant height and reduces individual plant productivity. In India when grown as a pure crop the average spacing reported to be used is 12–18 in. (30–45 cm) between the plants and 15–24 in. (37.5–60 cm) between the rows, although most cultivars, especially the spreading types are reported to give optimum yields when planted 24 in. (60 cm) apart, with 48 in. (120 cm) between the rows. In Puerto Rico, the plants are normally planted 60–72 in. (150–180 cm) apart with 72 in. (180 cm) between the plants; in E. Africa the recommended spacing for pure stands is 60 x 48 in. (150 x 120 cm) or 60 x 60 in. (150 x 150 cm), and Sri Lanka 36 x 24 in. (90 x 60 cm) is normally used. In Trinidad dwarf and semi-dwarf types are usually planted 42 x 42 in. (105 x 105 cm) and tall 72 x 72 in. (180 x 180 cm).

Pigeon pea

Seed-rate—in India the average seed-rate used when pigeon peas are intercropped is reported to be 1–5 lb/ac (1.12–5.6 kg/ha) and 12–20 lb/ac (13–22 kg/ha) when grown as a pure crop. In Sri Lanka when intercropped the average seed-rate is 1.5 to 3 lb/ac (1.7–3.4 kg/ha), and 8 lb/ac (9 kg/ha) when grown as a pure crop.

Pests and diseases

Pests—the major pests of the pigeon pea in India are the gram caterpillar, *Heliothis armigera* (*H. obsoleta*), the red gram plume moth, *Exelastis atomosa*, and the gram pod fly, *Melanagromyza* (*Agromyza*) *obtusa*, and so far no really effective control measures against these pests have been developed. Thrips, *Frankliniella sulphurea* and *Taeniothrips nigriconis* are reported to cause premature flower drop. In Sri Lanka the caterpillar of the spotted borer, *Maruca testulalis*, can cause serious damage, and the blue butterfly caterpillar, *Lampides boëticus* and the gram caterpillar, *Heliothis armigera*, can also be troublesome; spraying with dieldrin or endrin is reported to give reasonable control. In the W.I. the most important pests in the field are leafhoppers and pod borers. The former are of particular importance as they cause severe leaf curling and are probably the vector of a virus disease. In Trinidad the green leafhopper *Empoasca fabilis* is especially serious. Effective control is reported to be obtained by spraying with malathion. The pod borers, *Elasmopalpus rubedinellus*, *Ancylostomia stercorea* and *Heliothis virescens* are also serious pests in the W.I. In determinate types spraying with DDT, Dipterex or Gardona at peak flowering has given satisfactory control, but effective control is much more difficult with indeterminate, or all-season types, which have an extended flowering season. In Trinidad, heavy infestation with the black aphid, *Aphis craccivora*, has been reported recently. In Hawaii, a serious pest is the soft elongated flat scale, *Coccus elongatus*. In E. Africa the bollworm, or gram caterpillar, *Heliothis armigera*, and the spotted pod borer, *Maruca testulalis*, can cause considerable damage. In certain areas, eg Malawi, Hawaii and Puerto Rico, some cultivars suffer from attack from root-knot nematodes, principally *Meloidogyne javanica*, although in Puerto Rico, the presence of *Rotylenchus reniformis*, *Tylenchorhynchus* sp., *Criconemoides* sp., *Helicotylenchus* sp. and *Hoplolaimus* sp. has been reported.

At the post-harvest stage pulse beetles, *Callosobruchus chinensis*, are a very serious pest, particularly in India. Infestation can begin in the field and a pre-planting dusting with dieldrin, preferably mixed with fungicides such as captan or thiram, is frequently recommended. The dry seed may be protected by fumigation; carbon bisulphide is effective.

Diseases—the most important and widespread disease of pigeon peas is wilt, *Fusarium oxysporum* f. sp. *udum*, which is soil-borne and affects the plant at all stages of its development. Soil temperatures of between 62° and 84°F (17–20°C) favour pigeon pea wilt and it is usually more prevalent in the later part of the growing season. In India, where it is particularly serious, crop losses of 5 to 10 per cent are fairly commonplace, and in severe cases can amount to 50 per cent or more. A rotation of 3–5 years is required to free the soil of infection. Rotation of pigeon pea with tobacco has been suggested as a possible means of control through the adverse effect of the root exudates of tobacco on the pathogen. Mixed cropping with sorghum is reported to reduce the disease incidence. The best method of control is the cultivation of resistant cultivars and in India a number such as NP–15, NP–38, C–11, C–36 and T–17 have been released and proved to be fairly resistant.

Minor diseases of pigeon peas in India include a leaf spot caused by *Cercospora indica*, stem blight caused by *Phytophthora cajani*, and a stem canker, *Diplodia cajani*. In addition, a bacterial leaf spot and stem canker caused by *Xanthomonas cajani* is common during the summer in Maharashtra, when the relative humidity is 80–90 per cent and the temperature ranges from 75° to 88°F (24–31°C).

In the W.I. a collar rot and stem canker are serious, especially when pigeon peas are grown as a perennial crop. The causal organism is reported to be *Physalospora cajanae*, but it is similar to the disease that has also been ascribed to *Diplodia cajani* in India, and *Phoma cajani* in Puerto Rico. Rust, *Uredo cajani*, is also reported to be troublesome in the W.I. In addition, in Jamaica a root rot, *Rosellinia* sp., is sometimes a problem, especially when the crop is grown in soil where the roots of the previous crop remain. In Africa basal stem rot caused by *Macrophomina phaseoli* is of some economic importance, sometimes resulting in the loss of 2 to 4 per cent of plants in Uganda. In the Malagasy Republic the fungal disease *Phaeolus manihotis* is of economic importance. Leaf spots caused by *Cercospora* sp. and *Colletotrichum cajanae* and downy mildew, *Leveillula taurica*, also occur in many areas where pigeon peas are grown and frequently reduce productivity.

Several virus diseases affect pigeon peas. A sterility disease occurs in India and Burma and has been designated pigeon pea mosaic. It can cause almost complete crop failure. The vector was reported to be an eriophid mite, but recently it has been suggested that it may be transmitted by nematodes. No specific control measures are known. In the Caribbean witches broom disease can be troublesome. It is reported to be transmitted by leafhoppers, *Empoasca* spp. In Trinidad and Tobago, and Puerto Rico the pigeon pea crop is reported to be affected by cowpea mosaic.

Growth period

The growth period varies considerably from approximately 100 to 300 days, according to the cultivar, the growing location and the time of sowing. In India most late-maturing cultivars take about 240 to 280 days to produce a seed crop and many of the older early-maturing cultivars about 180 days. Recently improved short-duration cultivars have been developed, including Pusa ageta and Tenkasi red gram, which mature in 150 and 135 days, respectively, and UPAS–120, which produces seed in 110–117 days. In Zaire the average growth period is 158 days, in the Sudan 150 to 180 days, in Kenya 180 days and in Florida 120–150 days. In Trinidad the flowering of tall-types occurs during the period of short days and therefore depending upon the time of planting, flowering can occur as early as 125 days after sowing, or as late as 430 days.

Although the pigeon pea is a perennial and may be cropped for 2 or 3 years, seed yields usually fall very considerably after the first year so that it is best treated as an annual.

Harvesting and handling

Harvesting for dry seed is often by hand. The pods are picked as they mature, finally when most of the leaves have dried and been shed the plants are cut down to the ground level, tied into bundles, and left to dry for a few days, before being threshed. Sometimes the whole crop is cut down when about two-thirds or three-quarters of the pods are mature. The pods and leaves left after threshing are usually beaten with sticks and the seeds and chaff separated by winnowing. The unevenness of ripening has made mechanical harvesting difficult but combine harvesters have been used successfully in Florida and Guyana. In the latter country the pre-harvest application of a desiccant spray of diquate has been reported. After threshing the seeds are cleaned and thoroughly dried to a moisture content of approximately 10 per cent before being stored, usually in sealed earthenware or metal containers, often covered with a layer of sand, or in sacks which have been treated with malathion. Pigeon peas are susceptible to insect infestation during storage and periodic fumigation of the storage chamber with carbon bisulphide, phosphine, ethylene dibromide or methyl bromide is recommended. In the Bahamas pigeon peas are frequently stored by packing them firmly in oil drums which have been thoroughly steam-cleansed. Properly packed they are reported to keep in a reasonable condition for up to 5 years, although there is some loss of culinary quality.

In the Caribbean, where considerable quantities of pigeon peas are eaten in the green immature state as a vegetable, dwarf cultivars with a greater uniformity of pod maturity are often grown and cut by a mower and threshed in a conventional combine harvester.

Primary product

Seed—which vary greatly in shape, size and colour and may be spherical, oval, kidney-shaped or slightly rectangular—round or oval seed is normally about 0.15–0.3 in. (4–8 mm) in diameter. In India three grades have been distinguished: (i) small, 1 000 seeds weigh up to 2.4 oz (68.5 g); (ii) medium, 1 000 seeds weigh between 2.4 and 3.5 oz (68.5–100 g); (iii) bold, 1 000 seeds weigh more than 3.5 oz (100 g). The colour of the mature seed may be white or dirty greyish-white, various shades of brown, red, pinkish- or purplish-black, or mottled. The hilum is small and usually white, sometimes with two prominent ridges. In India, white seed tends to fetch premium prices.

Yield

In India yields of pigeon peas, when grown as a mixed crop, normally average between 200 and 800 lb/ac (225–900 kg/ha) and between 1 500 and 1 800 lb/ac (1 680–2 020 kg/ha) when grown as a pure crop with efficient management. Several of the recently released cultivars have a higher yield potential, eg Pusa ageta 2 400 lb/ac (2 700 kg/ha). In E. Africa yields normally average between 400 and 600 lb/ac (450–670 kg/ha), but with good husbandry may reach 1 000 lb/ac (1 120 kg/ha). In Egypt, on good soils, yields are reported to average 730 lb/ac (825 kg/ha).

In Jamaica the yield of green pigeon peas in the pod is reported to be 5 000–8 000 lb/ac (5 600–9 000 kg/ha), while in Puerto Rico the yield of green pods is reported to average less than 2 000 lb/ac (2 250 kg/ha), due to the primitive cultural methods used by many growers, and in Trinidad the yield of the green peas normally ranges from 80 to 490 lb/ac (90–550 kg/ha) depending upon the standard of crop husbandry.

Pigeon pea

Main use

The pigeon pea is an important protein food in many tropical areas. In India it is consumed mainly in the form of dhal. In Africa and Indonesia the mature seeds are usually soaked for several hours before being pounded and fried, or steamed, and eaten often in the form of a puree. The fresh green seeds are a very popular vegetable, particularly in the Caribbean area, where considerable quantities are processed.

Subsidiary uses

The ripe seeds may be germinated similarly to mung beans to produce sprouts. Small-seeded pigeon peas are sometimes used for poultry food, but crushing is advisable as the seeds are very hard. In Hawaii pigeon peas are used mainly for animal feeding. The green immature pods are sometimes eaten, boiled as a vegetable, similar to French beans, or as a constituent of curries. Pigeon peas are sometimes grown for forage. The average composition of the fresh green forage is: moisture 70.4 per cent; crude protein 7.1 per cent; crude fibre 10.7 per cent; N-free extract 7.9 per cent; fat 1.6 per cent; ash 2.3 per cent; and that of the whole plant, dried and ground into a meal: moisture 11.2 per cent; crude protein 14.8 per cent; crude fibre 28.9 per cent; N-free extract 39.9 per cent; fat 1.7 per cent; ash 3.5 per cent. Pigeon peas are also sometimes grown as a green manure or cover crop, as a windbreak, or to provide support or shade for crops such as vanilla.

Secondary and waste products

The dried stalks are sometimes used for thatching, for firewood or for the production of charcoal. The pods, husks, leaves remaining after threshing and the broken seeds obtained in the production of dhal are used for animal feeding. An approximate analysis of the pods and leaves left after threshing is as follows: moisture 6.2 per cent; crude protein 15.0 per cent; carbohydrate 54.0 per cent; fibre 14.4 per cent; ash 10.4 per cent. The leaves, flowers and roots are all used in traditional medicine in various countries. In Malagasy pigeon peas have been grown for rearing silkworms and in India were used as a host plant for lac insects during the last century.

Special features

A typical analysis of fresh, green pigeon peas is: moisture 67.4 per cent; protein 7.0 per cent; fat 0.6 per cent; carbohydrate 20.2 per cent; fibre 3.5 per cent; ash 1.3 per cent. The approximate composition of the mature dry seed is: moisture 7.1–10.35 per cent; crude protein 14.0–28.9 per cent, (av over 22 per cent); carbohydrate 36.0–65.8 per cent; fat 1.0–9.0 per cent (av 2.1 per cent); crude fibre 5.0–9.4 per cent; ash 3.8 per cent. Pigeon peas are an excellent source of vitamin B, average figures are: thiamine 500 mg/100 g of edible portion; riboflavin 150 mg/100 g; nicotinic acid 2.3 mg/100 g. Their average mineral content is calcium 154–194 mg/100 g and phosphorus 238–372 mg/100 g. The chief protein of the pigeon pea is the globulin cajanin, which accounts for 58 per cent of the total nitrogen, another globulin concajanin accounts for a further 8 per cent. Both are high in tyrosine and moderately rich in cystine, arginine and lysine. Pigeon peas have the following average amino acid composition (mg/gN): isoleucine 194; leucine 394; lysine 481; methionine 32; cystine 61; phenylalanine 517; tyrosine 126; threonine 182; valine 225; arginine 304; histidine 232; alanine 264; aspartic acid 600; glutamic acid 1 171; glycine 203; proline 247; serine 259. Thus, they are very low in the essential amino acids methionine and cystine.

Recently workers in Bangladesh have reported that local samples of pigeon peas contained 35.7 per cent starch and 3.4 per cent total sugars, of which 2.5 per cent is reducing sugar. Indian workers, however, have reported the following sugars present: sucrose 1.6 per cent; raffinose 1.1 per cent; stachyose 2.7 per cent; verbascose 4.1 per cent. The composition of the fat is: linolenic acid 5.7 per cent; linoleic 51.4 per cent; oleic 6.3 per cent; saturated fatty acids 36.6 per cent.

A trypsin inhibitor, which was found to be fairly heat stable and active over pH 2.5–10.1, has been isolated. The seeds are also reported to contain 200 mg/100 g of phytic-phosphorus, of which 75–89 per cent is phytin. In addition the enzyme urease and a single form of α–galactosidase with a relatively low molecular weight have been isolated from the seeds. The raw seeds are also reported to contain an unidentified soporific. The seed-coat, which is very hard and tough, constitutes about 10 per cent of the seed and is reported to contain an essential oil, having the odour of butter and containing α–, β–, and γ–selinenes, copaene and a mixture of eudesmols.

Processing

Dhal—is prepared by removing the seed-coat and splitting the seed into two cotyledons, it is accomplished by either the wet or dry method. In the dry method the seeds are dried in the sun for 3 or 4 days, and then split in a mill after they have been smeared with a vegetable oil to soften the husk. After milling all unsplit seeds are removed, treated with oil and left to dry in the sun before being remilled. This fractional milling is repeated three or four times, until all the seeds are split. The yield of clean dhal by this dry method is about 66 per cent. In the wet method, the seeds are first soaked in water for 6 to 10 hours, then mixed with sieved red earth, heaped up and left overnight. The seeds are then spread out, dried in the sun, and winnowed to remove the earth, before being milled. The split dhal is cleaned by repeated winnowing and heated with a small quantity of castor or gingelly oil to improve its appearance and preserve its quality. The yield of dhal by this method is about 75 to 80 per cent.

Dhal produced by the dry method is hemispherical in shape, and normally fetches a higher price because it softens more rapidly on cooking and has a good flavour. Dhal prepared by the wet method is normally flatter with a small depression in the centre. Its quality, particularly the time it takes to soften when cooked, is affected by the amount of mineral salts present in the soaking water.

The Central Food Technological Research Institute, Mysore, (CFTRI), has developed an improved method for producing dhal which reduces the time required for processing and the cost. In the improved process the seed is exposed to hot air for a predetermined time in specially designed units and equilibrated with gradual aeration to a critical moisture level in specially designed bins. The loosened skin-coat is then removed in an improved abrasion type machine. About 99.5 per cent removal of the skin-coat can be achieved by this improved method in one milling operation, if the seeds have been properly conditioned. There is considerable variation in the milling characteristics and the yield of dhal according to the cultivar, climate and crop management. The yield of dhal using the CFTRI improved process is usually 80–85 per cent. Recently a process has been developed in India by which the relatively expensive vegetable oil used in the traditional milling process is replaced by fermented sugar cane molasses (Sirka). The possibility of treating dhal with papain to reduce the time required for cooking is also being studied.

Dhal is graded according to the amount of broken, unripe and shrivelled grains present; that produced from pigeon peas is liable to be adulterated with other pulses such as lentils and peas.

Canning—considerable quantities of green, immature pigeon peas are canned in the Caribbean. A good quality final product, with an almost colourless brine of low turbidity, can be obtained provided that the pods are steam-scalded for 90 seconds before shelling to inactivate the peroxidases. The pre-heated pods are shelled, the peas are then cooled, washed, and blanched at 185°F (85°C) for 5 minutes. After thorough rinsing in cold water and a final sorting the peas are packed in plain cans with lacquered ends, a 2 per cent solution of brine at 195–200°F (91–93°C) is added, and A2 cans (307 x 408) closed, processed for 35 minutes at 240°F (116°C), and then immediately water-cooled to 100°F (38°C). Mature pigeon peas may be canned also. The peas should have a comparatively high moisture content (19 per cent) and are first soaked in water at 150°F (66°C) for 3 hours, then blanched for 3 minutes at 185°F (85°C). A 2 per cent brine solution is added and the sealed cans processed at 240°F (116°C) for 45 minutes.

Quick-freezing—the green peas are shelled and sorted as for canning then blanched for 5 minutes at 195°F (91°C), rinsed, hand-packed into plastic bags, or cardboard cartons, and frozen in a blast-freezer at −45°F (−43°C), followed by storage at −10°F (−23°C). Fully mature peas with an alcohol insoluble solids content of approximately 27 per cent give the best product.

Dehydration—attempts have been made to produce good quality dehydrated green pigeon peas, but the thickness of the seed-coat inhibits the dehydration process and water absorption during rehydration. Investigations suggest that an improved product can be obtained by soaking the seeds in 0.2 per cent sodium hydroxide for 4 hours before dehydration and packing in cans rather than polythene bags.

Production and trade

Production—pigeon peas constitute just over 4 per cent of the total reported global production of grain legumes. The reported world output averaged some 1 880 000 t/a for the period 1970–74, compared with 1 783 000 t/a for 1965–69, an increase of 5 per cent. In 1975 production was 1 960 000 t. About 95 per cent of the crop is produced in southern Asia, primarily in India. However, it is generally accepted that reported production figures for many African countries could be low by a factor of two or three times, since the crop is grown extensively in village compounds and kitchen gardens, at low and medium elevations throughout the humid to semi-arid areas of tropical Africa, but production is not reported.

Pigeon pea

Pigeon peas: Major producing countries

Quantity tonnes

	Annual average 1965–69	Annual average 1970–74	1975
India	1 662 000	1 740 000	1 818 000
Uganda	26 800	38 800	40 000
Dominican Repub.	21 600	27 800	29 000[1]
Burma	22 600	24 800	24 000

1 Unofficial figure.

Trade—pigeon peas are produced almost exclusively for domestic consumption and only very small quantities enter international trade. There is a small export trade from India, mainly to Indian nationals living abroad. Indian exports averaged 1 948 t/a for the years 1966–69 and 585 t/a for the period 1970–74, and in fact fell steadily from 1969, when they were 4 317 t, to 92 t in 1973, but rose to 402 t in 1974 and 878 t in 1975. At one time the bulk of these exports were consigned to Malaysia, the UK, Singapore and Bahrein, although shipments to individual countries fluctuated considerably from year to year. However, in more recent years the main destination has been Nepal.

Prices—average wholesale prices for pigeon peas in India for the period 1970–74 were £56.28/t, compared with £50.47/t for 1965–69, an increase of about 11 per cent. The average fob value of Indian exports was £121.00/t in 1975 and £110.00/t for the period 1970–74, compared with £82.75/t for the years 1966–69.

Major influences

Although a useful grain legume crop, the sensitivity of the pigeon pea to frost and waterlogging and the relatively long growth period of many of the established cultivars are factors which have tended to handicap the spread of its cultivation. Nevertheless, it is a crop which could have considerable potential over a wide range of tropical conditions from subhumid to semi-arid areas. Although currently average yields are between 400 and 1 000 lb/ac (450–1 120 kg/ha), experimental trials with improved cultivars indicate that the crop could have a high level of productivity. However, in India, by far the major producer, the development of cultivars with resistance to disease, especially fusarium, is of prime importance. While in the Caribbean the increasing demand for the green peas for processing, and the problem of hand-harvesting, have resulted in breeding programmes to develop high-yielding disease-resistant cultivars of a compact habit and determinancy of bearing, so that mechanical harvesting would be practicable. Recently, it has been shown that by planting dwarf cultivars at 12 x 6.5 in. (30 x 16.25 cm), which gives a density of approximately, 80 000 pl/ac (200 000 pl/ha), mechanical harvesting is economically practicable.

Bibliography

ABRAMS, R. 1975. Status of research on pigeon peas in Puerto Rico. *International workshop on grain legumes*. pp. 141–147. Hyderabad, India, Int. Crops Res. Inst. Semi-arid Trop., 350 pp.

ABRAMS, R. and JULIA, F. J. 1973. Effect of planting time, plant population and row spacing on yield and other characteristics of pigeon peas, *Cajanus cajan* (L.) Millsp. *J. Agric., Univ. P.R., 57*, 275–285.

ABRAMS, R. and JULIA, F. J. 1974. Effect of mechanical, cultural and chemical weed control on yield and yield components of pigeon peas, *Cajanus cajan* (L.) Millsp. *J. Agric., Univ. P.R., 58*, 466–472.

ACLAND, J. D. 1971. Pigeon peas. *East African crops*. pp. 140–141. London: Longman Group Ltd, 252 pp.

AKBAR, S., KHAN, N. A. and HUSSAIN, T. 1973. Amino acid composition and nutritive value of arhar (*Cajanus indicus*) grown in Peshawar region. *Pak. J. Sci. Ind. Res., 16*, 130–131.

AKINOLA, J. O. and WHITEMAN, P. C. 1972. A numerical classification of *Cajanus cajan* (L.) Millsp. accessions based on morphological and agronomic attributes. *Aust. J. Agric. Res., 23*, 995–1005.

AKINOLA, J. O. and WHITEMAN, P. C. 1975. Agronomic studies on pigeon pea (*Cajanus cajan* (L.) Millsp.): (i) Field responses to sowing time. *Aust. J. Agric. Res., 26*, 43–56.

AKINOLA, J. O. and WHITEMAN, P. C. 1975. Agronomic studies on pigeon pea (*Cajanus cajan* (L.) Millsp.): (ii) Responses to sowing density. *Aust. J. Agric. Res., 26*, 57–66.

AKINOLA, J. O. and WHITEMAN, P. C. 1975. Agronomic studies on pigeon pea (*Cajanus cajan* (L.) Millsp.): (iii) Responses to defoliation. *Aust. J. Agric. Res., 26*, 67–79.

Pigeon pea

AKINOLA, J. O., WHITEMAN, P. C. and WALLIS, E. S. 1975. The agronomy of pigeon pea (*Cajanus cajan*). *Commonw. Agric. Bur., Bur. Pastures & Field Crops, Hurley, Maidenhead, Berks, England, Rev. Ser.* 1, 57 pp.

ALVAREZ GARCIA, L. A. 1960. Phoma canker of pigeon peas in Puerto Rico. *J. Agric., Univ. P.R., 44*, 28–30.

AMIN, K. S., BALDEV, B. and WILLIAMS, F. J. 1976. Differentiation of *Phytophthora* stem blight from *Fusarium* wilt of pigeon pea by field symptoms. *FAO, Plant Prot. Bull., 24*, 123–124.

ANON. 1960. How to produce dhall. *Farm J., Dep. Agric. Brit. Guiana, 21* (3), 13–16; 20–22.

ANON. 1961. Recent advances in the production of arhar dhall. *Farm J., Dep. Agric. Brit. Guiana, 22* (2), 8.

ANON. 1972. Pulse varieties developed at IARI. *Indian Farming, 21* (10), 47.

ANON. 1975. *Cajanus cajan*—a new legume cover crop. *Planters Bull., Rubber Res. Inst. Malays.*, (137), pp. 51–52.

APONTE APONTE, F. 1963. El cultivo de grandules en Puerto Rico. [The cultivation of pigeon peas in Puerto Rico]. *Caribb. Agric., 1*, 191–197.

ARAULLO, E. V. 1974. Processing and utilization of cowpea, chick pea, pigeon pea and mung bean. *Proc. symp. interaction of agriculture with food science, Singapore*, 1974. (MacIntyre, R. ed.). pp. 131–142. Ottawa, Int. Dev. Res. Cent., IDRC–033e, 166 pp.

ARIYANAGAM, R. P. 1975. Status of research on pigeon peas in Trinidad. *International workshop on grain legumes*. pp. 131–139. Hyderabad, India, Int. Crops Res. Inst. Semi-arid Trop., 350 pp.

ARNON, I. 1972. Pulse or grain legumes. *Crop production in dry regions.* Vol. 2. pp. 217–260. London: Leonard Hill Books, 683 pp.

AYALA, A. 1962. Occurrence of the nematode *Meloidogyne javanica* on pigeon pea roots in Puerto Rico. *J. Agric., Univ. P.R., 46*, 154–156.

AYKROYD, W. R. and DOUGHTY, J. 1964. Legumes in human nutrition. *FAO Nutr. Stud.* 19, pp. 102; 115; 117. Rome: FAO, 138 pp.

BHARGAVA, R. N. 1975. Two new varieties of arhar for Bihar. *Indian Farming, 25* (1), 23.

BHATIA, B. S., RAMANATHAN, A., PRASAD, M. S. and VIJAYARAGHAVAN, P. K. 1967. Use of papain in the preparation of quick-cooking dehydrated pulses and beans. *Food Technol.*, *21*, 1395–1397.

CHOPRA, K. and SWAMY, G. 1975. *Pulses: An analysis of demand and supply in India. Institute for social and economic change; Monograph 2.* New Delhi: Sterling Publishers, PVT, Ltd, 132 pp.

CHOUDHURY, S. L. and BHATIA, P. C. 1971. Ridge-planted kharif pulses yield high despite waterlogging. *Indian Farming*, *21* (3), 8–9.

CHOWDHURY, S. L. and BHATIA, P. C. 1971. Profits triple when arhar is adequately fertilized. *Indian Farming*, *20* (12), 27–30.

COMMONWEALTH BUREAU OF PASTURES AND FIELD CROPS. nd. Pigeon peas (*Cajanus cajan*). Annotated bibliography 1253, 1959–70. *Commonw. Agric. Bur., Bur. Pastures & Field Crops, Hurley, Maidenhead, Berks, England*, 8 pp.

CROSS, L. A. 1970. The pigeon pea today. *Cajanus*, *3*, 103–109.

CRUZ, C. 1975. Observations on pod borer oviposition and infestation of pigeon pea varieties. *J. Agric., Univ. P.R.*, *59*, 63–68.

DAHIYA, B. S. and BRAR, J. S. 1976. The relationship between seed size and protein content in pigeon pea (*Cajanus cajan* (L.) Millsp.). *Trop. Grain Legume Bull.*, (3), pp. 18–19.

DAHIYA, B. S., BRAR, J. S. and KAUL, J. N. 1974. Changes in the growth habit of pigeon peas (*Cajanus cajan* (L.) Millsp.) due to late sowing. *J. Agric. Sci.*, (*Camb.*), *83*, 379–380.

DALAL, R. C. 1974. Effects of intercropping maize with pigeon peas on grain yield and nutrient uptake. *Exp. Agric.*, *10*, 219–224.

DART, P. J., ISLAM, R. and EAGLESHAM, A. 1975. The root nodule symbiosis of chickpea and pigeonpea. *International workshop on grain legumes*. pp. 63–83. Hyderabad, India, Int. Crops Res. Inst. Semi-arid Trop., 350 pp.

DAVIES, J. C. and LATEEF, S. S. 1975. Insect pests of pigeonpea and chickpea in India and prospects for control. *International workshop on grain legumes*. pp. 319–323. Hyderabad, India, Int. Crops Res. Inst. Semi-arid Trop., 350 pp.

DE, D. N. 1974. Pigeon pea. *Evolutionary studies in world crops: Diversity and change in the Indian subcontinent*. (Sir Joseph Hutchinson ed.). pp. 79–87. London: Cambridge University Press, 175 pp.

Pigeon pea

Derieux, M. 1971. Quelques données sur le comportement du pois d'Angole en Guadeloupe (Antilles françaises). [Some data on the behaviour of pigeon pea in Guadeloupe (French West Indies)]. *Ann. Amélior. Plant., 21*, 373–407.

Elías, J. G., Hernández, M. and Bressani, R. 1976. The nutritive value of precooked legume flours processed by different methods. *Nutr. Rep. Int., 14*, 385–403.

Food and Agriculture Organization of the United Nations. 1970. Amino acid content of foods and biological data on proteins. *FAO Nutr. Stud. 24*, pp. 56–57. Rome: FAO, 286 pp.

Göhl, B. 1975. *Cajanus cajan* (L.) Millsp. (*C. indicus* Spreng.). *Tropical feeds: Feeds information summaries and nutritive values*. pp. 165; 511; 527. Rome: FAO, 661 pp.

Gooding, H. J. 1962. The agronomic aspects of pigeon peas. *Field Crop Abstr., 15*, 1–5.

Gupta, G. L., Nigam, S. S., Sastry, S. D. and Chakravarti, K. K. 1969. Investigations on the essential oil from *Cajanus cajan* (Linn.) Millsp. *Perfum, Essent. Oil Rec., 60*, 329–336.

Hammerton, J. L. 1975. Effects of defoliation on pigeon peas (*Cajanus cajan*). *Exp. Agric., 11*, 177–182.

Hammerton, J. L. 1975. Effects of growth regulators on pigeon pea (*Cajanus cajan*). *Exp. Agric., 11*, 241–245.

Hammerton, J. L. 1976. Effects of planting date on growth and yield of pigeon pea (*Cajanus cajan* (L.) Millsp.). *J. Agric. Sci.*, (*Camb.*), *87*, 649–660.

Hammerton, J. L. and Pierre, R. E. 1971. *Cajanus cajan* the pigeon or gungo pea. *Cajanus, 4*, 81–88.

Henderson, T. H. 1965. Some aspects of pigeon pea farming in Trinidad. *Trinidad, Univ. W.I., Dep. Agric. Econ. Farm Manage. Occas. Ser.* 3, 40 pp.

Hulse, J. H. 1975. Problems of nutritional quality of pigeonpea and chickpea and prospects of research. *International workshop on grain legumes*. pp. 189–207. Hyderabad, India, Int. Crops Res. Inst. Semi-arid Trop., 350 pp.

International Crops Research Institute for the Semi-arid Tropics. 1975. The pulses—Pigeon pea improvement. *ICRISAT, Annu. Rep. 1973/74*. pp. 35–40. *Hyderabad, India, Int. Crop Res. Inst. Semi-arid Trop.*, 87 pp.

INTERNATIONAL INSTITUTE OF TROPICAL AGRICULTURE. 1975. Grain legume program: Lima beans and pigeon peas. *IITA Annu. Rep. 1974*. pp. 83–85. *Ibadan, Nigeria, Int. Inst. Trop. Agric.*, 199 pp.

IRVINE, F. R. 1969. Pigeon pea (*Cajanus cajan*). *West African agriculture*. Vol. 2. 3rd ed. pp. 205–207. London: Oxford University Press, 272 pp.

JADHAV, P. S., JAIN, T. C. and PRASANNALAKSHMI, S. 1975. *Sorghum, millets and peas: A bibliography of the Indian literature*, 1969–73. pp. 72–73. India, Hyderabad, Int. Crops Res. Inst. Semi-arid Trop., 116 pp.

JANARTHANAN, R., SATHIABALAN, S. G., SUBRAMANIAN, K. S., NAVANEETHAN, G. and KANDASAMY, T. K. 1973. A report on the survey of sterility mosaic disease incidence on red gram in Tamilnadu. *Madras Agric. J.*, *60*, 41–44.

JESWANI, L. M. 1975. Varietal improvement of seed legumes in India. *Food protein sources*. (Pirie, N. W. and Swaminathan, M. S. eds.). pp. 9–18. London: Cambridge University Press, 260 pp.

JOHNSON, R. M. and RAYMOND, W. D. 1964. The chemical composition of some tropical food plants: Pigeon peas and cowpeas. *Trop. Sci.*, *6*, 68–70.

KANWAR, J. S. 1974. Improvement of crops and their relationship to nutrition and food science technology in the semi-arid tropics. *Proc. symp. interaction of agriculture with food science, Singapore*, 1974. (MacIntyre, R. ed.). pp. 53–64. Ottawa, Int. Dev. Res. Cent., IDRC–033e, 166 pp.

KANWAR, J. S. and SINGH, K. B. (eds.). 1974. Pigeon pea (*Cajanus cajan*). *Guide for field crops in the tropics and subtropics*. (Litzenberger, S. C. ed.). pp. 146–153. Washington, Off. Agric. Tech., Assist. Bur., Agency Int. Dev., 321 pp.

KASASIAN, L. 1971. Vegetable crops. *Weed control in the tropics*. pp. 163–175. London: Leonard Hill Books, 307 pp.

KHAN, T. N. and ASHLEY, J. M. 1975. Factors affecting plant stand in pigeon pea. *Exp. Agric.*, *11*, 315–322.

KHAN, T. N. and RACHIE, K. O. 1972. Preliminary evaluation and utilization of pigeon pea germplasm in Uganda. *East Afr. Agric. For. J.*, *38*, 78–82.

KILLINGER, G. B. 1968. Pigeon peas (*Cajanus cajan* (L.) Druce), a useful crop for Florida. *Proc. Soil Crop Sci. Soc., Florida*, *28*, 162–167.

Pigeon pea

KRAUSS, F. G. 1932. The pigeon pea (*Cajanus indicus*) its improvement, culture and utilization in Hawaii. *US Dep. Agric., Hawaii Agric. Exp. Stn. Bull.* 64, 46 pp.

KRISHNAMURTHY, K., GIRISH, G. K., RAMASIVAN, T., BOSE, S. K., SINGH, K. and TOMER, R. P. S. 1972. A new process for the removal of husk of red gram using sirka. *Bull. Grain Technol., 10*, 181–186.

KURIEN, P. P., DESIKACHAR, H. S. R. and PARPIA, H. A. B. 1972. Processing and utilization of grain legumes in India. Symposium on food legumes. pp. 225–236. *Tokyo, Jpn., Minist. Agric. & For., Trop. Agric. Res. Cent., Trop. Agric. Res. Ser.* 6, 253 pp.

KURIEN, P. P. and PARPIA, H. A. B. 1968. Pulse milling in India: (i) Processing and milling of tur arhar (*Cajanus cajan* Linn.). *J. Food Sci. Technol., 5*, 203–207.

LAURENCE, G. A. 1971. Insect pests of pigeon pea and their control. *J. Agric. Soc. Trinidad & Tob., 71*, 501–504.

MAHTA, D. N. and DAVE, B. B. 1933. Studies in *Cajanus indicus. Mem. Dep. Agric., India, Bot. Ser., 19*, 1–25.

MALHOTRA, R. S. and SINGH, K. B. 1973. Genetic variability and genotype-environment interaction in Bengal-gram. *Indian J. Agric. Sci., 43*, 914–917.

MATHUR, H. G. 1958. Marketing of pulses in India. *Nagapur, India, Agric. Mark. Gov. India, Advis. Agric. Mark. Ser.* AMA 102, 182 pp.

MODI, J. D. and KULKARNI, P. R. 1976. Studies on the starches of ragi and red gram. *J. Food Sci. Technol., 13*, 9–10.

MOHAMED-HANIFA, A., BALASUBRAMANIAM, G., LEELA, D. A. and SUBRAMANIAM, T. R. 1974. Granular insecticides for the control of pod borers in red gram. *Madras Agric. J., 61*, 970–972.

MOODY, K. 1973. Weed control in tropical grain legumes. *Proc. 1st Int. Inst. Trop. Agric., Grain legume improvement workshop.* pp. 162–183. Ibadan, Nigeria, Int. Inst. Trop. Agric., 325 pp.

MORTON, J. F. 1976. The pigeon pea (*Cajanus cajan* Millsp.), a high-protein tropical bush legume. *Hortscience, 11*, 11–19.

MUÑOZ, A. M. and ABRAMS, R. 1971. Inheritance of some quantitative characters in pigeon peas (*Cajanus cajan* (L.) Millsp.). *J. Agric., Univ. P.R., 55*, 23–43.

NICHOLS, R. 1964. Mineral deficiency symptoms of pigeon peas (*Cajanus cajan*). *J. Agric. Soc. Trinidad, & Tob.*, *64*, 27–29.

OKE, O. L. 1969. Sulphur nutrition of legumes. *Exp. Agric.*, *5*, 111–116.

PALIWAL, K. W. and MALIWAL, G. L. 1973. Salt tolerance of some arhar (*Cajanus indicus*) and cowpea (*Vigna sinensis*) varieties at germination and seedling stages. *Ann. Arid Zone*, *12*, 135–144.

PATHAK, G. N. 1970. Red gram (*Cajanus cajan* (L.) Millsp.). *Pulse crops of India*. (Kachroo, P. and Arif, M. eds.). pp. 14–53. New Delhi: Indian Counc. Agric. Res., 334 pp.

PAVGI, M. S. and SINGH, R. A. 1965. Some parasitic fungi on pigeon pea from India. *Mycopathol. Mycol. Appl.*, *27*, 97–106.

PIETRI, R., ABRAMS, R. and JULIA, F. J. 1971. Influence of fertility level on the protein content and agronomic characters of pigeon peas in an oxisol. *J. Agric., Univ. P.R.*, *55*, 474–477.

PRASAD, S., PRAKASH, R. and HASSAN, M. A. 1972. Natural crossing in pigeon pea (*Cajanus cajan* (L.) Millsp.). *Mysore J. Agric. Sci.*, *6*, 426–429.

PURSEGLOVE, J. W. 1968. *Cajanus cajan* (L.) Millsp. *Tropical crops: Dicotyledons*. Vol. 1. pp. 236–241. London: Longmans, Green and Co Ltd, 332 pp.

PUSHPAMMA, P. 1975. Evaluation of nutritional value, cooking quality and consumer preferences of grain legumes. *International workshop on grain legumes*. pp. 213–220. Hyderabad, India, Int. Crops Res. Inst. Semi-arid Trop., 350 pp.

RACHIE, K. O., NANGJU, D., RAWAL, K., LUSE, R. A., WILLIAMS, R. J., SINGH, S. R. and WIEN, H. C. 1975. Descriptions of CITA–1, CITA–2 and CITA–4 pigeon peas (GP–1). *Trop. Grain Legume Bull.*, (1), pp. 12–15.

RACHIE, K. O. and ROBERTS, L. M. 1974. Grain legumes of the lowland tropics. *Adv. Agron.*, *26*, 1–132.

RACHIE, K. O. and SILVESTRE, P. 1977. Grain legumes. *Food crops of the lowland tropics*. (Leakey, C. L. A. and Wills, J. B. eds.). pp. 41–74. Oxford: Oxford University Press, 345 pp.

RAHMAN, A. R. 1961. The effect of chemical pretreatment on the quality of dehydrated pigeon peas. *J. Agric., Univ. P.R.*, *45*, 172–181.

RAHMAN, A. R. 1964. Effect of storage and packaging on the quality of dehydrated and dehydrofrozen pigeon peas. *J. Agric., Univ. P.R.*, *48*, 318–326.

RAHMAN, Q. N., AKHTAR, N. and CHOWDHURY, A. M. 1974. Proximate composition of foodstuffs in Bangladesh: (i) Cereals and pulses. *Bangladesh J. Sci. Ind. Res.*, *9*, 129–133.

RAMANUJAM, S. 1973. Grain legume improvement in India. *Proc. 1st Int. Inst. Trop. Agric., Grain legume improvement workshop*. pp. 37–41. Ibadan, Nigeria, Int. Inst. Trop. Agric., 325 pp.

RATHNASWAMY, R., VEERASWAMY, R. and PALANISWAMY, G. A. 1973. Studies in red gram (*Cajanus cajan* (L.) Millsp.) seed characters, cooking quality and protein content. *Madras Agric. J.*, *60*, 396–398.

RAWAT, R. R. and JAKHMOLA, S. S. 1967. Estimation of losses in grain yield in different varieties of tur (*Cajanus cajan*) by pod fly (*Melanagromyza obtusa*), plume moth (*Exelastis atomosa*), pulse beetle (*Bruchus* sp.) and other means. *Madras Agric. J.*, *54*, 601–602.

REGUPATHY, A. and RATHNASWAMY, R. 1970. Studies on comparative susceptibility of seeds of certain red gram (*Cajanus cajan* (L.) Millsp.) varieties to pulse beetle, *Callosobruchus chinensis* L. (Bruchidae: Coleoptera). *Madras Agric. J.*, *57*, 106–109.

RIOLLANO, A. 1964. Effects of photoperiodism and other factors on the improvement of pigeon pea varieties. *J. Agric., Univ. P.R.*, *48*, 232–235.

RIOLLANO, A., PÉREZ, A. and RAMOS, C. 1962. Effects of planting date, variety, and plant population on the flowering and yield of pigeon pea (*Cajanus cajan* (L.). *J. Agric., Univ. P.R.*, *46*, 127–134.

ROYES, W. VERNON. 1975. Amino acid profiles of *Cajanus cajan* protein. *Nutritional improvement of food legumes by breeding*. (Milner, M. ed.). pp. 193–196. New York/London: John Wiley and Sons, 399 pp.

ROYES, W. VERNON. 1975. Testa pigments in *Cajanus cajan* canning quality and flavour. *Nutritional improvement of food legumes by breeding*. (Milner, M. ed.). pp. 297–298. New York/London: John Wiley and Sons, 399 pp.

ROYES, W. VERNON. 1976. Pigeon pea. *Evolution of crop plants*. (Simmonds, N. W. ed.). pp. 154–156. London: Longman Group Ltd, 339 pp.

ROYES, W. V. and FINCHAM, A. G. 1975. Grain quality in *Cajanus* and *Cicer*. *International workshop on grain legumes*. pp. 209–211. Hyderabad, India, Int. Crops Res. Inst. Semi-arid Trop., 350 pp.

RYAN, J. G., OPPEN, M. VON, SUBRAHMANYAM, K. V. and ASOKAN, M. 1974. Socio-economic aspects of agricultural development in the semi-arid tropics. *ICRISAT, International workshop on farming systems*. pp. 389–431. Hyderabad, India, Int. Crop. Res. Inst. Semi-arid Trop.

SÁNCHEZ NIEVA, F. 1961. The influence of degree of maturity on the quality of canned pigeon peas. *J. Agric., Univ. P.R., 45*, 217–231.

SÁNCHEZ NIEVA, F. 1963. Application of shear press to determine the degree of maturity of pigeon peas. *J. Agric., Univ. P.R., 47*, 212–216.

SÁNCHEZ NIEVA, F., GONZÁLEZ, M. A. and BENERO, J. R. 1961. The freezing of pigeon peas for market. *J. Agric., Univ. P.R., 45*, 205–216.

SÁNCHEZ NIEVA, F., GONZÁLEZ, M. A. and BENERO, J. R. 1961. The effects of some processing variables on the quality of canned pigeon peas. *J. Agric., Univ. P.R., 45*, 232–258.

SÁNCHEZ NIEVA, F., GONZÁLEZ, M. A., BENERO, J. R. and HERNÁNDEZ, I. 1963. The brine grading of pigeon peas. *J. Agric., Univ. P.R., 47*, 14–23.

SÁNCHEZ NIEVA, F., RODRIGUEZ, A. J. and BENERO, J. R. 1961. Improved methods of canning pigeon peas. *Univ. P.R., Rio Piedras, Agric. Exp. Stn. Bull.* 157, 26 pp.

SAXENA, M. C. and YADAV, D. S. 1975. Some agronomic considerations of pigeon peas and chickpeas. *International workshop on grain legumes*. pp. 31–61. Hyderabad, India, Int. Crops Res. Inst. Semi-arid Trop., 350 pp.

SELLSCHOP, J. and MULLER, H. 1953. The pigeon pea or dhal bean. *Farming South Afr., 28*, 159–160.

SEN GUPTA, P. K. 1974. Diseases of major pulse crops in India. *PANS, 20*, 409–415.

SENEWIRATNE, S. T. and APPADURAI, R. R. 1966. Dhal. *Field crops of Ceylon*. pp. 153–160. Colombo: Lake House Investments Ltd, 376 pp.

SHARMA, Y. K., TIWARI, A. S., RAO, K. C. and MISHRA, A. 1972. Studies on chemical constituents and their influence on cookability in pigeon pea. *J. Food Sci. Technol., 14*, 38–40.

SHAW, F. J. F., KHAN, A. R. and SINGH, H. 1933. Studies in Indian pulses: (iii) The types of *Cajanus indicus* Spreng. *Indian J. Agric. Sci., 3*, 1–36.

SHERIFF, N. M. and RAJAGOPALAN, C. K. 1971. A comparative study of the intensity of infestation of the pod fly, *Melangromyza* (*Agromyza*) *obtusa* Mall, on different varieties of red gram (*Cajanus cajan* Linn.). *Madras Agric. J., 58*, 842–843.

Pigeon pea

SIEGEL, A. and FAWCETT, B. 1976. *Food legume processing and utilization (with special emphasis on application in developing countries)*. Ottawa, Int. Dev. Res. Cent., IDRC–TS1, 88 pp.

SINGH, B. B., GUPTA, S. C. and SINGH, B. D. 1974. Note on 'UPAS 120' an early-maturing mutant of pigeon pea. *Indian J. Agric. Sci.*, *44*, 233–234.

SINGH, K. B. 1973. Punjab can take on arhar in a big way. *Indian Farming*, *22* (10), 19.

SINGH, L., MAHESHWARI, S. K. and SHARMA, D. 1971. Effect of date of planting and plant population on growth, yield, yield components and protein content of pigeon pea (*Cajanus cajan* (L.) Millsp.). *Indian J. Agric. Sci.*, *41*, 535–538.

SINGH, L., SHARMA, D., DAODHAR, A. D. and SHARMA, Y. K. 1973. Variation in protein, methionine, tryptophan, and cooking period in pigeon pea (*Cajanus cajan* (L.) Millsp.). *Indian J. Agric. Sci.*, *43*, 795–798.

SINGH, L., SINGH, N., SHRIVASTAVA, M. P. and GUPTA, A. K. 1977. Characteristics and utilization of vegetable types of pigeon peas (*Cajanus cajan* (L.) Millsp.). *Indian J. Nutr. Diet.*, *14*, 8–10.

SINGH, N. D. 1975. Evaluation of nematode population in the pigeon pea. *Tropical diseases of legumes*. (Bird, J. and Maramorosch, K. eds.). pp. 147–149. New York/San Francisco/London: Academic Press, 171 pp.

SMARTT, J. 1976. *Cajanus* D.C. *Tropical pulses*. pp. 54–56. London: Longman Group Ltd, 348 pp.

SPENCE, J. A. 1975. The importance of diseases in relation to the grain legume research program in the eastern Caribbean. *Tropical diseases of legumes*. (Bird, J. and Maramorosch, K. eds.). pp. 151–155. New York/San Francisco/London: Academic Press, 171 pp.

SRIKANTIA, S. G. 1975. Chickpea and pigeonpea: some nutritional aspects. *International workshop on grain legumes*. pp. 221–223. Hyderabad, India, Int. Crops Res. Inst. Semi-arid Trop., 350 pp.

SUARD, C. and DEGRAS, L. 1975. Études pour la conservation des semences du pois d'Angole (*Cajanus cajan*) et du niébé (*Vigna sinensis*). [Studies on the conservation of pigeon pea (*Cajanus cajan*) and cowpea (*Vigna sinensis*)]. *Nouv. Agron. Ant. Guy.*, (1), pp. 92–97.

SUAREZ, J. J. and HERRERA, D. 1971. Response of pigeon pea (*Cajanus cajan* Millsp.) at different populations submitted to different heights of cutting. *Rev. Cubana Cienc. Agric.* (*Engl. ed.*), *5*, 71–75.

SWAMINATHAN, M. S. and JAIN, H. K. 1975. Food legumes in Indian agriculture. *Nutritional improvement of food legumes by breeding*. (Milner, M. ed.). pp. 69–82. New York/London: John Wiley and Sons, 399 pp.

TARA, M. R. and RAMA RAO, M. V. 1972. Changes in the essential amino acid content of arhar dhal (*Cajanus cajan*) on dehydration. *J. Food Sci. Technol.*, *9*, 76–79.

TARA, M. R. and RAMA RAO, M. V. 1975. Changes in the free amino acids of arhar dhal (*Cajanus cajan*) on processing. *J. Food Sci. Technol.*, *12*, 71–74.

TARA, M. R., RAWAL, T. N. and RAMA RAO, M. V. 1972. Effect of processing on the proteins of arhar dhal (*Cajanus cajan*). *Indian J. Nutr., Diet.*, *9*, 208–212.

TIWARI, A. S., YADAN, L. N., SINGH, L. and MAHADIK, C. N. 1977. Spreading plant type does better in pigeon pea. *Trop. Grain Legume Bull.*, (7), pp. 7–9.

VEERASWAMY, R., RAJASEKARAN, V. P. A., SELVAKUMARI, G. and SHERIFF, N. M. 1972. Effect of phosphoric acid and organic manure on red gram (*Cajanus cajan* (L.) Millsp.). *Madras Agric. J.*, *59*, 304–305.

VEERASWAMY, R. and RATHNASWAMY, R. 1972. Red gram Co.1—an improved short-term strain from Tamil Nadu. *Madras Agric. J.*, *59*, 177–179.

VEERASWAMY, R., RATHNASWAMY, R., SELVAKUMARI, G. and BADRI NARAYANAN, P. 1972. Studies on the spacing of red gram—(*Cajanus cajan* (L.) Millsp.). *Madras Agric. J.*, *59*, 435–436.

WALLIS, E. S., WHITEMAN, P. C. and AKINOLA, J. O. 1975. Pigeon pea (*Cajanus cajan* (L.) Millsp.) research in Australia. *International workshop on grain legumes*. pp. 149–166. Hyderabad, India, Int. Crops Res. Inst. Semi-arid Trop., 350 pp.

WESTPHAL, E. 1974. *Cajanus cajan* (L.) Millsp. Pulses in Ethiopia, their taxonomy and agricultural significance. pp. 64–72. *Wageningen, PUDOC, Cent. Agric. Publ. Doc., Agric. Res. Rep.* 815, 276 pp.

YEGNA NARAYAN AIYER, A. K. 1966. Togare (*Cajanus indicus*). *Field crops of India*. 6th ed. pp. 103–107. Bangalore City: Bangalore Printing and Publishing Co Ltd, 564 pp.

Common names RICE BEAN, Red bean[1].

Botanical name *Vigna umbellata* (Thunb.) Ohwi & Ohashi, syn. *Phaseolus calcaratus* Roxb.

Family Leguminosae.

Other names Anipay (Philipp.); Bamboo bean, Climbing mountain bean, Crab-eye bean; Dungay (Philipp.); Frijol arroz (Sp.); Gai-kalai (Beng.); Ghurush (Pun.); Gurounsh, Gurush (Ind.); Haricot (grain) de riz (Fr.); Judia de arroz (Sp.); Kachang sepalit (Malays.); Kalipan (Philipp.); Katjang otji (Indon.); Kilkilang (Philipp.); Lazy-man pea (As.); Linay (Philipp.); Mambi bean (Cuba); Mangulasi (Philipp.); Meth (Ind.); Mungo-lising (Philipp.); Mu-tsa (China); Pagapay, Pagsei, Paksai (Philipp.); Pau maia (Ind.); Pè-gin (SE. As.); Pè-yin (Burm.); Pois jaune, P. pigeon[2], P. zombi (Ant.); Reisbohne (Ger.); Shiltong (Ind.); Sem or Sim[3] (Assam); Sita-mas, Sut(a)ri (Ind.); Take-azuki (Jpn.); Taklauo, Tapilan (Philipp.); Tsuru-adsuki (Jpn.).

Botany

The rice bean is a short-term perennial, usually grown as an annual. It shows great diversity, and may be erect, semi-erect or twining, and normally attains a height of 12 to 40 in. (30–100 cm), but some forms can reach 80 in. (200 cm).

It has a very extensive root-system; the tap-root can be 40–60 in. (100–150 cm) in length and bears small nodules and numerous fine-rooting branches. The stems are grooved and in some forms covered with short, fine, white hairs. The leaves are trifoliolate, the leaflets 2.3–3.5 in. (6–9 cm) long, sometimes entire, but not infrequently three-lobed. The infloresence is an erect, axillary

[1] Frequently also used for the adzuki bean, *Vigna angularis*.
[2] Also used for the pigeon pea, *Cajanus cajan*, in some francophone countries.
[3] Also frequently used in India for the hyacinth bean, *Lablab purpureus*, and for the runner bean, *Phaseolus coccineus*.

raceme, 2–3 in. (5–7.5 cm) long, with 10–20 self-fertile, bright-yellow flowers. The pod is 3–5 in. (7.5–12.5 cm) long and approximately 0.15–0.24 in. (4–6 mm) broad, somewhat curved, with a prominent beak. It usually contains 6–10 oblong seeds, which may be green, yellow, brown, maroon, black, mottled or straw coloured.

The rice bean shows considerable variation, and four types have been recognized: (i) *glaber*, which has smooth stems and leaves; (ii) *major*, which has larger flowers; (iii) *rumbaiya*, which has shorter stems; (iv) *gracilis*, a wild form found in India, which has slender, smooth stems and narrow leaflets.

Origin and distribution

The rice bean is found growing wild in India, central China and Malaysia. It is now widely distributed throughout Asia and is also cultivated to a limited extent in other tropical areas such as, Mauritius, E. Africa, the W.I., Queensland and the USA.

Cultivation conditions

Temperature—essentially a tropical crop and very susceptible to frost, it is usually found growing in areas with an average temperature of between 64° and 86°F (18–30°C). Some cultivars are reported to be tolerant of high temperatures, although in W. Bengal the maximum and minimum temperatures of flower initiation was found to be 77–79°F (25–26°C) and 50–54°F (10–12°C), respectively.

Rainfall—although fairly tolerant of drought conditions, for optimum yields the rice bean requires an annual rainfall of 40–60 in. (1 000–1 500 mm).

Soil—the rice bean can be grown on a wide range of soil types, but for optimum yields requires fertile loams. The application of superphosphate 44–53 lb/ac (50–60 kg/ha) prior to planting has been recommended in India.

Altitude—in the Western Himalayas the rice bean is found growing at altitudes up to 6 000 ft (1 800 m) and in the Khasi hills, Assam, up to 5 000 ft (1 500 m).

Day-length—the rice bean is a short-day legume. Recent investigations have shown that the photoperiod is very important for flower initiation and that the day-length threshold for this species is less than 12 hours.

Planting procedure

Material—seed is used and germination is hypogeal.

Method—in India the rice bean is frequently sown broadcast, after two or three ploughings, and receives little after care.

Seed-rate—in India, when grown for seed, the average seed-rate is 36–45 lb/ac (40–50 kg/ha); when grown as a catch crop for fodder a slightly higher seed-rate of at least 53 to 62 lb/ac (60–70 kg/ha) has been recommended. In Burma the average seed-rate is somewhat lower, 19 lb/ac (21 kg/ha).

Pests and diseases

The rice bean appears to be remarkably free from most of the pests and diseases which affect food legume crops. It is, however, reported to be susceptible to root-knot nematodes, *Meloidogyne* spp., which may account for the fact that it grows very successfully on soils which have been flooded for rice, thus controlling the nematodes.

Growth period

This varies according to the cultivar and the day-length. When grown in Angola (6°–17°S latitude) the rice bean is reported to require 60 days from seeding to maturity, but in W. Bengal in the northern hemisphere, the crop requires about 130 days to produce an economic yield of seed. The time of sowing has been shown to be critical in W. Bengal, and must take place during June, July or early August. If delayed to September the seed yield decreases to less than 50 per cent and when sown in October or November the plants remain in a vegetative stage until the flowering time (December) of the following year. When grown for fodder a crop may be obtained in a minimum of 70–80 days from sowing, but considerably higher yields are obtained after 120–130 days.

Harvesting and handling

The pods shatter easily and crop losses are heavy even when they are picked by hand, although harvesting in the early morning when the pods are moist does reduce seed losses. The seeds are dried and threshed by hand. Reports suggest that they can be stored safely without any treatment since they are not normally affected by the common storage insects which usually infest grain legumes.

Primary product

Seed—these are approximately 0.24–0.3 in. (6–8 mm) in length, rounded at the ends, with a raised, concave hilum. They may be greenish-yellow, yellow, brown, maroon, black, mottled or straw coloured. The greenish-yellow seeds usually become yellowish-brown during storage. These, and the brownish-yellow types, are reported to be the most nutritious. One hundred seeds weigh approximately 0.28–0.4 oz (8–12 g).

Yield

Yields of the seed are reported to average between 180 and 270 lb/ac (200–300 kg/ha), although in W. Bengal with good crop management yields up to 2 000 lb/ac (2 240 kg/ha) are reported to be attainable. In Burma yields are reported to average between 375 and 750 lb/ac (420–840 kg/ha). When grown for green fodder average yields can range from 1 960 to 3 110 lb/ac (2 200–3 500 kg/ha).

Main use

The rice bean is a nutritious pulse crop, the beans are frequently cooked with, or instead of, rice, often in soups and stews. It is not popular in India, because it cannot be processed into dhal.

Subsidiary uses

The seeds after crushing and soaking, are sometimes used as a livestock food. In some parts of Asia, the seeds are utilized, similarly to mung beans, for the production of bean sprouts.

Secondary and waste products

The green immature seed pods and the young leaves may be boiled and eaten as a vegetable. The rice bean is sometimes cultivated as a fodder or green manure crop. When grown for fodder it should be harvested when the pods are half developed, since the leaves drop easily as the plant reaches maturity. An approximate analysis of the fodder has been given as: *Vegetative stage:* dry matter 16.0 per cent; crude protein 18.0 per cent; fat 1.1 per cent; crude fibre 31.5 per cent; N-free extract 39.9 per cent; ash 9.5 per cent; calcium 1.4 per cent; phosphorus 0.35 per cent; *Flowering stage:* dry matter 24.0 per cent; crude protein 14.5 per cent; fat 1.0 per cent; crude fibre 32.1 per cent; N-free extract 41.6 per cent; ash 10.8 per cent; calcium 1.2 per cent; phosphorus 0.4 per cent.

Special features

The approximate composition of rice beans from Zambia has been reported as: moisture 11.0–13.8 per cent; protein 19.1–22.7 per cent; fat 0.6–1.2 per cent; total carbohydrate 60.7–65.4 per cent; crude fibre 4.0–5.8 per cent; ash 4.2–4.3 per cent; calcium 142–257 mg/100 g; phosphorus 301–480 mg/100 g; iron 7.2–10.9 mg/100 g; thiamine 0.39–0.57 mg/100 g; riboflavin 0.08–0.21 mg/100 g; niacin 2.2–2.4 mg/100 g; vitamin A 30 iu. Analysis of dry mature Indian seeds (kernel 90.2–91.1 per cent) gave the following values: moisture 3.2–4.5 per cent; carbohydrate 56.7–59.8 per cent; protein 22.0–23.0 per cent; fat 3.5–5.0 per cent; fibre 4.9–5.3 per cent; ash 3.0–4.1 per cent.

The amino acid content (expressed as mg/gN) has been given as: arginine 462; histidine 380; leucine 606; isoleucine 387; lysine 769; methionine 769; cystine 44; phenylalanine 325; tyrosine 262; threonine 294; valine 394.

The seeds contain an off-white, fibrous, mucilage which on hydrolysis yields xylose, arabinose, galactose and galacturonic acid. The decorticated seed on extraction with petroleum-ether yields a bright yellow, odourless and flavourless oil having the following characteristics: $SG^{25°C}$ 0.91–0.92; $N_D^{25°C}$ 1.40–1.46; acid val. 4.1–4.2; sap. val. 182.2–185.4; iod. val. (Wijs') 66.5–68.8; unsaponifiable matter 1.80–2.35 per cent. The fatty acid composition of the oil is: myristic 6.3–7.3 per cent; palmitic 5.6–6.0 per cent; stearic 2.1–4.4 per cent; behenic 4.6–5.8 per cent; arachidic 3.0–3.9 per cent; lignoceric 2.9–3.2 per cent; linoleic 7.5–9.7 per cent; oleic 61.3–68.0 per cent. The unsaponifiable matter is reported to contain β-sitosterol.

The rice bean is reported to be free of cyanogenic glycosides and to be a relatively good source of calcium compared with other food legumes.

Production and trade

Separate statistics are not available for the production of rice beans. Most of the crop is consumed locally, but Burma, Thailand and China export supplies surplus to their own requirements, mainly to Japan. Japan is the major market for rice beans and dominates the international trade in this commodity.

Rice beans: Japanese imports

Quantity tonnes

	1969	Annual average 1970–74	1975
Total	6 695	11 645	12 235
of which from:			
Burma	5 080	4 548	2 091
People's Repub. China	772	3 531	3 123
Thailand	843	3 402	7 021

The average value of Japanese imports was £79.80/t for the years 1970–74, compared with £52.00/t for 1969. Values have shown a steady upward trend since 1970 and reached £112.00/t in 1974 and £125.00/t in 1975.

Major influences

The rice bean is a minor pulse crop whose development is handicapped by the fact that average yields are low and the seed-pods shatter easily, making economic harvesting difficult. Until very recently very little work has been done on improving the rice bean, but recently the Indian Agricultural Research Institute, New Delhi, has started a breeding programme in an attempt to develop early-maturing types suitable for cultivation on the plains. There has been increasing interest in India in the development of this bean as a fodder crop, and also in West Africa, where it is less susceptible than many other grain legumes to diseases and pests.

Bibliography

ALLARD, H. A. AND ZAUMEYER, W. J. 1944. Responses of beans (*Phaseolus*) and other legumes to length of day. *US Dep. Agric., Tech. Bull.* 867, 24 pp.

CANADA, DEPARTMENT OF INDUSTRY, TRADE AND COMMERCE. 1974. *World pulses market survey*. pp. 8; 105–107. Ottawa, Dep. Ind. Trade Commer., Agric. Fish Food Prod. Branch, 123 pp.

CHADA, Y. R. (ed.). 1976. *V. umbellata* (Thunb.) Ohwi & Ohashi. *The wealth of India: Raw materials*. Vol. 10 (Sp–W). pp. 496–497. New Delhi: Indian Counc. Sci. Ind. Res., Publ. Inf. Dir., 591 pp.

Rice bean

CHAUDHURI, A. P. AND PRASAD, B. 1972. Flowering behaviour and yield of rice bean (*Phaseolus calcaratus* Roxb.) in relation to date of sowing. *Indian J. Agric. Sci.*, *42*, 627–630.

CROWTHER, P. C. AND GREENWOOD-BARTON, L. H. G. 1962. The nutritional value of the rice bean, *Phaseolus calcaratus* Roxb. from Northern Rhodesia. *Trop. Sci.*, *4*, 163.

HERKLOTS, G. A. C. 1972. Red bean. *Vegetables in south-east Asia*. pp. 247–248. London: George Allen and Unwin Ltd, 525 pp.

MACKENZIE, D. R., HO, L., LIU, T. D., WU, H. B. F. AND OYER, E. B. 1975. Photoperiodism of mung bean and four related species. *Hortscience*, *10*, 486–487.

MAJUMDAR, B. R., SEN, S. AND ROY, S. R. 1968. Raise rice bean for rich nutritious fodder. *Indian Farming*, *18* (6), 29–30.

PIPER, C. V. 1914. Five oriental species of beans. pp. 13–16. *US Dep. Agric. Bull.* 119, 32 pp.

PURSEGLOVE, J. W. 1968. *Phaseolus calcaratus* Roxb. *Tropical crops: Dicotyledons*. Vol. 1. pp. 294–295. London: Longmans, Green and Co Ltd, 332 pp.

RACHIE, K. O. AND ROBERTS, L. M. 1974. Grain legumes of the lowland tropics. *Adv. Agron.*, *26*, 1–132.

SASAKI, T. 1967. What is so called Take-azuki? *Jpn. J. Trop. Agric.*, *10*, 209–214. (In English and Japanese).

SEVILLA-EUSEBIO, J., MAMARIL, J. C., EUSEBIO, J. A. AND GONZALES, R. R. 1968. Studies on Philippine leguminous seeds as protein foods: (i) Evaluation of protein quality in some local beans based on their amino acid patterns; (ii) Effect of heat on the biological value of munggo, paayab, tapilan and kadyos beans. *Philipp. Agric.*, *52*, 211–217; 218–232.

SINGH, H. B., JOSHI, B. S. AND THOMAS, T. A. 1970. Rice bean: *Phaseolus calcaratus* Roxb. *Pulse crops of India*. (Kachroo, P. and Arif, M. eds.). pp. 158–160. New Delhi: Indian Counc. Agric. Res., 334 pp.

THOMPSTONE, E. AND SAWYER, A. M. 1914. The peas and beans of Burma. pp. 52–54. *Burma, Dep. Agric. Bull.* 12, 107 pp.

Common names	**RUNNER BEAN[1], Multiflora bean, Scarlet runner bean[2].**
Botanical name	*Phaseolus coccineus* L., syn. *P. multiflorus* Lam.
Family	Leguminosae.
Other names	Ayecote, Ayocote, Botol (C. Am.); Benibanaigen (Jpn.); Butter bean[3] (Ken.); Caraota Florida (Venez.); Caseknife bean[4]; Chiapas (Mex.); Fagiolini rampicanti, Fagiolo di Spagna (It.); Feuerbohne (Ger.); Frijol ayocote (Mex.); Hanasasage (Jpn.); Haricot à fleurs rouges (Mor.); H. d'Españe (Fr.); Judia escarlata[2], J. d'Espagna (Sp.); Lubya kusais (Ar.); Pole bean[5]; Poroto de Espagna; P. pallar (Arg.); Pronkboon (Dut.); Sem[6] (Ind.); Sheuit aduma (Is.); Sim[7] (Assam); Stick or String bean[8]; Urahi[9], Uri[9] (Ind.); Vanlig blomsterböna (Sw.).

Botany

A climbing and branching, slightly pubescent perennial, which is frequently grown as an annual. The runner bean can reach a height of 8–10 ft (2.4–3.0 m), although bushy, semi-erect, dwarf forms have been developed to facilitate harvesting. It has thick, fleshy, branched, tuberous roots. The stems are twisting, slightly ribbed. The leaves are trifoliolate, with ovate leaflets, 3–5 in. (7.5–12.5) cm) long. The inflorescence is an axillary raceme and bears

---

1 Sometimes used for climbing types of the haricot bean, *Phaseolus vulgaris*.
2 Used for red-flowered types.
3 Used for a white-seeded cultivar in Kenya, but more generally applied to the white large-seeded lima bean, *Phaseolus lunatus*, or in India for the hyacinth bean, *Lablab purpureus*.
4 Usually used for white-seeded cultivars.
5 Sometimes used in the UK, but more usually used for climbing forms of the haricot bean, *Phaseolus vulgaris*, or the lima bean, *P. lunatus*, or in India for the hyacinth bean, *Lablab purpureus*.
6 Also used in India for the lima bean, *Phaseolus lunatus*, and the hyacinth bean, *Lablab purpureus*.
7 Also used for the hyacinth bean, *Lablab purpureus* and the rice bean, *Vigna umbellata*.
8 Used for climbing types only.
9 Also commonly used in parts of India for the hyacinth bean, *Lablab purpureus*.

several, (approximately 12), flowers, 0.7–1 in. (1.8–2.5 cm) long, on long axillary peduncles. They are normally bright-scarlet, but white forms also exist. The pods are usually 4–12 in. (10–30 cm) in length, but can reach 18 in. (45 cm), often slightly curved, plump, glabrous or slightly pubescent, with a stout beak and containing 6–10 very large, broadly oblong seeds, which are usually purplish-black, with red mottling, but occasionally may be white.

Some authorities consider that *Phaseolus coccineus* includes three botanical varieties, var. *rubronanus*, a red-flowered bushy type, var. *albus*, the White Dutch Runner, and var. *albonanus*, a bush form with white seeds. Popular cultivars of the red-flowered types in the UK include Achievement, Enorma, Kelvedon Marvel, Princess and Streamline, and of the white-flowered types, Czar, Desiree, Emergo and Prizewinner.

Origin and distribution

The runner bean is considered to have originated in Mexico, where it occurs wild at altitudes of about 6 000 ft (1 800 m). Perennial forms are still cultivated on a limited scale in parts of Latin America. It is now widely distributed in temperate areas and is very popular in the UK, where it is grown as an annual.

Cultivation conditions

Temperature—although the runner bean is a plant of the humid tropical highlands it is more tolerant of relatively cool temperatures than most other *Phaseolus* species. There is evidence that high temperatures, above 77°F (25°C), inhibit seed-setting. It is, however, susceptible to damage at temperatures below 41°F (5°C) and is very susceptible to frost damage at all stages of growth. For this reason it can be grown only in temperate areas with a frost-free period of 120–130 days.

Rainfall—a well-distributed rainfall throughout the growing period is required as the runner bean is extremely susceptible to drought conditions and requires a relatively high humidity to set seeds. In the UK the crop is often grown with supplementary irrigation, from the green-bud stage onwards, normally from June to August. On a coarse sandy loam with an available water supply (A.W.S.) of not more than 1.6 in. (4 cm), 1 in. (2.5 cm) of water is required when a 1 in. (2.5 cm) soil moisture deficit (S.M.D.) has been reached. On a loamy sand to clay soil with an A.W.S. of more than 1.6 in. (4 cm) and less than 2.5 in. (6.25 cm), 2 in. (5 cm) of water is required when 2 in. (5 cm)

S.M.D. has been reached. On a very fine sandy loam to a peaty loam where the A.W.S. is more than 2.5 in. (6.25 cm), 2 in. (5 cm) of water is required when there is a 3 in. (7.5 cm) S.M.D. In Ethiopia runner beans are grown successfully at higher altitudes in areas with an average annual rainfall of approximately 60 in. (1 500 mm).

Soil—deep, well-prepared, well-drained, loamy soils, of a light to medium texture are essential for optimum yields. Runner beans cannot tolerate cold, heavy, badly drained soils and the pH should be preferably between 6.0 and 7.0. Although a degree of acidity may be tolerated, soil nitrification is not so effective at levels below pH 6.0 and the availability of phosphate is reduced; for this reason the application of lime is recommended on soils with a pH below 6.5. On moderately fertile soils in the UK the application of 90 units of nitrogen, 60 units of phosphoric acid and 120 units of potash is suggested. On highly fertile soils this may be reduced to, nitrogen 60 units, phosphoric acid 40 units, potash 90 units. Where farmyard manure is applied fertilizer dressings may be reduced by up to 30 units of nitrogen, 40 units of phosphoric acid and 75 units of potash, but at no time should the base dressings be less than 40 units of phosphoric acid and 60 units of potash. Recently it has been suggested that the application of as much as 240 units of nitrogen, split over two or three applications could be beneficial.

Altitude—in the tropics the runner bean can be grown successfully at elevations above about 4 000 ft (1 200 m), below this it will not set seed. In Kenya, it is grown as a minor pulse crop at elevations between 6 500 and 8 500 ft (1 950–2 550 m) below 6 000 ft (1 800 m) seed-set is affected adversely.

Day-length—most cultivars are day-neutral, but short-day types exist.

Planting procedure

Material—seed is usually used, preferably obtained from early or mid-season crops; it should be disease-free. Germination is hypogeal. Tubers with a small piece of stem attached can also be used, provided that they have been kept at temperatures above 41°F (5°C). At one time this was a popular practice amongst small growers and gardeners in the UK.

Method—the seeds are sown either by drill or by hand, about 2 in. (5 cm) deep, in deep, well-prepared, weed-free soil. As the vines are very susceptible to wind damage the crop should not be planted in an exposed position. In the

UK there are two methods of growing runner beans. The cheapest consists of maintaining the plants in a dwarf or bushy condition by pinching out the leading shoots as they develop. This method has the disadvantage that the pods are borne close to the ground and are more susceptible to disease during wet weather, moreover, there is a greater tendency for the pods to grow curved. In the other method the beans are provided with sticks, canes, or wires and strings for support, and allowed to climb naturally. In the UK beans grown by this method are often referred to as 'Stick' or 'String' beans. Runner beans should be kept free from weeds during the early stages of growth and this is achieved by hoeing or the use of herbicides. Contact pre-emergence sprays of a paraquat diquat mixture have been recommended, also dinosebamine or dinoseb in oil applied 4-7 days prior to crop emergence. More recently the use of the herbicides diphenamid and trifluralin combined with bentazon has been suggested.

In C. America the runner bean is often interplanted with maize, and sprouts from the tuberous roots take over the maize fields for the first year or so of the fallow period. The crop is also sometimes planted as a perennial in small plots near to the houses.

The presence of insects is essential for pollination and in the UK young colonies of bees are sometimes placed in the crop soon after it starts to flower.

Field-spacing—in the UK, for the pinched crop, the seeds are normally planted 6 in. (15 cm) apart, in single rows 27–30 in. (67.5–75 cm) apart. For the supported crop, research has indicated that on 36 in. (90 cm) path width, twin-rows 24 in. (60 cm) apart and canes 12 in. (30 cm) apart within the row, with two plants per cane, is likely to give the optimum profit per unit area; but in commercial practice the path width is usually about 42 in. (105 cm) to facilitate picking, and the distance between the twin-rows is reduced to 18 in. (45 cm). Single rows 48 in. (120 cm) wide are also sometimes used.

Seed-rate—in the UK the seed-rate for the pinched crop is approximately 112 lb/ac (125 kg/ha), and for the supported crop is varied between about 84 and 112 lb/ac (94–125 kg/ha).

Pests and diseases

Pests—in the UK several insect pests are troublesome. Black bean aphids, *Aphis fabae*, may colonize the flowers and young shoots resulting in stunted and distorted pods. The use of disulphoton granules at planting is reported to be a very effective control measure, although it requires considerable care. As a curative treatment, the systemic insecticides demeton-S-methyl and oxydemeton-methyl can be used up to 3 weeks before picking, and the contact insecticides mevinphos and malathion up to 2 days before picking. Capsid bugs, *Lygus* spp., can cause serious pod distortion and are visible as foliage damage before bloom; treatment with DDT has been suggested for effective control. Bean seed fly larvae, *Hylemya platura*, can destroy newly-sown seeds or damage them so that stunted plants result; treating the seed with a dressing containing dieldrin or BHC has been found to be effective against this pest. The spider mite, *Tetranychus urticae*, can sometimes be troublesome and some measure of control can be obtained by the use of S-methyl or oxydemeton-methyl applied up to 3 weeks before picking. Care must be taken with the use of these so as not to kill bees active on the crop. In the UK sparrows frequently cause considerable damage by pecking the flowers and in certain parts of Europe considerable crop losses due to flower biting by the bumble bee, *Bombus terrestris*, have been reported.

Diseases—the seed-borne bacterial disease *Pseudomonas phaseolicola* can reduce yields seriously and can result in a complete crop failure. It causes spotting of the leaves and pods, wilting, and in severe cases the death of the plant. Mild outbreaks can be controlled by spraying with copper oxychloride. As the disease is spread by rain-splashing, irrigation should cease as soon as the first symptoms are noticed. In cases of serious infection it is sometimes necessary to plough in the whole crop. Wilt caused by the fungus *Fusarium oxysporum* is a soil-borne disease which can sometimes be troublesome, as can bean rust, *Uromyces phaseoli*. The former may cause yellowing of the margins of the primary leaves and the whole leaf to turn dry and brittle. On the compound leaves the margins of the leaflets roll inwards and later the leaflets drop. Leaves are affected progressively higher up the stem until eventually the whole plant may wither and die. The fungus is carried over in the soil from season to season, and rotation, combined with efficient field sanitation, is the only satisfactory control. Bean rust produces numerous brown powdery pustules on the leaves and pods which become black as they develop and make the pods unsaleable. In the UK the fungus spores can survive the winter on poles, wires and twine. After an outbreak of the disease the wires and twine used to support the affected beans should be destroyed and the support poles or canes disinfected with a 2 per cent solution of formaldehyde.

Runner bean

In the tropics the runner bean is also reported to be affected by anthracnose, *Colletotrichum lindemuthianum*, and fusarium wilt, *Fusarium solani*, f. *phaseoli*. The runner bean is of interest to plant breeders as it is resistant to most of the root rot organisms which attack the haricot bean, *Phaseolus vulgaris*, and can be crossed with it and the lima bean, *P. lunatus*.

Growth period

The green immature pods are normally ready for harvesting from 80 to 90 days after sowing, and the mature seeds in approximately 120 days.

Harvesting and handling

The green pods are picked when they are succulent and non-fibrous and the seeds are still undeveloped. Picking is usually at 4 to 5 day intervals and continues over a period of 60 to 90 days in temperate countries, depending upon the incidence of frost. After picking, the beans should be stored in a cool packing shed otherwise they deteriorate rapidly. In the UK runner beans are sorted, size-graded and are usually marketed in 8 lb (3.5 kg) chip baskets at the beginning of the season, but the bulk of the crop is marketed in bushel, 28–30 lb (13–14 kg), or half bushel, 14 lb (6 kg) boxes.

When grown for mature seed, the plants are also picked over several times by hand since the seeds do not ripen uniformly. After picking they are handled similarly to haricot beans.

Primary product

Pod—the runner bean is grown in many countries for the production of the immature edible pod. Normally the pods vary from 4 to 12 in. (10–30 cm) in length and are usually slightly curved and plump, with a pronounced beak. The pod wall is rough, with small, oblique ridges. The pods normally contain 1–5 seeds, separated by varying amounts of sept-like tissue. The seeds are large, broadly oblong, flattened, about 1 in. (2.5 cm) in length, and usually shiny black, mottled with red or purple, although a number of cultivars produce white seeds.

Yield

In the UK yields of runner bean pods are reported to average between 3.5 and 5.5 T/ac (8.75–13.75 t/ha), although with efficient crop management yields can reach as high as 15 T/ac (37.5 t/ha). In Kenya yields of dry mature seeds from smallholder production have been estimated to be about 800 to 1 000 lb/ac (900–1 120 kg/ha).

Main use

The immature pods are a nutritious vegetable. In the UK runner beans are produced mainly for the fresh vegetable market and are also a popular vegetable in kitchen gardens. When boiled as a vegetable the sliced beans are succulent and well flavoured and are regarded by many British consumers as being superior to French beans. In Latin America and certain European countries the runner bean is grown mainly for the production of the mature seeds, which are eaten fresh as a vegetable, or dried and eaten as a pulse.

Subsidiary uses

In certain countries, eg the USA, the runner bean is grown as an ornamental climbing plant.

Secondary and waste products

In C. America the tuberous roots are sometimes dug up and used as a starchy foodstuff, after they have been boiled and the cooking liquor discarded. Their flavour is said to resemble that of the potato, although the texture is more woody.

Special features

An analysis of the edible portion (59 per cent) of the pods gave the following values: moisture 58.3 per cent; protein 7.4 per cent; fat 1.0 per cent; carbohydrate 29.8 per cent; fibre 1.9 per cent; ash 1.6 per cent; calcium 50 mg/100 g; phosphorus 160 mg/100 g; iron 2.6 mg/100 g; thiamine 0.34 mg/100 g; riboflavin 0.19 mg/100 g; vitamin C 27 mg/100 g; vitamin A 57 iu/100 g. The fresh shelled immature seeds are reported to have the following approximate composition: moisture 34.2 per cent; carbohydrate 47.9 per cent; fat 0.3 per cent; crude fibre 12.2 per cent; protein 2.6 per cent; ash 2.8 per cent; calcium 60.6 mg/100 g; phosphorus 276.6 mg/100 g; iron 4.1 mg/100 g; carotene 0.034 mg/100 g; thiamine 0.538 mg/100 g; riboflavin 0.138 mg/100 g; niacin 2.3 mg/100 g; ascorbic acid 0.2 mg/100 g. Gibberellins A_1, A_5, A_6 and A_8 have been detected in the immature seeds. An approximate composition of the mature seeds has been given as: moisture 12.0 per cent; protein 20.0 per cent; fat 1.5 per cent; total carbohydrate 63.0 per cent; fibre 5.0 per cent; ash 3.5 per cent; calcium 120 mg/100 g; iron 10.0 mg/100 g; thiamine 0.3 mg/100 g; riboflavin 0.1 mg/100 g; niacin 2.0 mg/100 g. The fatty oil has the following characteristics: SG $^{15°C}$, 0.920; N_D $^{40°C}$, 1.476; sap. val. 190; iod. val. 141.2; solidifying point – 12°C; unsaponifiable matter 1–2 per cent. The tubers are reported to contain 18.6 per cent starch, amylose content 27 per cent, and to have a crude protein content of 4.2 per cent. According to some authorities a poisonous bitter principle is present.

Processing

The immature pods can be sliced and canned or quick-frozen similar to French beans, *Phaseolus vulgaris*. At one time considerable quantities were processed in the UK, but they are no longer popular with the vegetable processing industry, because of the labour involved in their production in the field and the necessity to remove 'strings' from older pods before processing.

Production and trade

The runner bean is of minor importance as a food legume and there is very little information relating to production and trade. The dry mature beans are mainly produced and consumed in Latin America and certain European countries. Production in Argentina, the leading producer, was estimated to be 40 400 t in 1973, compared with 36 000 t in 1972 and 29 200 t in 1971. There is a considerable production of the immature green pods for consumption as a vegetable in the UK. Production, however, has shown a downward trend in recent years and averaged 33 500 t/a for the years 1971–75, compared with 43 300 t/a for 1966–70, a decrease of 23 per cent. Later figures are not available.

The dry mature beans are of negligible importance in international trade. There is, however, a specialized demand for the large, white-seeded types, known as 'elephant' or 'soissons' beans, in certain European countries, notably Switzerland, where they fetch premium prices, (approximately £250.00/t in 1972). Argentina is understood to be the principal source of supply.

Major influences

Although the immature, green pods of the runner bean are a very popular vegetable in the UK, commercial production is likely to continue to decline because the crop is capital and labour intensive. Dwarf forms have been developed to reduce the capital and labour required for producing runner beans, but these, like the climbing forms, are not suitable for mechanical harvesting, owing to uneveness of maturity. The high cost of production and problem of 'string' which forms as the pods mature and can effectively be removed only by hand has made their use by the vegetable processing industry uneconomic.

Although suitable for cultivation at higher altitudes in the tropics, the runner bean is unlikely to become an important grain legume crop in these areas because of the need to provide supports and the uneveness of maturity.

Bibliography

AYKROYD, W. R. AND DOUGHTY, J. 1964. Legumes in human nutrition. *FAO Nutr. Stud.* 19, pp. 109; 115. Rome: FAO, 138 pp.

BLACKWALL, F. L. C. 1969. Effects of weather, irrigation, and pod-removal on the setting of pods and marketable yield of runner beans (*Phaseolus multiflorus*). *J. Hortic. Sci.*, *44*, 371–384.

BLACKWALL, F. L. C. 1971. A study of the plant/insect relationships and pod-setting in the runner bean (*Phaseolus multiflorus*). *J. Hortic. Sci.*, *46*, 365–379.

BLACKWALL, F. L. C. 1971. Pod-setting and yield in the runner bean (*Phaseolus multiflorus*). *J. R. Hortic. Soc.*, *96*, 121–130.

BLEASDALE, J. K. A. 1968. Effects of plant spacing on the yield and profitability of the scarlet runner bean (*Phaseolus multiflorus* Willd.). *Hortic. Res.*, *8*, 155–169.

BRIEN, R. M., CHAMBERLAIN, E. E., COTTIER, W., CRUICKSHANK, I. A. M., DYE, D. W., JACKS, H. AND REID, W. D. 1955. Diseases and pests of peas and beans in New Zealand and their control. pp. 34–68. *N.Z. Dep. Sci. Ind. Res. Bull.* 114, 91 pp.

CANADA, DEPARTMENT OF INDUSTRY TRADE AND COMMERCE. 1974. *World pulses market survey*. p. 7. Ottawa, Dep. Ind. Trade Commer., Agric., Fish, Food Prod. Branch, 123 pp.

CRAMP, K. W., EWAN, J. W., HUME, W. G., SHEARD, G. F. AND FINCH, C. G. 1962. Beans. pp. 12–17. *Lond., Minist. Agric., Fish. Food, Bull.* 87, 24 pp.

GANE, A. J., KING, J. M., GENT, G. P., BIDDLE, A. J., HANDLEY, R. D. AND BINGHAM, R. J. B. 1975. *Pea and bean growing handbook*. Vol. 2. *Beans*. Peterborough, England, Processors and Growers Res. Organ., 7 sections, (loose leaf).

HIORTH, G. 1942. Eiweissreiche Wurzelknollen bei niedrigen Feuerbohnen. [Tuberous roots rich in protein in dwarf scarlet runners]. *Zuchter*, *14*, 43–47. (*Plant Breed. Abstr.*, *12*, 892).

Runner bean

KAPLAN, L. 1965. Archeology and domestication in American *Phaseolus* (beans). *Econ. Bot., 19*, 358–368.

KRISHNAMURTHI, A. (ed.). 1969. *P. coccineus* Linn. syn. *P. multiflorus* Lam. *The wealth of India: Raw materials*. Vol. 8 (Ph–Re). p. 4. New Delhi: Publ. Inf. Dir. Counc. Sci. Ind. Res., 394 pp.

MINISTRY OF AGRICULTURE, FISHERIES AND FOOD. 1969. Scarlet runner bean. *Lond., Minist. Agric. Fish. Food, Advis. Leafl.* 502, 6 pp.

OLDHAM, C. H. 1950. Runner beans. *Vegetable growers guide*. pp. 124–129. London: Crosby Lockwood and Son Ltd, 472 pp.

PURSEGLOVE, J. W. 1968. *Phaseolus coccineus* L. *Tropical crops: Dicotyledons*. Vol. 1. pp. 295–296. London: Longmans, Green and Co Ltd, 332 pp.

ROBERTS, H. A. 1974. Encouraging work on weed control in runner beans. *Commer. Grow.*, (4091), pp. 929–930.

SCHREIBER, K., WEILAND, J. AND SEMBDNER, G. 1970. Isolierung von gibberellin A_8-0(3)-β-D-glucopyranosid aus früchten von *Phaseolus coccineus*. [Isolation of gibberellin A_8-0(3)-β-D-glucopyranoside from *Phaseolus coccineus* fruits]. *Phytochemistry, 9*, 189–198.

SUTTIE, J. M. 1969. The butter bean (*Phaseolus coccineus* L.) in Kenya. *East Afr. Agric. For. J., 35*, 211–212.

WESTPHAL, E. 1974. *Phaseolus coccineus* L. Pulses in Ethiopia, their taxonomy and agricultural significance. pp. 135–140. *Wageningen, PUDOC, Cent. Agric. Publ. Doc., Agric. Res., Rep.*, 815, 278 pp.

WILLIAMS, I. H. AND FREE, J. B. 1975. The pollination and set of the early flowers of runner bean (*Phaseolus coccineus* L.). *J. Hortic. Sci., 50*, 405–413.

WRIGHT, W. J. 1959. Flower biting bees may ruin the bean crop. *Commer. Grow.*, (3308), p. 1294.

Common name SWORD BEAN[1].

Botanical name *Canavalia gladiata* (Jacq.) DC.

Family Leguminosae.

Other names Abai[1] (Ind.); Avarakai (Malays.); Babricorn bean (W.I.); Bengal butter bean (Aust.); Caraota grande (Venez.); Cut-eye bean[1]; Fève Jack[1] (Haiti); Gotani bean[1] (Rhod.); Haba de burro[1] (Sp.); Habas[1] (Philipp.); Haricot-rouge (Guad.); Horse bean[2] (Ind., Jam.); Kachang parang (Malays.); Kaos bakol, K. bebedogan, K. parasaman, K. pedang (Sud.); Kara bedog, K. we-dung, Krandang (Indon.); Lubia elfil[1] (Sud.); Magtambokau (Philipp.); Makendal rouge (Mart.); Nata-mame (Jpn.); 'One-eye' bean[1]; Pataning-españa (Philipp.); Pearson bean (USA); Pois gagne rouge (Guad.); P. sabre (rouge) (Fr.); Sabre bean[1], Scimitar bean (Aust.); Too-a lund tow, Tua pra (Thai.); Tum bekai (Ind.); Valavarai (Malays.); Wonder bean (USA).

Botany

A vigorous, perennial climber, often cultivated as an annual, and sometimes reaching 15 to 32 ft (4.5–10 m) in height, but showing considerable variation in form, particularly in the degree of twining, the size of the seed pods and the number and colour of the seeds. In some areas semi-erect forms are found. The vegetative portions and the flowers are very similar to those of the jack bean, *Canavalia ensiformis*, with which it is often confused. The taxonomy of the two species has already been discussed in the section on the jack bean. The sword bean, however, has larger pubescent leaflets, 4–7.2 x 2.3–5.6 in. (10–18 x 6–14 cm) and larger flowers, approximately 1.6 in. (4 cm) long, white, pink, reddish or purplish in colour. The seed-pods are usually broader than those of the jack bean, more curved and with strongly developed ridges.

---

[1] Also frequently used for the jack bean, *Canavalia ensiformis*.
[2] More commonly used for the broad bean, *Vicia faba*, especially in Europe; also used for the jack bean, *Canavalia ensiformis*.

Sword bean

They are about 8–16 in. (20–40 cm) long and 1.4–2 in. (3.5–5 cm) broad, and contain on average 8 to 12 seeds. In India the following types of sword bean are recognized: (i) flowers and seeds red; (ii) flowers white, seeds red; (iii) flowers and seeds white.

Origin and distribution

The sword bean is considered by some authorities to have originated in Asia, where it is sometimes found growing wild in forests. It has spread throughout the tropics and is now cultivated on a limited scale throughout Asia, the W.I., Africa and S. America. It has also been introduced into tropical areas of Australia.

Cultivation conditions

Temperature—like the jack bean, the sword bean requires fairly high temperatures, 59–86°F (15–30°C).

Rainfall—the sword bean requires a moderately high, evenly distributed rainfall, about 36 to 60 in./a (900–1 500 mm/a); although some cultivars are fairly resistant to drought when once established.

Soil—it requires reasonably fertile soil and many cultivars are susceptible to waterlogging.

Altitude—it is reported to be found at elevations up to 3 000 ft (900 m) in Asia.

Planting procedure

Material—sound healthy seed is used and germination is epigeal.

Method—the seed is often sown in drills about 2–3 in. (5–7.5 cm) deep and support, such as bamboo trellis, must be provided. In Hong Kong, where the sword bean is grown as an annual vegetable, the seed is sometimes sown singly in 4 in. (10 cm) pots and after 4 to 8 weeks, the seedlings are transplanted into open ground.

Field-spacing—usually the plants are planted 18 to 24 in. (45–60 cm) apart, with 30 to 36 in. (75–90 cm) between the rows.

Seed-rate—in Angola a seed-rate of 36 lb/ac (40 kg/ha) is reported to be used.

Pests and diseases

The sword bean is relatively resistant to attack from pests and diseases.

Spodoptera frugiperda is sometimes troublesome. Like the jack bean, the sword bean is susceptible to a root rot, *Colletotrichum lindemuthianum*. In Asia the crop is also reported to suffer from scab, *Elsinoe canavaliae*.

Growth period

A crop of mature seed is produced in about 180 to 300 days. The immature green pods, which are sometimes utilized as a vegetable, are ready for harvesting in 90 to 150 days. Although a perennial the sword bean is often treated as an annual and it is recommended that the plants should not be kept for more than 2 years.

Harvesting and handling

The mature seeds are harvested by hand as soon as they are ripe, otherwise losses are heavy due to the pods shattering. For use as a vegetable the green pods are normally hand-picked when they are about 5 or 6 in. (12.5–15 cm) long, before they swell and become tough.

Primary product

Seed—the mature seed may be bright pink, red, reddish-brown or white. White seed is usually considered to have a better flavour and texture. The seed is usually larger than that of the jack bean, at least 1 in. (2.5 cm) long, strongly compressed, with a dark brown hilum extending almost the entire length. The seed-coat is very tough and thick.

Yield

Average yields ranging from 640 to 1 330 lb/ac (720–1 500 kg/ha) have been reported.

Main use

The mature dry beans may be cooked and eaten as a foodstuff; but care is required in their preparation because of the toxic principles present. The sword bean is not a popular grain legume because of its strong flavour, poor texture and thick, tough seed-coat. In Indonesia, the beans are usually boiled twice in water to soften, then washed in clean water, after which the seed-coat is removed and the decorticated beans are soaked in water for 2 days, drained and then fermented for 3 to 4 days. In other parts of Asia the beans are often soaked in water overnight, then boiled until soft, in water to which a small quantity of sodium bicarbonate has been added, after which they are rinsed, boiled again in fresh water, and finally pounded and used in curries, or as a substitute for mashed potato.

The immature green pods are widely utilized in Asia as a vegetable similar to French, or snap beans, *Phaseolus vulgaris.* Sometimes the immature, white-seeded types are cooked and used as a substitute for broad beans, *Vicia faba.*

Subsidiary uses
The sword bean is sometimes grown as a cover, green manure, or forage crop.

Secondary and waste products
The seeds and herbage are occasionally used for animal feeding. The seeds, particularly the pink coloured ones, are sometimes employed in traditional Chinese medicine. The roasted beans have been used as a coffee substitute.

Special features
The approximate composition of the fresh beans is: moisture 88.6 per cent; protein 2.7 per cent; carbohydrate 6.4 per cent; fat 0.2 per cent; fibre 1.5 per cent; ash 0.6 per cent; vitamin A 40 iu/100 g. The seed-coat constitutes approximately 29 per cent of the total weight of the mature dry beans. An approximate analysis of the edible portion of the beans is: moisture 14.9 per cent; protein 27.1 per cent; fat 0.6 per cent; carbohydrate 42.2 per cent; fibre 11.6 per cent; ash 3.6 per cent. The beans are also reported to contain 0.00972 per cent hydrocyanic acid and a toxic saponin. The immature beans contain gibberellins.

Production and trade
No reliable statistical data are available.

Major influences
The sword bean is a very minor grain legume and its use for human or animal feeding requires caution because of its toxicity, which requires further investigation. The presence of toxic principles combined with a fibrous texture and a very thick, tough seed-coat, constituting almost one-third of the total weight of the seed, makes it unlikely that this bean will increase in importance.

Bibliography

BURKILL, I. H. 1935. *Canavalia. A dictionary of the economic products of the Malay Peninsula*. Vol. 1 (A–H). pp. 432–435. London: Crown Agents for the Colonies, 1220 pp.

CHARAVANAPAVAN, C. 1943. The utilization of the sword-bean and jack bean as a food. *Trop. Agric.*, (*Ceylon*), *99*, 157–159.

FOOD AND AGRICULTURAL ORGANIZATION OF THE UNITED NATIONS. 1959. *Tabulated information on tropical and subtropical legumes*. pp. 76–77. Rome: FAO, Plant Prod. Prot. Div., 367 pp.

HERKLOTS, G. A. C. 1972. Sword or jack beans. *Vegetables in south-east Asia*. pp. 233–236. London: George Allen and Unwin Ltd, 525 pp.

JENKINS, A. E. 1931. Scab of *Canavalia* caused by *Elsinoe canavaliae*. *J. Agric. Res.*, *42*, 1–12.

LUCIE-SMITH, M. N. 1933. Photography as a help in the examination of cattle foods: Structure of the pod and seeds of *Canavalia* spp. *J. Southeast Coll. Agric.*, (*Wye*), (32), pp. 42–48.

MUROFUSHI, N., TAKAHASHI, N., YOKOTA, T., KATO, J., SHIOTANI, Y. and TAMURA, S. 1969. Gibberellins in immature seeds of *Canavalia:* (i) Isolation and biological activity of gibberellins A_{21} and A_{22}; (ii) Structure of gibberellins A_{21} and A_{22}. *Agric. Biol. Chem.*, *33*, 592–597; 598–609.

PIPER, C. V. 1920. The jack bean. *US Dep. Agric.*, *Dep. Circ.* 92, 12 pp.

PURSEGLOVE, J. W. 1968. *Canavalia gladiata* (Jacq.) DC. *Tropical crops: Dicotyledons*. Vol. 1. p. 245. London: Longmans, Green and Co Ltd, 332 pp.

QUISUMBING, E. 1965. *Canavalia* in the Philippines. *Araneta J. Agric.*, *12*, 1–7.

SASTRI, B. N. (ed). 1950. *Canavalia gladiata* (Jacq.) D.C. *The wealth of India: Raw materials*. Vol. 2 (C). pp. 56–57. Delhi: Indian Counc. Sci. Ind. Res., 427 pp.

SMARTT, J. 1976. *Canavalia gladiata* (Jacq.) D.C. (Sword bean). *Tropical pulses*. p. 58. London: Longman Group Ltd, 348 pp.

Sword bean

STEHLÉ, H. 1953. Étude botanique et agronomique des legumineuses autochtones et exotiques des genres: *Canavalia*, *Clitoria* et *Crotalaria* aux Antilles françaises. [A botanical and agronomic study of native and introduced legumes of the genera *Canavalia*, *Clitoria* and *Crotalaria* in the French Antilles]. *Rev. Int. Bot. Appl., Trop. Agric.*, *33*, 490–517.

TIHON, L. 1946. Á propos de deux *Canavalia* rencontrés au Congo Belge. [An account of two *Canavalia* spp. encountered in the Belgian Congo]. *Bull. Agric. Congo Belge*, *37*, 156–162.

WESTPHAL, E. 1974. Pulses in Ethiopia, their taxonomy and agricultural significance. pp. 72–77. *Wageningen, PUDOC, Cent. Agric. Publ. Doc., Agric. Res. Rep.* 815, 278 pp.

WHITE, C. T. 1943. The sword bean or scimitar bean and the jack bean. *Qd. Agric., J.*, *57*, 25–27.

Common names — **TEPARY BEAN, Rice haricot bean.**

Botanical name — *Phaseolus acutifolius* Gray var. *latifolius* Freem.

Family — Leguminosae.

Other names — Dinawa[1] (Afr.); Frijol trigo (Chile); Garbancillo bolando (Mex.); Haricot riz (Alg.); H. Sudan (Sen.); Texan bean (USA).

Botany

A glabrous, or slightly hairy, sub-erect annual, normally 6–12 in. (15–30 cm) high, which develops a bushy form when grown on poor land, but where luxuriant growth is possible it may be trailing or twining, with stems up to 78 in. (2 m) long. The first leaves are simple, but the mature ones are alternate, trifoliolate. The leaflets are ovate, about 2 in. (5 cm) long and 1.5 in. (4 cm) broad, with prominent veins and carried on petioles which may reach 4 in. (10 cm) in length. The inflorescences are axillary, with up to five small, white, pink or pale-lilac, self-fertilized flowers; approximately 0.6 in. (1.5 cm) in length. The flat, papery seed-pods can be straight or slightly curved, usually 2–3.5 in. (5–9 cm) long and about 0.5 in. (1.2 cm) broad. They are covered with long silky hairs when immature and have a prominent beak and a fibrous rim at the margins. They contain up to 7 (av 5) small, roundish or oblong seeds of various colours. Forty-seven different cultivars, showing considerable variation in the colour of the seeds, have been isolated in Arizona.

Origin and distribution

The tepary bean occurs wild in Arizona, and NW. Mexico where it was cultivated by the Aztecs. At one time it was of considerable commercial importance along the Pacific coasts of California and northern Mexico, but has been largely replaced by the haricot bean, *Phaseolus vulgaris*. Attempts to introduce the tepary bean into the more arid areas of Africa, Asia, Australia and S. America have had varying degrees of success and nowadays it is confined mainly to its native habitat, largely to experimental stations.

Cultivation conditions

Temperature—the tepary bean is well adapted to hot, dry conditions, requiring high temperatures and bright sunshine. It can be grown successfully in areas

[1] Also used for the cowpea, *Vigna unguiculata*.

where other beans will fail to flower, or set seed. It is intolerant of frost and cannot be grown in areas where night temperatures fall below 46°F (8°C).

Rainfall—this bean is particularly suited to arid conditions and can be grown in areas with a rainfall of 20 to 24 in./a (500–600 mm/a), or even less with irrigation. It thrives in arid areas which receive heavy but infrequent rains. With an average rainfall of about 40 in./a (1 000 mm/a) the plant usually makes excessive vegetative growth at the expense of seed yield. When grown under irrigation it is usual for the plant to receive about 3 or 4 irrigations during the early stages of growth, prior to flowering.

Soil—light, well-drained soils are preferred, but reasonable yields can be obtained on poor sandy soils, and the tepary bean is reported to be moderately tolerant of saline and alkaline soils. It cannot stand waterlogging, and heavy clay soils are unsuitable. Little is known of its manurial requirements, but it is reported to respond to applications of nitrogen and potash, the latter at the rate of 67 lb/ac (75 kg/ha). Experience in Algeria suggests that the use of fertilizers at planting is not advisable. There, the tepary bean has been found to do well as a second crop after the soil has received a previous application of superphosphate.

Altitude—in Mexico and Arizona the tepary bean is usually grown at medium altitudes. When grown in coastal areas in Algeria it was found to take longer to reach maturity.

Day-length—according to Allard and Zaumeyer (1944), the tepary bean requires a short day-length, with day-lengths in excess of 12 hours it becomes recumbent and weakly twining. However, recently Hartman (1969) reports the existence of day-neutral lines.

Planting procedure

Material—seed is used, germination is epigeal. The seed absorbs water very easily, in moist soils the seed-coat wrinkles within 5 minutes.

Method—occasionally the seed is broadcast, but more usually it is planted in rows, at a depth of 1 to 4 in. (2.5–10 cm), depending upon the type of soil and the availability of moisture. Sometimes 3 or 4 seeds are planted on mounds 18 in. (45 cm) high. Efficient weed control particularly during the early stages of growth is essential. It has been recommended that the seedlings are hoed two or three times to control weeds, except when the seed is planted at the end of the rainy season, when weeding is not usually necessary.

Field-spacing—the seeds are reported to be planted at distances varying from 3 to 10 in. (7.5–25 cm) in rows 24 to 36 in. (60–90 cm) apart.

Seed-rate—with a spacing of 6 x 36 in. (15 x 90 cm) the seed-rate averages 10–15 lb/ac (11–17 kg/ha); if broadcast the average seed-rate is 25 to 30 lb/ac (28–34 kg/ha), and if grown as a forage crop a much higher seed-rate of approximately 60 lb/ac (67 kg/ha) has been suggested.

Pests and diseases

The tepary bean is reported to be moderately resistant to insect pests and diseases in the field. In Algeria the crop is reported to suffer attack from the black bean aphid, *Aphis fabae*, in some seasons, but this pest can be controlled successfully by spraying with nicotine. When stored the tepary bean is usually immune to attack from the bean weevil, *Acanthoscelides obtectus*, but in Uganda stored tepary beans are reported to become infested with the rice weevil, *Sitophilus oryzae*. When introduced into Burma the tepary bean was found to be susceptible to attack from a root rot, *Rhizoctonia* spp.

Growth period

In the tropics short-duration types can mature in about 60 days, but most types have a growth period of 70 to 90 days and in cooler regions, such as the coastal region of Algeria, the growth period averages 120 days.

Harvesting and handling

In the field individual plants can show a range of 14 days in ripening, so that harvesting is usually by hand. The pods shatter freely when ripe and are normally picked as soon as they begin to change colour, occasionally the whole plant is pulled up when about 80 or 90 per cent of the pods are ripe. Normally the pods are left to dry for a few days before the beans are threshed, cleaned and stored, similarly to haricot beans.

Primary product

Seed—the tepary bean is small, rounded or oblong, sometimes flattened, approximately 0.3 x 0.24 in. (8 x 6 mm), with a dull, matt seed-coat, which can be white, yellow, brown, deep-violet, self-coloured or marbled, and sometimes, with radiating lines similar to those on the lima bean, *Phaseolus lunatus*. In the USA only white-seeded types are encountered commercially. The interior of the seed has a hard, waxy appearance. One hundred seeds weigh about 0.5 oz (15 g).

Yield

In the USA under dry conditions the average yield of dry seed is reported to range from 450 to 700 lb/ac (500–780 kg/ha) and with irrigation from 800 to 1 500 lb/ac (900–1 680 kg/ha). In Uganda the average yield is between 400 and 600 lb/ac (450–670 kg/ha). In Algeria, with efficient cultivation, an average yield of between 3 560 and 4 450 lb/ac (4 000–5 000 kg/ha) is reported to be attainable.

Main use

The dry seeds may be eaten similar to haricot beans, but become very hard on storage and take a very long time to cook. In Uganda they are usually boiled and then coarsely ground before being added to soup. Tepary beans have a strong flavour and odour and are less palatable than haricot beans. In Mexico the beans are sometimes soaked in water to produce a gelatinous extract used in the preparation of soups.

Subsidiary uses

The tepary bean has been grown occasionally for fodder or green manure in the USA. Dry hay yields of between 2 and 4 T/ac (5–10 t/ha) have been recorded. An analysis of the hay has been given as: moisture 6.6 per cent; protein 9.9 per cent; fat 1.9 per cent; N-free extract 43.1 per cent; fibre 29.3 per cent; ash 9.2 per cent.

Secondary and waste products

The pods and haulms remaining after harvesting may be used for animal feed. The approximate composition of the pods has been given as: moisture 8.0 per cent; protein 4.1 per cent; fat 0.5 per cent; N-free extract 43.6 per cent; fibre 37.0 per cent; ash 6.8 per cent.

Special features

The approximate composition of the mature seed is: moisture 8.6–9.5 per cent; total carbohydrate 59.3–69.2 per cent; protein 21.0–22.2 per cent; fat 1.3–1.4 per cent; fibre 3.4–5.2 per cent; ash 3.3–4.2 per cent. The amino acid content (mg/gN) has been reported as: isoleucine 280; leucine 480; lysine 410; phenylalanine 330; tyrosine 200; methionine 60; cystine 90; threonine 250; valine 360. The absence of alkaloids and cyanogenic glycosides has been reported.

Production and trade

No statistical data are available regarding the production of tepary beans, but nowadays they are understood to be grown in N. America on a very limited scale in California, Arizona and northern Mexico. They are also reported to be grown on prison farms in Uganda. Attempts have also been made to cultivate them in Algeria and eastern Africa, as far south as Lesotho and Botswana, and in Australia.

Major influences

The tepary bean, although a nutritious foodstuff well adapted as a quick catch crop suitable for arid tropical areas, has never become a commercially popular food legume, probably because it is laborious to harvest and cook, has a strong flavour, and according to some authorities an objectionable odour.

Bibliography

ALLARD, H. A. and ZAUMEYER, W. J. 1944. Responses of beans (*Phaseolus*) and other legumes to length of day. *US Dep. Agric., Tech. Bull.* 867, 24 pp.

ANON. 1916. Edible beans from Burma: Tepary bean. *Bull. Imp. Inst.*, *14*, 154–156.

AYKROYD, W. R. and DOUGHTY, J. 1964. Legumes in human nutrition. *FAO Nutr. Stud.*, 19, pp. 108; 116; 118. Rome: FAO, 138 pp.

BUSSON, F. 1965. *Phaseolus acutifolius* Gray. *Plantes alimentaires de l'ouest Africain: Étude botanique, biologique et chimique*. pp. 242; 252–254. Marseille: Leconte, 568 pp.

CARNE, W. M. 1915. The tepary bean. *Agric. Gaz. N.S.W.*, *26*, 979–980.

COBLEY, L. S. 1956. Tepary bean: *Phaseolus acutifolius. An introduction to the botany of tropical crops*. pp. 145–146. London: Longmans, Green and Co Ltd, 357 pp.

Tepary bean

DEUEL, H. J. 1924. The digestibility of tepary beans. *J. Agric. Res.*, *29*, 205–208.

FREEMAN, G. F. 1912. South western beans and teparies. *Univ. Arizona Agric. Exp. Stn., Bull.* 68, 44 pp.

HARTMANN, R. W. 1969. Photoperiod responses of *Phaseolus* plant introductions in Hawaii. *J. Am. Soc. Hortic. Sci.*, *94*, 437–440.

HENDRY, G. W. 1921. Bean culture in California. *Univ. Calif., Agric. Exp. Stn. Berkeley, Bull.* 294, 70 pp.

KAPLAN, L. 1965. Archeology and domestication in American *Phaseolus* beans. *Econ. Bot.*, *19*, 358–368.

ONABA, G. R. 1970. Tepary bean: *Phaseolus acutifolius* Gray. *Agriculture in Uganda.* (Jameson, J. D. ed.). 2nd ed. p. 247. London: Oxford University Press, (on behalf of Uganda Minist. Agric. & For.), 395 pp.

PURSEGLOVE, J. W. 1968. *Phaseolus acutifolius* Gray var. *latifolius* Freem. *Tropical crops: Dicotyledons.* Vol. 1. pp. 287–289. London: Longmans, Green and Co Ltd, 332 pp.

RACHIE, K. O. and ROBERTS, L. M. 1974. Grain legumes of the lowland tropics. *Adv. Agron.*, *26*, 1–132.

SMARTT, J. 1976. *Phaseolus acutifolius* A. Gray. (Tepary bean). *Tropical pulses.* p. 70. London: Longman Group Ltd, 348 pp.

THEAU, A. 1951. Le haricot riz. [The rice haricot]. *Rev. Hortic. Alger.*, *55*, 8–10.

THOMPSON, H. C. and KELLY, W. C. 1957. Tepary bean. *Vegetable crops.* 5th ed. pp. 433–434. New York/Toronto/London: McGraw-Hill Book Co Inc, 611 pp.

WENHOLZ, H. 1930. Tepary beans (*Phaseolus acutifolius* var. *latifolius*). *Agric. Gaz. N.S.W.*, *41*, 618.

Common names **URD (BEAN), Black gram[1], Mash[1].**

Botanical name *Vigna mungo* (L.) Hepper, syn. *Phaseolus mungo* L.

Family Leguminosae.

Other names Adad (Ind.); Ambérique[1] (haricot) (Fr.); Balatong (Philipp.); Biri (Ind.); Chiroko, Choroko (E. Afr.); Dâu-muoi (Viet.); Grâo de pulha (Ang.); Haricot mungo[1] (Fr.); H. velu (Zar.); Illundu (Sri La.); Kachang hijau[1], K. hitam (Malays.); Kalai[2] (Ind.); Kambulu (Zar.); Kifudu[1] (Ug.); Mahasha, March, Mash kalai, Mate mah, Matikalai (Ind.); Mât-pè (Burm.); Minumulu (Malays.); Muñggo[1] (Philipp.); Tikari (Beng.); Udad, Uddulu, Udid, Uhunnu, Ulundu (Ind.); Undu (Sri La.); Urad, Urid (kai) (Ind.); Woolly pyrol (W.I.).

Botany

A hairy bushy annual, usually about 1–3 ft (30–90 cm) in height, sometimes spreading or trailing. The plants have a well-developed tap-root and many laterals covered with nodules. Like the mung bean the root-system of the urd is of two types: (i) mesophytic or shallow; (ii) xerophytic or deep-rooted. The stem is diffuse, furrowed, much branched from the base, and often covered with rough, reddish-brown hairs. The leaves are trifoliolate, with large leaflets 2–4 in. (5–10 cm) long, entire-ovate to rhombic-ovate in outline and acuminate. The leaflets are membranous and light-green in colour. The inflorescence is axillary with terminal clusters of 5–6 flowers on a short, usually hairy, peduncle, which elongates as the pods develop. The petiole is long, hairy, channelled and often diffused with purple. The flowers are small, self-pollinated, and may be pale-yellow, lemon, or a bright golden-yellow, according to the cultivar, sometimes there are diffuse red spots at the back of the standard.

The pods are slender, somewhat cylindrical, about 1.6 to 2.3 in. (4–6 cm) long, erect or sub-erect, with a short, hooked beak, normally covered with hairs,

---

[1] Also sometimes used for the mung bean, *Vigna radiata*.
[2] Also sometimes used for the horse gram, *Macrotyloma uniflorum*.

and varying in colour from buff to light or dark-brown when fully mature. They contain 4–10 small, oblong seeds which are generally black, but may be dark olive-green, grey, brown, or mottled. Two principal types of urd are grown in Asia: (i) the larger-seeded, early-maturing, black-seeded types; (ii) the smaller-seeded, later-maturing types, which can have brown, olive-green, grey, or mottled seeds.

As has already been discussed, urd, *Vigna mungo*, is morphologically very similar to the mung bean, *Vigna radiata*, but should be regarded as a separate species. The two food legume crops may be distinguished by the fact that the mung bean has spreading or reflexed pods, with short hairs and globose seeds with flat hilums, whereas urd has erect or sub-erect pods, usually with long hairs, and larger oblong seeds, with a concave hilum.

Attempts have been made in India during the past 50 years to improve local strains of urd and a number of improved cultivars such as: Mashi–1, Mash–48, T–27 and T–9, Khargone–3, Sindh Kheda and ADT–1 have become available. More recently early-maturing cultivars with a high degree of tolerance to virus diseases have been developed, including Pusa–1, H–10 and G–1.

Origin and distribution
Urd originated in the SE. Asian sub-continent, where it is a highly prized grain legume and is cultivated in most areas. It has now spread to other tropical areas in Asia, Africa and America. In the W.I. it is grown mainly as a green manure crop, under the name woolly pyrol.

Cultivation conditions
Temperature—urd is essentially a tropical crop, resistant to high temperatures, normally grown in areas with an average temperature of between 77° and 95°F (25–35°C) and plenty of sunshine, since prolonged cloudy weather is detrimental to growth. It cannot withstand frost.

Rainfall—it is relatively drought resistant, more so than the mung bean, and is usually grown in areas with an average annual rainfall of less than 36 in. (900 mm). In areas of heavier rainfall it is grown as a dry season crop, when one or two irrigations are often necessary. Rain at flowering time has a very adverse effect upon seed yield.

Soil—heavy soils that are water retentive are preferred and in India optimum results are usually obtained on clay, or black cotton soils. Urd, however, is cultivated on other types of soils, including red or light loams and paddy soils, provided that they are not shallow. The precise fertilizer requirements of urd have not been studied in detail, but it is known to respond significantly to the application of phosphate. There is also an indication that the application of nitrogen could increase the protein content of the seed. The basal application of 112 lb/ac (125 kg/ha) of superphosphate and 28 lb/ac (31 kg/ha) of potassium muriate is reported to give a good response. In many areas in India the practice of applying wood ash and dung is fairly widespread. Urd is reported to be liable to chlorosis in certain soils, but responds to applications of gypsum.

Altitude—in the SE. Asian sub-continent urd is grown from sea level up to elevations of 5 000–6 000 ft (1 500–1 800 m).

Day-length—most cultivars are short-day plants.

Planting procedure

Material—seed, which can retain its viability for at least 2 years, and normally has a germination rate of about 90 per cent, is used. Germination is epigeal. Inoculation with the appropriate rhizobium culture has been found to increase seed yield and the nitrogen status of the soil.

Method—urd is quick growing and in the SE. Asian sub-continent is grown as an early, mid-season or late crop, but seldom as the sole crop of the year. It is sometimes grown in pure stands, but is frequently grown as a mixed crop, subordinate to cotton, maize or sorghum, when it is normally broadcast and ploughed in, and receives the same cultural attention as the main crop. When grown as a pure stand it is usually sown in rows, and when grown during the rainy season planting on ridges is recommended to avoid the possibility of waterlogging. Urd does not require a very well-prepared seed-bed, a rough tilth is satisfactory, otherwise the plant makes excessive vegetative growth at the expense of seed production. Efficient weed control, particularly during the early stages of growth, is essential and usually two or three inter-cultivations are given; the first 14 days after sowing, the second 28 days, and the third, if required, after 42 days. The pre-emergence application of herbicides such as: chlorthal, trifluralin, diphenamid, alachlor and nitrofen is reported to give effective weed control.

Urd

In India in recent years defective setting of pods and incomplete grain filling have become a problem on occasions. Hormone spraying at the opening of the first flowers and then again at full bloom, particularly with α–naphthoxyacetic acid, p–chlorophenoxyacetic acid, alone, or mixed, and gibberellic acid and 1–naphthaleneacetic acid, has been found to increase pod-set significantly.

Field-spacing—in India when grown as a pure crop, urd is usually sown in rows 10–12 in. (25–30 cm) apart, with about 4–8 in. (10–20 cm) between the plants. Recent investigations suggest that for the rain-fed crop a spacing of 12 x 4 in. (30 x 10 cm) is best. In E. Africa the recommended spacing is 18 x 6 in. (45 x 15 cm).

Seed-rate—in India the average seed-rate normally ranges from 12 to 20 lb/ac (13–22 kg/ha) when urd is grown as a pure crop and 8 to 12 lb/ac (9–13 kg/ha) when grown in mixtures. Recent research suggests that a seed-rate of 27 lb/ac (30 kg/ha) would be beneficial for late sown crops. In E. Africa the average seed-rate is reported to be 10–15 lb/ac (11–17 kg/ha).

Pests and diseases

In India urd is subject to attack by many insects, as a result of which the average annual loss in yield is estimated to be about 10 to 15 per cent. Bean fly, *Ophiomyia* (*Melanagromyza*) *phaseoli*, the common pulse beetle fly, *Madurasia obscurella*, and white fly, *Bemisia tabaci*, are of major importance in Uttar Pradesh. The soil application of the granular insecticides, aldicarb, phorate and monocrotophos is reported to give effective control during the early stages of growth. The pod borer, *Apion ampulum*, and the aphid, *Aphis craccivora*, are often troublesome in many parts of Asia, in addition to the hairy caterpillar, *Diacrisia obliqua*. In E. Africa the crop is reported to suffer from attacks by the American bollworm, *Heliothis armigera*, and the bean aphid, *Aphis fabae*. Like most grain legumes urd suffers from insect attack during storage, the cowpea weevil, *Callosobruchus chinensis*, being particularly troublesome. The use of systemic insecticides such as parathion, have been suggested to control this pest in the field and so considerably reduce infestation during storage.

Urd is subject to a number of fungal and bacterial diseases, none of which normally cause appreciable crop losses. The more important diseases are: powdery mildew, *Erysiphe polygoni*, leaf spot, *Cercospora cruenta*, dry root rot, *Macrophomina phaseoli*, rust, *Uromyces appendiculatus*, angular leaf spot,

Protomycopsis phaseoli, blight, *Ascochyta phaseolorum*, seedling blight, *Phomopsis* sp., bacterial leaf spot, *Xanthomonas phaseoli*, halo blight, *Pseudomonas phaseolicola*, and anthracnose, *Colletotrichum lindemuthianum*.

The occurrence of a damping-off disease due to *Phythium aphanidermatum* has recently been reported on crops in India. In addition, urd is susceptible to a number of virus diseases which can cause considerable yield losses. The principal virus diseases infecting urd in the SE. Asian sub-continent are: yellow mosaic, leaf crinkle, mosaic mottle and leaf curl. Infection with yellow mosaic or leaf crinkle reduces pod production. Mosaic mottle virus causes phyllody. Infection with leaf curl often results in the death of the plants; this virus is transmitted through sap, seed and by the aphid, *Aphis craccivora*. The most effective control measures against these diseases are the use of healthy, disease-resistant seed and the complete destruction of all diseased plant material.

Growth period
Urd is a quick-growing legume and produces a seed crop within 75 to 130 days, according to the cultivar. For example, the improved cultivar T–9 will produce a seed crop in 75 to 80 days and Pusa–1 in about 85 days.

Harvesting and handling
The low habit of urd makes harvesting a tedious operation and the plants are often pulled up by the roots, stacked to dry for 3–7 days, then threshed by beating with sticks or trampling under the feet of oxen. When grown in rice fields the pods are normally hand-picked, the plants grazed for a day or two and then ploughed in. The seed is thoroughly dried in the sun to a moisture content of 11 per cent or less, cleaned and stored, usually in straw baskets, or earthenware or metal containers, at the rural level. Urd, like most grain legumes, is susceptible to insect infestation during storage and these containers are usually sealed with a layer of sand, earth, ash or cow dung, in an attempt to protect the seed from infestation. Dried neem leaves, *Azadivachta indica*, are sometimes burnt in the storage receptacles, as a precaution against insect infestation.

Primary product
Seed—urd has small, oblong seed, approximately 0.12 in. (3 mm) long, often with almost flattened ends, but at times rounded. The surface is usually dull and rough, although some cultivars produce smooth, shiny seeds. The hilum

is broad and concave, covered with a dense white pad which is grooved down the centre. There are two main types of urd recognized in India, the black which is common throughout the country and the olive-green type which is more common in Uttar Pradesh and the drier areas of the Punjab. The seeds are often classified by Indian traders as follows: (i) *small*—1 000 seeds weigh 0.75–0.96 oz (21.4 to 27.5 g); (ii) *medium*—1 000 seeds weigh between 0.97 and 1.6 oz (27.6–47.0 g); (iii) *bold*—1 000 seeds weigh more than 1.6 oz (47.0 g).

Yield

Yields of seed in India can range between 300 and 500 lb/ac (340–560 kg/ha) when unimproved types are grown as a pure crop, and even less when grown as a subordinate crop. Average yields of the improved, quick-maturing cultivar T–9 are reported to be 830 lb/ac (930 kg/ha) and of the more recently released Pusa–1, 1 330 lb/ac (1 500 kg/ha).

Main use

Urd is a very nutritious grain legume and is popular in Asia, where it is eaten whole or split, husked and unhusked, or parched. In India, approximately 50 per cent of the crop is used for the production of dhal. The seeds may also be ground into a flour and utilized in the preparation of various food products such as papads, a fried dough product, spiced balls, and fermented products such as dosa and idli.

Subsidiary uses

The seed is sometimes used as an animal feedingstuff. In Assam it is usual to boil the seeds in water to produce a broth before they are fed to cattle.

Secondary and waste products

The green pods are occasionally eaten boiled as a vegetable. Urd is sometimes grown as a forage or green manure crop. The vines or haulms left after harvesting are also used for animal feeding, but are reported to be inferior to the waste from mung beans. An approximate analysis of the haulms (percentage of dry matter) has been given as: crude protein 8.9 per cent; crude fibre 28.6 per cent; ash 12.6 per cent; fat 2.8 per cent; N-free extract 47.1 per cent; calcium 1.7 per cent; phosphorus 0.2 per cent. The husks and splits remaining after the preparation of dhal are also used for animal feeding. Urd is also used in traditional medicine; the root is reported to be a narcotic.

Special features

Urd resembles the mung bean very closely in its major constituents. The seed-coat constitutes between 12.1 and 14.6 per cent of the whole seed. An approximate analysis of husked seed has been given as: moisture 10.9 per cent; protein 23.9 per cent; fat 1.4 per cent; ash 3.4 per cent; carbohydrate 60.4 per cent; calcium 0.2 per cent; phosphorus 0.4 per cent; vitamin A 65 iu/100 g; vitamin B 140 iu/100 g. The presence of a trypsin inhibitor and haemagglutinins has been reported.

The carbohydrate portion consists of sucrose 1.6 per cent, raffinose 0.51 per cent, stachyose 1.8 per cent and verbascose 3.7 per cent. The starch present consists of oval granules of medium size (4.6–11.2μ long x 8.3μ diameter); it contains 30.5 per cent amylose and 69.5 per cent amylopectin.

The protein content has been found to vary considerably according to cultivar, climate, and soil conditions, etc, values ranging from 17.3 to 25.95 per cent have been reported. The average amino acid composition, expressed as a percentage of the total protein content has been reported as: aspartic acid 12.7; threonine 3.5; serine 5.1; glutamic acid 18.6; proline 4.3; glycine 4.4; alanine 4.5; valine 5.8; methionine 1.6; isoleucine 4.8; leucine 8.7; tyrosine 2.9; phenylalanine 6.0; lysine 7.2; histidine 3.0; arginine 6.9. The recently developed Indian cultivars are reported to exhibit considerable variability with regard to methionine content. The thiamine content of stored seed has been found to decrease with insect infestation. Recently, the presence of the peptides γ–glutamylmethionine and its sulphoxide has been reported in urd, but not in the mung bean.

A sample of Pakistani seeds was found to contain 2.1 per cent of oil with the following fatty acid composition: palmitic 14.1 per cent; behenic 9.3 per cent; lignoceric 3.8 per cent; oleic 20.8 per cent; linoleic 16.3 per cent and linolenic 35.7 per cent.

A protein surfactant of the nature of a globulin and a closely associated arabogalactan type polysaccharide have been isolated from the seed. The polysaccharide confers stability on the foam formed by the protein and prevents its disruption by heat. The proteins of urd compare favourably, as regards foam-forming activity, to those of egg white and it has been suggested that flour made from urd might be used as a substitute for eggs in cake mixtures and other baked products.

Processing
Urd is processed into dhal by wet or dry methods. In the dry method the slippery wax seed-coat is partially removed by an initial pitting treatment in a rough roller-mill. The seed is next given a coating of vegetable oil (1–2 g/100 g of seed) and left overnight. It is then spread out in the sun to dry for 3–6 hours, after which it is sprayed with water and then left to dry for 3–4 days. The seed is then passed through a roller or other suitable machine. In parts of S. India, for example, a rice-huller-type of machine is sometimes used for dehusking and splitting. In the wet method, after sorting and cleaning, the seed is soaked in water, this softens the seed-coat, which is removed easily by passing it through a roller. The yield of dhal at the rural level is reported to be of the order of 69 per cent, 71 per cent by conventional commercial methods, and 82–85 per cent by the improved method developed by the Central Food Technology Research Institute, (CFTRI), Mysore.

Production and trade
Urd is produced throughout SE. Asia, but India is by far the most important producer. Although in the past 20 years there has been a steady decrease in the area under urd, total production in India has remained fairly steady, and for the period 1970–74 averaged 588 000 t/a. Average production in Bangladesh for the period 1968–72 averaged 36 600 t/a, and for 1973–74 was 36 670 t and for 1974–75 41 700 t. Urd is produced almost entirely for local consumption and there is little trade in this food legume, apart from interstate trade in India.

Major influences
Urd is a valuable pulse crop in Asia, particularly in the SE. Asian subcontinent where it is often the most prized of all the grain legumes. It would seem to offer considerable potential for development as a legume crop for the semi-arid or subhumid tropics, because of the high nutritional quality of the seed. However, currently average yields are very low and there is need to develop high-yielding, disease-resistant cultivars.

Bibliography
ACLAND, J. D. 1971. Grams. *East African crops*. p. 117. London: Longman Group Ltd, 252 pp.

AYKROYD, W. R. and DOUGHTY, J. 1964. Legumes in human nutrition. *FAO Nutr. Stud.* 19, pp. 109; 116; 118. Rome: FAO, 138 pp.

AZIZ, M. A. and SHAH, S. S. 1966. Improvement of pulses in the former Punjab. *Agric. Pak.*, *17*, 267–287.

BALASUBRAHMANYAM, N., SHURPALEKAR, S. R. and VENKATESH, K. V. L. 1973. Moisture sorption behaviour and packaging of papads. *J. Food Sci. Technol.*, *10*, 20–24.

BOSE, R. D. 1932. Studies in Indian pulses: (v) Urid or black gram (*Phaseolus mungo* Linn. var. *roxburghii* Prain). *Indian J. Agric. Sci.*, *2*, 625–637.

BOSE, R. D. and JOGLEKAR, R. G. 1933. Studies in Indian pulses (vi) The root systems of green and black grams. *Indian J. Agric. Sci.*, *3*, 1045–1056.

CHADA, Y. R. (ed.). 1976. *V. mungo* (Linn.) Hepper. *The wealth of India: Raw materials*. Vol. 10 (Sp–W). pp. 476–484. New Delhi: Counc. Sci. Ind. Res., Publ. Inf. Dir., 591 pp.

CHANDRA, S., SAGAR, P. and SINGH, B. P. 1974. We can break the yield barrier in pulses. *Indian Farming*, *24* (1), 11–13.

CHOPRA, K. and SWAMY, G. 1975. *Pulses: An analysis of demand and supply in India. Institute for social and economic change; Monograph 2.* New Delhi: Sterling Publishers, PVT, Ltd, 132 pp.

CHOUDHURY, S. L. and BHATIA, P. C. 1971. Ridge-planted kharif pulses yield high despite waterlogging. *Indian Farming*, *21* (3), 8–9.

COMMONWEALTH BUREAU OF PASTURES AND FIELD CROPS. nd. *Phaseolus mungo*. Annotated bibliography 1241, 1962–70. *Commonw. Agric. Bur., Bur. Pastures & Field Crops, Hurley, Maidenhead, Berks, England*, 4 pp.

DAYANAND, and MAHAPATRA, I. C. 1973. It pays to grow moong and urid in summer. *Indian Farming*, 23 (2), 23–25.

GÖHL, B. 1975. *Tropical feeds. Feeds information: summaries and nutritive values*. p. 206. Rome: FAO, 661 pp.

HAWARE, M. P. and PAVGI, M. S. 1976. Field reaction of black gram and green gram to angular black-spot. *Indian J. Agric. Sci.*, *46*, 280–282.

INDIAN COUNCIL OF AGRICULTURAL RESEARCH. 1969. Black gram. *Handbook of agriculture*. 3rd ed. pp. 182–183. New Delhi: Indian Counc. Agric. Res., 911 pp.

JAGANATHAN, T., NARAYANASAMY, P., PALANISAMY, A. and RANGANATHAN, K. 1974. Studies on the damping-off disease of blackgram (*Phaseolus mungo* L.). *Madras Agric. J., 61*, 156–159.

JESWANI, L. M. 1975. Varietal improvement of seed legumes in India. *Food protein sources*. (Pirie, N. W. and Swaminathan, M. S. eds.). pp. 9–18. London: Cambridge University Press, 260 pp.

KADKOL, S. B., DESIKACHAR, H. S. R. and SRINIVASAN, M. 1961. The mucilaginous principles in black gram (*Phaseolus mungo*) dhal. *J. Sci. Ind. Res.*, (*India*), *20C*, 252–253.

KURIEN, P. P., DESIKACHAR, H. S. R. and PARPIA, H. A. B. 1972. Processing and utilisation of grain legumes in India. Symp. food legumes. pp. 225–236. *Tokyo, Jpn., Min. Agric. & For., Trop. Agric. Res. Cent., Trop. Agric. Res. Ser.* 6, 253 pp.

LUKOKI-LUYEYE. 1975. Distinction entre *Vigna radiata* et *Vigna mungo*. [Distinction between *Vigna radiata* and *Vigna mungo*]. *Bull. Rech. Agron.*, (*Gembloux*), *10*, 372–373.

MACKENZIE, D. R., HO, L., LIU, T. D., WU, H. B. and OYER, E. B. 1975. Photoperiodism of mung bean and four related species. *Hortscience, 10*, 486–487.

MATHUR, H. G. 1958. Marketing of pulses in India. *Nagapur, India, Agric. Mark. Advis., Gov. India, Agric. Mark. Ser. AMA* 102, 182 pp.

MEHROTRA, O. N., SAXENA, H. K., ROY, A. N. and NATH, S. 1968. Effects of growth regulators on fruiting and yield of black gram (*Phaseolus mungo* Roxb.) in India. *Exp. Agric., 4*, 339–344.

MOHAMED ALI, A., BALAKRISHNAN, V. K., SANKARAN, S., RAJAN, A. V. and MORACHAN, Y. B. 1974. Chemical weed control in black gram (*Phaseolus mungo* Roxb.). *Madras Agric. J., 61*, 785–786.

MOODY, K. 1973. Weed control in tropical legumes. *Proc. 1st Int. Inst. Trop. Agric., Grain improvement workshop*. pp. 162–183. Ibadan, Nigeria, Int. Inst. Trop. Agric., 325 pp.

NARAYANASAMY, P. and JAGANATHAN, T. 1974. Effects of virus infection on the yield components of black gram. *Madras Agric. J., 61*, 451–456.

NARESH, J. S. and THAKUR, R. P. 1972. Efficacy of systemic granular and spray insecticides for the control of insect pests of black gram (*Phaseolus mungo* Roxb.). *Indian J. Agric. Sci.*, *42*, 732–735.

OTOUL, E. and MARÉCHAL, R. 1975. Des dipeptides soufres differencient settement *Vigna radiata* de *Vigna mungo*. [Clearly differentiated sulphur-dipeptides of *Vigna radiata* and *Vigna mungo*]. *Phytochemistry 14*, 173–179.

PARPIA, H. A. B. 1975. Utilisation problems in food legumes. *Nutritional improvement of food legumes by breeding*. (Milner, M. ed.). pp. 281–295. New York/London: John Wiley and Sons, 399 pp.

PIPER, C. V. 1914. Five oriental species of beans. pp. 26–28. *US Dep. Agric., Bull.* 119, 32 pp.

PURSEGLOVE, J. W. 1968. *Phaseolus mungo* L. *Tropical crops: Dicotyledons*. Vol. 1. pp. 301–304. Longmans, Green and Co Ltd, 332 pp.

RAJAGOPALAN, C. K., VEERASWAMY, R. and MOHAMED SHERIFF, N. 1972. A note on studies on optimum spacing for black gram. *Madras Agric. J.*, *59*, 655–656.

RAJAGOPALAN, C. K., VENUGOPAL, K. and MOHAMED SHERIFF, N. 1970. An economic manurial schedule for dryland black gram. *Madras Agric. J.*, *57*, 271–273.

RAJENDRAN, K., SIVAPPAH, A. N. and KRISHNAMOORTHY, K. K. 1974. Effect of fertilisation on yield and nutrient concentration of black gram (*Phaseolus mungo* L.). *Madras Agric. J.*, *61*, 447–450.

RAMANUTAM, S. 1973. Grain legume improvement in India. *Proc. 1st Int. Inst. Trop. Agric., Grain legume improvement workshop*. pp. 37–41. Ibadan, Nigeria, Int. Inst. Trop. Agric., 325 pp.

RAO, M. V. L., SUSHEELAMMA, N. S. and SRINIVASAN, M. R. 1974. Texture principles in the black gram (*Phaseolus mungo*). *Madrid, 4th Int. Congr. Food Sci. & Technol., Work Doc. 1a, Chemical constituents of foods in relation to flavour, colour and texture, Pap. 17*, pp. 17–18.

SAHU, S. K. 1973. Effect of rhizobium inoculation and phosphate application on black gram (*Phaseolus mungo*) and horse gram (*Dolichos biflorus*). *Madras Agric. J.*, *60*, 989–993.

SATTAR, A. and HAFIZ, A. 1952. Diseases of pulses. Researches on plant diseases of the Punjab. pp. 108–110. *Lahore, Pak. Assoc. Adv. Sci., Monogr.* 1, 158 pp.

SAXENA, R. C., SHARMA, M. M. and SINGH MALIK, N. P. 1972. Effect of systemic insecticides on the germination and subsequent growth of urid (*Phaseolus mungo* L.) seed. *Madras Agric. J., 59*, 272–275.

SEN GUPTA, P. K. 1974. Diseases of major pulse crops in India. *PANS, 20*, 409–415.

SENEWIRATNE, S. T. and APPADURAI, R. R. 1966. Black gram. *Field crops of Ceylon.* pp. 170–172. Colombo: Lake House Investments Ltd, 376 pp.

SHARMA, B. M. 1972. Effect of dates of sowing, seed rates and spacings on the grain yield of black gram (*Phaseolus mungo*). *Indian Agric., 16*, 13–16. (*Field Crop Abstr., 28*, 368).

SHURPALEKAR, S. R., PRABHAKAR, J. V., VENKATESH, K. V. L., VIBHAKAR, S. and AMLA, B. L. 1972. Some factors affecting the quality of black gram (*Phaseolus mungo*) papads. *J. Food Sci. Technol., 9*, 26–29.

SINGH, H. B., JOSHI, B. S. and THOMAS, T. A. 1970. The *Phaseolus* group: Black gram, *Phaseolus mungo* L. *Pulse crops of India.* (Kachroo, P. and Arif, M. eds.). pp. 149–158. New Delhi: Indian Counc. Agric. Res., 334 pp.

SINGH, U. P., SINGH, U. and SINGH, P. 1975. Estimate of variability, heritability and correlations for yield and its components in urd (*Phaseolus mungo* L.). *Madras Agric. J., 62*, 71–72.

SIVAPRAKASAM, K., PILLAYARSAMY, K., ARAJAMANI, and SOUMINIRA-JAGOPLAN, C. K. 1974. Evaluation of black gram (*Phaseolus mungo* L.) and green gram (*P. aureus* Roxb.) varieties for resistance to yellow mosaic virus of green gram. *Madras Agric. J., 61*, 1021–1022.

SRIVASTAVA, K. M., VERMA, G. S. and VERMA, H. N. 1969. A mosaic disease of black gram (*Phaseolus mungo*). *Sci. Cult., 35*, 475–476.

SUSHEELAMMA, N. S. and RAO, M. V. L. 1974. Surface-active principle in black gram (*Phaseolus mungo*) and its role in the texture of leavened foods containing the legume. *J. Sci., Food Agric., 25*, 665–673.

SWAMINATHAN, M. S. and JAIN, H. K. 1975. Food legumes in Indian agriculture. *Nutritional improvement of food legumes by breeding.* (Milner, M. ed.). pp. 69–82. New York/London: John Wiley and Sons, 399 pp.

TOURNEUR, M. 1958. L'ambérique et le mungo ne sont pas des *Phaseolus*. [Urid and mung are not *Phaseolus*]. *Riz Rizic.*, *4*, 131–148.

VENKATRAO, S., NUGGEHALLI, R. N., PINGALE, S. V., SWAMINATHAN, M. and SUBRAHMANYAN, V. 1960. Effect of insect infestation on stored field bean (*Dolichos lablab*) and black gram (*Phaseolus mungo*). *Food Sci.*, *9*, 79–82.

VERDCOURT, B. 1970. Studies in the *Leguminosae-Papilionoideae* for the 'Flora of tropical East Africa': IV. *Kew Bull.*, *24*, 507–569.

WATT, E. E. and MARÉCHAL, R. 1977. The difference between mung and urid beans. *Trop. Grain Legume Bull.*, (7), pp. 31–33.

YEGNA AIYER NARAYAN, A. K. 1966. Black gram (*Phaseolus mungo*). *Field crops of India*. 6th ed. pp. 118–120. Bangalore City: Bangalore Printing and Publishing Co Ltd, 564 pp.

Common name | **VELVET BEAN.**

Botanical name | *Mucuna pruriens* (L.) DC. var. *utilis* (Wall. ex Wight) Baker ex Burck.

Family | Leguminosae.

Other names | Banana stock pea (Aust.); Dolique de Floride (Fr.); Fluweelboontjie (S. Afr.); Frijol terciopelo (Mex.); Haba de terciopelo (P.R.); Haricot velouté (Fr.); Makhmali sem (Ind.); Mauritius bean[1]; Ojo de venado (Sp.); Pois mascate (Fr.); Poroto aterciopelado (Arg.); Stizolobia (It.).

Botany

An annual, or perennial, herbaceous, vigorous climbing vine, which can reach 60 ft (18 m) in length when grown on supports and even on the ground can attain a length of 18 ft (5.5 m), although bushy forms also exist. The roots are fleshy, usually well nodulated and produced near the soil surface. The long trailing stems are rather slender, sparsely pubescent. They bear numerous, alternate, trifoliolate leaves on short, hairy, fleshy petioles, with large, ovate leaflets. The infloresence is axillary and the flowers, usually 5–30, are showy, and purple, red or greenish-yellow in colour. The pods are normally 2–6 in. (5–15 cm) long, slightly ridged and densely covered with black, white or grey hairs, which give them a velvety appearance. The nearly globular seeds (3–5) are normally speckled or marbled, brown or black, although black, grey, or white forms sometimes occur.

There has long been some uncertainty as to whether the velvet beans should be segregated from *Mucuna* as the genus *Stizolobium*, and although not currently fashionable, the latter name has been used in much of the literature, especially in the USA. Only two species are commonly grown, the true velvet bean *M. pruriens* var. *utilis*, which has medium-sized seeds, and the horse bean, *M. sloanei* Fawcett & Rendle (syn. *M. urens* auctt.), which has larger seeds with an extremely hard seed-coat. There are, however, numerous cultivars of *M. pruriens*, some of which were formerly treated as separate species. The principal ones are:

---

[1] Also used for the winged bean, *Psophocarpus tetragonolobus*.

(i) Deering, Florida or Georgia velvet bean, frequently designated as *Stizolobium deeringianum* or *Muncuna deeringiana*. It is of some importance as a cattle fodder, particularly in the USA and parts of S. America.

(ii) Bengal velvet bean, frequently designated as *M. utilis* or *S. utile*, and grown in India.

(iii) Mauritius velvet bean, often designated as *M. aterrima* or *S. aterrimum*, and grown in Mauritius, Australia, Brazil and the W.I., often as a rotation crop with sugar cane or as a drought resistant cover crop.

(iv) Yokohama velvet bean, usually designated as *S. hassjoo*. This is an early-maturing type and is a less vigorous grower than the other velvet beans, it is thought to have originated in Japan, where it is known under the names 'Osharuka-mame' and 'Hasshomame'.

(v) Lyon bean, designated *M. nivea*, *S. niveum*, *M. cochinchinensis* or *Carpogon niveum*, which is sometimes cultivated as a vegetable for its immature pods in the Philippines and SE. Asia.

Origin and distribution

Velvet beans are thought to have originated in Asia and to have been introduced into the western hemisphere via Mauritius. They are now cultivated in many tropical and subtropical areas, and through breeding, certain cultivars tolerant of more temperate conditions have been developed.

Cultivation conditions

Temperature—a warm equable temperature of between 68° and 86°F (20–30°C) throughout the growing period is preferred; night temperatures of 70°F (21°C) are reported to stimulate flowering. Velvet beans are susceptible to frost and require a frost-free period of 180 to 240 days. Exposure to temperatures below 41°F (5°C) for periods as short as 24 or 36 hours is reported to be fatal to young Florida velvet bean plants.

Rainfall—velvet beans are often grown in the tropics and subtropics in areas with an average rainfall of between 45 and 60 in./a (1 200–1 500 mm/a), or more. The Mauritius velvet bean, some types of the Lyon bean and the Florida velvet bean are tolerant of drought conditions and show considerable promise for dryland farming.

Velvet bean

Soil—a wide range of soil types are suitable, including heavy clays, provided that they are well drained, since velvet beans cannot stand waterlogging. They are tolerant of fairly acid soils, but for optimum yields light sandy loams, with a pH of between 5 and 6.5, are required. Little is known of the crop's precise manurial requirements, but it responds to applications of phosphate, and in the USA the application of superphosphate 100–200 lb/ac (112–225 kg/ha) has been recommended.

Altitude—this crop can be grown from sea level up to elevations of between 6 000 and 7 000 ft (1 800–2 100 m) in the tropics.

Day-length—velvet beans have been reported to be a short-day legume, but in experiments in S. Africa the plants did not show a significant response to day-length.

Planting procedure
Material—seed is used, germination is hypogeal and usually occurs within 5–7 days, with a rate of between 90 and 95 per cent. Inoculation of the seed is not usually necessary if velvet beans are grown in areas where the cowpea has been cultivated, otherwise inoculation with a commercial cowpea inoculum may be necessary.

Method—velvet beans are frequently grown mixed with other vigorous growing crops such as sugar cane or maize. When grown for the production of seed, planting is usually in rows, but when grown as a fodder or green manure crop, the seeds are often broadcast. For optimum results the seed-bed should be well prepared to a depth of at least 6 in. (15 cm) and completely free from weeds. Adapted maize planters with thick plates and enlarged holes are sometimes used for planting in countries such as the USA, S. Africa and Australia. In Asia the crop normally receives little attention, but the suppression of weeds during the early stages of growth and the provision of supports for the vines to climb is beneficial.

Field-spacing—when grown for seed, velvet beans are frequently planted in rows 3 to 6 ft (90–180 cm) apart, with 0.5 to 3 ft (15–90 cm) between the plants. When grown as a mixed crop with cereals, such as maize or millet, it is usual to plant every alternate or third row with velvet beans.

Seed-rate—in India the average seed-rate is reported to vary between 25 and 45 lb/ac (28–50 kg/ha), in most other countries it normally varies from

10–20 lb/ac (11–22 kg/ha), when the crop is planted in rows, and from 40 to 80 lb/ac (45–90 kg/ha), when sown broadcast.

Pests and diseases

Velvet beans are relatively free from serious attacks by pests and diseases. In the USA the velvet bean caterpillar, *Anticarsia gemmatilis*, can sometimes cause serious defoliation of the plant and occasionally the crop is reported to be affected by the root-knot nematode, *Heterodera radicola*.

In India velvet beans are reported to be infected by a bacterial leaf-spot, *Xanthomonas stizolobiicola*, a leaf spot disease, *Cercospora stizolobii*, and a rust, *Uromyces mucanae*. All three are reported to be controlled effectively by the complete destruction of diseased plant debris. In Queensland a wilt disease, *Phytophthora drechsleri*, is reported to cause considerable crop losses on the Mauritius velvet bean given wet weather conditions.

Growth period

When grown for seed in the tropics many cultivars are normally ready for harvesting approximately 180 to 270 days after planting. In the USA, Florida velvet beans take about 240–270 days to reach maturity, but early-maturing cultivars, which produce seed in 110–130 days, have been developed. The Yokohama velvet bean normally matures within 110–120 days. When grown for forage, velvet beans are usually harvested between 90 and 120 days after sowing.

Harvesting and handling

The pods should be thoroughly ripe before being harvested. As they are liable to shatter, they are often hand-picked and are usually left to dry for several days before being threshed. Threshing can be carried out by hand or machine, but whichever method is used only fully mature, dry pods can be threshed without difficulty. Velvet beans are reported to be relatively resistant to insect attack during storage.

Primary product

Seed—the velvet bean is globular, approximately 0.5 x 0.5 in. (1.2 x 1.2 cm), often a mottled brown or black colour, sometimes with a pale-grey background; a few cultivars produce pure grey, white or black seed. The hilum is about 0.15 in. (4 mm) in length and is surrounded by a distinctive white aril. One hundred seeds weigh approximately 3.8 oz (109 g).

Yield

In India seed yields are reported to average between 650 and 1 000 lb/ac (730–1 120 kg/ha), yields of hay to average between 2 500 and 3 200 lb/ac (2 800–3 580 kg/ha), and yields of fodder from 8 200 to 16 400 lb/ac (9 180–18 370 kg/ha). Yields of seed in the USA are reported to average between 1 500 and 2 000 lb/ac (1 680–2 240 kg/ha), although with efficient cultivation yields of around 3 000 lb/ac (3 360 kg/ha) are by no means unusual. At Grafton, New South Wales, average yields have been reported to be as low as 500 lb/ac (560 kg/ha).

Main use

Velvet beans are a nutritious animal feedingstuff and are used mainly for grazing, although the mature seeds are also used in the manufacture of compound feedingstuffs, or fed direct to the animals, when they are often soaked in water for 24 hours, or ground into a meal. They are used mainly for feeding cattle or sheep and can only be fed to pigs if they constitute less than 25 per cent of the diet. They are considered unsuitable for poultry.

Subsidiary uses

Velvet beans can be used as a human foodstuff but require considerable care in their preparation, because of the toxic principle they contain. In many parts of Africa and Asia they are regarded as a famine food. The toxic principle can be removed by boiling and soaking the seeds in several changes of water. In parts of Asia, the seeds are sometimes roasted before being eaten. In other parts of Asia, notably Java, the seeds are sometimes boiled, the seed-coat removed, and the decorticated seeds soaked in water, after which they are chopped, steamed, and left to ferment to produce a bean cake, 'tempe benguk', which resembles tempe produced from soyabeans. The immature pods and leaves are occasionally boiled and eaten as a vegetable.

Secondary and waste products

Velvet beans are also grown as a green manure or cover crop and are particularly valuable as an anti-erosion crop. The possibility of utilizing the seed as a source of industrial starch has been investigated in Brazil and results indicate that a starch with a high viscosity, similar to that obtained from cowpeas, and suitable as a thickening agent for food products, or as an adhesive in the paper and textile industries, could be obtained. The possibility of utilizing the seeds as a commercial source of L-dopa, used in the treatment of Parkinsons' disease, has also been investigated recently.

Special features

The approximate composition of the green forage of velvet beans, on a dry weight basis, has been given as: protein 15.1 per cent; fat 2.1 per cent; N-free extract 48.6 per cent; fibre 19.3 per cent; ash 14.9 per cent. Digestible protein 10.7 per cent; digestible carbohydrate 49.6 per cent; total digestibile nutrients 63.4 per cent; nutritive ratio 4:9. The composition of the whole, dry pods has been reported as: moisture 10.0 per cent; protein 18.1 per cent; fat 4.4 per cent; N-free extract 50.3 per cent; fibre 13.0 per cent; ash 4.2 per cent. The approximate composition of the mature seeds is: moisture 10.0 per cent; protein 23.4 per cent; fat 5.7 per cent; N-free extract 51.5 per cent; fibre 6.4 per cent; ash 3.0 per cent; calcium 0.18 per cent; phosphorus 0.99 per cent; potassium 1.36 per cent; vitamin A 50 iu/100 g; thiamine 0.50 mg/100 g; riboflavin 0.20 mg/100 g; niacin 1.7 mg/100 g. The amino acids present (mg/gN) are: isoleucine 300; leucine 475; lysine 388; methionine 75; cystine 56; phenylalanine 300; tyrosine 319; threonine 250; valine 344; arginine 494; histidine 131; alanine 219; aspartic acid 794; glutamic acid 763; glycine 288; proline 369; serine 306.

The oil present in the seeds has been found to be highly unsaturated with 47.2 per cent linoleic acid; 14.2 per cent oleic acid; 3.8 per cent linolenic and 0.5 per cent palmitoleic acid. The saturated fatty acids are: palmitic 19.5 per cent; stearic 12.6 per cent; arachidic 2.2 per cent. In feeding trials with rats it was shown that the toxic principle occurs in the protein fraction of the seed and not in the oil. The toxic principle L-dopa, 3-(3, 4-dihydroxyphenyl)-L-alanine, is present mainly in the seed embryo and has been isolated in amounts equivalent to 1.5 per cent of the whole seed weight. It has been suggested that the presence of free L-dopa is the reason that velvet beans are relatively immune to attack from insects and small mammals. In addition to L-dopa, a new amino acid (-)-1-methyl-3-carboxy-6, 7-dihydroxy-1, 2, 3, 4-tetrahydroisoquinoline, has been isolated recently from velvet beans.

Production and trade

No statistical data have been traced.

Major influences

Velvet beans are a minor legume crop, many cultivars of which are suitable for cultivation in the more humid regions of the tropics, while others, eg the Mauritius velvet bean, are suitable for dryland farming. They can be grown successfully on soils unsuitable for cowpeas, but have the disadvantage of a longer growth period and are more difficult to thresh. Moreover, the seed is not highly valued for human or animal feeding, because of the prolonged soaking and, or, boiling required before it can be consumed safely. The leaves and vines make an excellent fodder and efforts are being made in India to develop improved early-maturing, high-yielding strains of indeterminate growth habit, for use in multi-cropping programmes in areas of low or uncertain rainfall.

The possibility of utilizing velvet beans as a commercial source of L-dopa, which is relatively expensive to produce synthetically, has received attention in recent years. Yields of around 4.8 per cent are reported to be attainable from the decorticated seed meal. The residual cake has a protein content of 15–20 per cent and could be used for livestock feeding.

Bibliography

ABDEL KADER, M. M., EL-KIRDASSY, Z. H. M., SHOEB, Z. E. AND EISSA, M. H. 1973. Chemical and nutritional evaluation of Egyptian velvet beans *Stizolobium deeringianum* Bort. *Fette Seifen Anstrichm.*, *75*, 25–27.

ANON. 1957. Velvet beans as a grazing crop. *Qd. Agric. J.*, *83*, 488.

BELL, E. A. AND JANZEN, D. H. 1971. Medical and ecological considerations of L-dopa and 5-HTP in seeds. *Nature*, *229*, 136–137.

DABADGHAO, P. M. AND GANDHI, R. T. 1954. Mucuna the new green fodder. *Indian Farming*, *4*(6), 16–17.

DAKO, D. Y. AND HILL, D. C. 1977. Chemical and biological evaluation of *Mucuna pruriens* (*utilis*) beans. *Nutr. Rep. Int.*, *15*, 239–244.

DAXENBICHLER, M. E., KLEIMAN, R., WEISLEDER, D., VANETTEN, C. H. AND CARLSON, K. D. 1972. A new amino acid, (-)-1-methyl-3-carboxy-6, 7-dihydroxy-1, 2, 3, 4-tetrahydroisoquinoline, from velvet beans. *Tetrahedron Lett.*, (18), pp. 1801–1802.

DAXENBICHLER, M. E., VANETTEN, C. H., EARLE, F. R. AND TALLENT, W. H. 1972. L-dopa recovery from mucuna seed. *J. Agric. Food Chem.*, *20*, 1046–1048.

FRENCH, M. H. 1940. Some recent observations on the feeding value of some local animal feedingstuffs. *East Afr. Agric. J.*, *6*, 87–90.

GANJAR, I. AND SLAMET, D. S. 1974. The nutrient content of fermented *Mucuna pruriens* seeds *1st ASEAN Workshop on grain legumes. Bogor, Minist. Agric. Indonesia, ASEAN 74 FA/Wrks, GLI/WOP-11*, 11 pp.

GILLET, J. B., POLHILL, R. M. AND VERDCOURT, B. 1971. 51. *Mucuna. Flora of tropical East Africa: Leguminosae* (*Part 4*), *Subfamily Papilionoideae* (2). pp. 566–567. London: Crown Agents for Overseas Governments and Administrations, 1108 pp.

GÖHL, B. 1975. *Stizolobium* spp. (*Mucuna* spp.). *Tropical feeds: Feeds information summaries and nutritive values*. pp. 217–218. Rome: FAO, 661 pp.

GREEN, N. K., MOFFATT, J. R. AND PARKINSON, S. T. 1931. Identification of commercial species of pea and bean seeds. *J. Southeast Coll. Agric.*, (*Wye*), *28*, 21–40.

KING, N. J., MUNGOMERY, R. W. AND HUGHES, C. G. 1965. Velvet beans. *Manual of cane-growing*. pp. 123–124. New York: American Elsevier Publishing Co Inc, 375 pp.

MAGOON, M. L., SINGH, A. AND MEHRA, K. L. 1974. Improved velvet beans for increased forage production. *Indian Farming*, *23* (12), 23–27.

MEAD, K. J. 1959. Velvet beans. *Agric. Gaz. N.S.W.*, *70*, 248–251.

MES, M. G. 1959. The influence of night temperature and day-length on the growth, nodulation, nitrogen assimilation and flowering of *Stizolobium deeringianum* (Velvet bean). *South Afr. J. Sci.*, *55*, 35–39.

PIPER, C. V. AND MORSE, W. J. 1938. The velvet bean. *US Dep. Agric. Farmers' Bull.* 1276, 21 pp.

PIPER, C. V. AND SHULL, J. M. 1917. Structure of the pod and the seed of the Georgia velvet bean, *Stizolobium deeringianum. J. Agric. Res.*, *11*, 673–675.

POLLOCK, N. A. R. 1934. The velvet bean. *Qd. Agric. J.*, *42*, 136–142.

PONS, J. E. 1942. Less well-known legumes. *Farming South Afr.*, *17*, 769–774.

QUEENSLAND DEPARTMENT OF AGRICULTURE AND STOCK. 1962. Velvet bean. *The Queensland agricultural and pastoral handbook*. 2nd ed. Vol. 1. *Farm crops and pastures*. (Hockings, E. T. ed.). pp. 447–449. Brisbane, Queensland, Dep. Agric. & Stock, 583 pp.

REHR, S. S., JANZEN, D. H. AND FEENY, P. P. 1973. L-dopa in legume seeds: A chemical barrier to insect attack. *Science*, *181*, 81–82.

SASTRI, B. N. (ed.). 1962. *Mucuna* Adans. (Leguminosae). *The wealth of India: Raw materials*. Vol. 6 (L-M). pp. 439–444. New Delhi: Indian Counc. Sci. Ind. Res., 483 pp.

SELLSCHOP, J. P. F. AND SALMON, S. C. 1928. The influence of chilling above the freezing point on certain crop plants. *J. Agric. Res.*, *37*, 315–338.

SINGH, S. 1954. Let velvet beans be in your rotation. *Indian Farming*, *4* (9), 13.

TOLMASQUIM, E., CORRÊA, A. M. N. AND NAKAMURA, T. 1970. Studies on new starches (ii): A study of the properties of four varieties of mucuna bean (*Stizolobium*). *Die Stärke*, *22*, 313–317.

VERDCOURT, B. 1970. Studies in the *Leguminosae-Papilionoideae* for the 'Flora of tropical East Africa': II *Mucuna* Adans. *Kew Bull.*, *24*, 286–287.

VISSER, H. C. 1957. The velvet bean. *Handbook for farmers in South Africa*. Vol. 2. *Agronomy and horticulture*. p. 199. Pretoria, Dep. Agric., 882 pp.

WESTPHAL, E. 1974. *Mucuna pruriens* (L.) DC. Pulses in Ethiopia, their taxonomy and agricultural significance. pp. 121–129. *Wageningen, PUDOC, Cent. Agric. Publ. Doc., Agric. Res. Rep.* 815, 278 pp.

WHYTE, R. O., NILSSON-LEISSNER, G. AND TRUMBLE, H. C. 1953. *Stizolobium:* Velvet beans. Legumes in agriculture. *FAO Agric. Stud.* 21, pp. 323–324. Rome: FAO, 367 pp.

Common names | **WINGED BEAN[1], Asparagus bean, or pea[2], Four-angled[1], or Four-cornered bean[1], Goa bean[3], Manila bean[4], Mauritius bean[5].**

Botanical name | *Psophocarpus tetragonolobus* L.

Family | Leguminosae.

Other names | Amali, Batong-baimbing (Philipp.); Burma haricot; Calamismis (Philipp.); Chara-koni-sem (Beng.); Chaudhaari-phali (Hind.); Chavdhari-ghevda (Bom.); Chichipir, Chipir (Indon.); Cigarillas (Philipp.); Dara-d(h) ambala (Sri La.); Dâu cau (Viet.); Dragon bean; Fava de cavallo (Port); Flügelbohne (Ger.); Garbanso (Philipp.); Haricot dragon (Fr.); Kachang bêlimbing, K. botol, K. botor (Malays.); K. embing (Indon.); K. kélisah, K. kotor (Malays.); Kalamismis (Philipp.); Katchang botor (Malays.); Katjeper, Kêchipir, Kêtjeepir (blinger), Kêtjeper (Indon.); Lakar-sem (Beng.); Morisuavarai, Murukavarai (Tam.); Pallang, Parupa-gulung (Philipp.); Pè-myît, Pè-saung-sa, (ya or za) (Burm.); Pois ailé, P. carré[6] (Fr.); Princess pea; Sabidokong (Philipp.); See-kok-tau (China); Segidilla, Seguidilla, Sequidilla, Sererella (Philipp.); Sesquidilla (Sp.); Shambe kayi (Ind.); Sigarilya (Philipp.); Tjeepir bee-bas, T. we-loo, Tjeetjeepir (Indon.); Too-a-poo, Tua pu (Thai.); Winged pea[1].

---

[1] Also applied to other legumes with four-winged, edible pods, including *Psophocarpus palustris*, which is similar and occasionally cultivated in Africa, and *Lotus tetragonolobus* (syn. *Tetragonolobus purpureus*), which is grown in S. Europe and the Mediterranean region.

[2] More usually used for the asparagus bean, *Vigna unguiculata* ssp. *sesquipedalis*.

[3] Also used for the jack bean, *Canavalia ensiformis*.

[4] Also used for the bambara groundnut, *Voandzeia subterranea*.

[5] Also used for the velvet bean, *Mucuna pruriens* var. *utilis*.

[6] Also used for marrowfat peas, *Pisum sativum*, and occasionally used for the grass pea, *Lathyrus sativus*.

Botany

A climbing perennial, producing new growth annually, from shallow, persistent roots, but for optimum results the winged bean is treated as an annual. The fibrous roots are numerous with the main laterals running horizontally near the soil surface; after a few months they usually become thickened and tuberous, although certain strains are unable to form tubers. The roots are normally heavily nodulated. Plants in Malaysia may carry up to 440 large nodules each and their fresh weight can reach 700 lb/ac (780 kg/ha). A single nodule may weigh 0.02 oz (0.6 g) and have a diameter of up to 0.5 in. (1.2 cm). The stem is moderately thick, slightly ridged and grooved, and can reach 10–12 ft (3–3.6 m) in height, if given support. The leaves are trifoliolate, on long, stiff petioles; the leaflets are ovate, 3–6 in. (7.5–15 cm) long, the terminal one is usually longer than the laterals and attached to the petiole by a marked pulvinus. The infloresence is borne on an axillary raceme, up to 6 in. (15 cm) in length, and bearing 2–10 flowers, which may be blue, white or lilac. It has been reported that pollination in some species is by bees, and in their absence pod-set is very low. The pods are four-sided, with characteristic serrated wings running down the four corners. They contain 5–20 seeds which can vary in colour from white, through varying shades of yellow and brown to black, and may also be mottled.

There are many different local strains of the winged bean. The species is not found growing wild although it has been noted growing as an escape in Burma and the Philippines. There are four closely related species found wild in Africa, of which, *P. palustris* Desv. and *P. scandens* (Endl.) Verdc. (syn. *P. longipedunculatus* Hassk.) are occasionally cultivated.

Origin and distribution

The winged bean is thought to have originated in Africa (Malagasy or Mauritius) and to have spread to Asia. It is now cultivated usually as a market garden crop in S. India, Burma, Malaysia, New Guinea, Indonesia, the Philippines, Vietnam and Thailand, and to a lesser extent in Africa, mainly in Ghana and Nigeria, and in the W.I.

Cultivation conditions

Temperature—a tropical crop resistant to high temperatures, grown between 20° N and 10° S latitude in Asia.

Rainfall—it requires a well-distributed rainfall in excess of 60 in./a (1 500 mm/a), and thrives in areas with an annual rainfall of 100 in. (2 500 mm), or more. It can be grown as a dry season crop, provided that there is adequate irrigation and the water does not remain on the soil, as this tends to reduce the yield of roots. Despite its perennial nature and extensive root-system, it does not survive prolonged drought.

Soil—the winged bean is not very demanding in its soil requirements, provided that there is adequate drainage. It cannot tolerate waterlogging or salinity. Well-cultivated, rich, sandy loams are best for optimum yields of pods; on clay soils the tubers are frequently small and lacking in flavour. It is frequently grown successfully in nitrogen-poor soils because of its exceptional ability to nodulate. Although recently in Nigeria, experimentally grown winged beans have made comparatively slow growth accompanied by markedly chlorotic, light-green foliage, which suggests that the rate of nodulation is very dependent upon the availability of the most effective rhizobial strains, probably of the cowpea group. The manurial requirements of the winged bean have not been studied in detail, but it responds favourably to nitrogen fertilization. When grown as a vegetable the routine application of standard NPK fertilizer at intervals of 14 to 21 days has been recommended.

Altitude—it can be grown at elevations up to 7 000 ft (2 100 m) in the tropics.

Day-length—the winged bean requires short days for normal flower induction, since when grown under a long photoperiod there is excessive vegetative growth at the expense of flowers.

Planting procedure

Material—seed, which is viable for approximately one year, is normally used. Problems handicapping the future development of this crop are the lack of adequate commercial supplies of seeds and the genetic variability of existing supplies. In certain areas of Burma and the Philippines the crop is treated as a perennial and the tubers are left in the ground to produce fresh plants.

Method—the winged bean is often interplanted with sweet potatoes, taro, bananas, sugar cane, or other vegetable crops. For pod and seed production planting is usually on the flat and the seeds are dibbed in holes about 1 in. (2.5 cm) deep, at the beginning of the rainy season. It is usual to provide the winged bean with supports, bamboo poles arranged singly, or in tripods, are often used. When poles are used the plants may grow so tall that picking is difficult and the use of a trellis or wire fence 4 ft (1.2 m) high has been recommended recently. In Burma, where the crop is grown on a field-scale for the production of tubers, the seeds are normally planted 2–3 in. (5–7.5 cm) deep on ridges, and earthed up, stakes are frequently dispensed with, and the plants are left to ramble over the ground reaching a height of 12 in. (30 cm). Seedlings make slow growth for the first 3 to 5 weeks and efficient weed control is usually very necessary until they are well established.

Field-spacing—in Ghana, for seed production, a spacing of 24 x 24 in. (60 x 60 cm) with 3 seeds per hole has been recommended. For pod production, reported spacings range from 18 to 24 in. (45–60 cm) between the plants and 24 to 48 in. (60–120 cm) between the rows. In Burma, for tuber production on a field-scale, the seeds are reported to be planted 3–6 in. (7.5–15 cm) apart, on ridges 24 in. (60 cm) wide and 8–10 in. (20–25 cm) high. In the Philippines the winged bean is planted in hills 3.2–6.5 ft (1–2 m) apart in rows 6.5 ft (2 m) apart, although some growers use a 13 x 13 ft (4 x 4 m) spacing. Usually 2 or 3 seeds are placed in each hill which requires about 2.7 lb/ac (3 kg/ha) of seed using 13 x 13 ft (4 x 4 m) spacing.

Pests and diseases

As it is currently grown in mixed market garden culture or shifting agriculture the winged bean is generally free from serious pests and diseases. In Ghana, the flowers and pods are sometimes eaten by caterpillars. Occasionally crops are damaged by leaf miners, grasshoppers and spider mites. A root-knot nematode has been reported in Papua New Guinea. False rust, *Synchytrium psophocarpi*, has recently been reported to be a major fungus disease in Papua New Guinea, and also to occur in other parts of Asia and W. Africa. In the last area it is reported to have been controlled experimentally by the application of copper fungicides.

Growth period

Pods—the first green pods are usually ready for picking about 42 to 70 days after sowing and each plant continues to produce about 25 pods every 5 or 6 days for several weeks, depending upon the fertility of the soil and the availability of soil moisture. It has been suggested that production normally declines to an uneconomic level after several months, although in the Philippines the crop is sometimes treated as a perennial and is reported to produce an economic yield of pods for 5 years or more.

Seed—mature fully ripe seed is normally produced 180 to 270 days after planting. In Ghana certain improved early-maturing strains are reported to produce seed about 114 days after sowing.
Tubers—when grown for the tubers harvesting normally occurs 120 to 240 days after sowing.

Harvesting and handling
The immature pods are usually picked by hand about 2 weeks after fertilization and before they have become fibrous. In the Philippines they are usually size-graded, washed, tied into bundles and packed in baskets for local sale. When grown for the edible seed, the pods are left on the plant until they are fully ripe when they are picked by hand, threshed, and the seeds dried before being packed in airtight containers to prevent insect infestation. The tubers are usually harvested when they reach 1–2 in. (2.5–5 cm) in diameter and 3–5 in. (7.5–12 cm) in length. They are normally dug out by hand with a fork to avoid damage, and sometimes the ground is flooded to facilitate digging, and to reduce the possibility of mechanical injury.

Primary product
Pod—the winged bean is currently grown in most tropical countries for the production of the immature edible pod. The pods vary in length from 2 to 14 in. (5–35 cm) and are approximately 1 in. (2.5 cm) wide. They are four-sided with characteristic serrated papery wings about 0.12–0.2 in. (3–6 mm) broad. The pod length appears to be genetically determined with long- and short-pod strains. The pods are normally green, although purple forms sometimes occur. They contain 5–20 shiny seeds, which can vary in shape from nearly globular to almost conical, and may be white, yellow, brown, black, or mottled. One hundred seeds normally weigh approximately 1 oz. (28–30 g).

Yield
There is little information on the yield of pods, probably because cropping is irregular and the winged bean is grown mainly on a small-scale in market gardens. When grown experimentally in Malaysia a yield of green pods of 31 620 lb/ac (35 530 kg/ha) has been reported. In Ghana seed yields ranging from 730 to 1 230 lb/ac (820–1 380 kg/ha) have been obtained and recently yields of 2 140 lb/ac (2 400 kg/ha) have been recorded in agronomic trials at the International Institute of Tropical Agriculture (IITA), Ibadan, Nigeria. In Malaysia considerably higher yields of approximately 4 090 lb/ac

(4 580 kg/ha) have recently been reported. Tuber yields are reported to vary from 1 to 2.4 T/ac (2.5–6 t/ha) in different regions of Burma and from 0.9 to 2 T/ac (2.25–5 t/ha) in Malaysia, where it has been observed that plots grown for green pod production give the highest yields of tubers.

Main use

The pods of the winged bean are a nutritious vegetable and are eaten sliced and boiled similarly to French beans, *Phaseolus vulgaris*; sometimes the very young pods are eaten raw in salads. However, compared with the potentialities of the other products of the crop, discussed in the section *Secondary and waste products*, the pods are likely to become of considerably less economic importance in the future.

Subsidiary uses

Because of its capacity for nodulation, the winged bean is occasionally grown as a green manure or cover crop, and as a restorative fallow crop. In certain areas of Burma, sugar cane following a crop of the winged bean is reported to increase its productivity by about 50 per cent.

Secondary and waste products

Tubers—the tuberous roots may be peeled, eaten raw, or boiled and eaten as a vegetable similar to potatoes. They are reported to have the consistency of an apple and to taste slightly sweet. In Burma they are regarded as a delicacy and are often eaten as a snack. They are reported to be most suitable for eating when they are about the thickness of a thumb; as they mature they become stringy and insipid.

Seeds—the unripe seeds are sometimes eaten as a vegetable similar to peas, when they are often used as an ingredient for soups and curries. The mature dried seeds are reported to cook with difficulty and to be rather indigestible. They are sometimes roasted and eaten like groundnuts; in Indonesia they are usually parched before eating. The seeds are similar to soyabeans in composition and it has been suggested that they could be utilized for the production of protein foodstuffs, culinary oil and soap. The residual cake after oil extraction could be used for human or livestock feeding. It has been demonstrated that a flour made from the seeds is suitable for use as a milk substitute in the treatment of children suffering from kwashiorkor.

Foliage—the young leaves and flowers are sometimes eaten raw, or steamed, or added to fish soups. In Papua New Guinea the flowers are sometimes fried in oil and have a taste resembling that of mushrooms. The stems and leaves have possibilities for use as forage. Their nutritive value has been reported as follows: moisture 78.9 per cent; protein 6.3 per cent; digestible protein 4.8 per cent; fat 1.0 per cent; carbohydrate 7.9 per cent; fibre 4.1 per cent; ash 1.8 per cent; calcium 0.37 per cent; phosphorus 0.12 per cent.

Special features

Pods—the edible portion amounts to about 96 per cent, the proximate composition is: moisture 76.0–92.0 per cent; protein 1.9–2.9 per cent; fat 0.2–0.3 per cent; carbohydrate 3.1–3.8 per cent; fibre 1.2–2.6 per cent; ash 0.4–1.9 per cent. The mineral and vitamin content of Philippine pods, per 100 g of edible portion, has been given as: calcium 42 mg; phosphorus 46 mg; iron 0.9 mg; sodium 5.0 mg; potassium 230 mg; thiamine 0.12 mg; riboflavin 0.13 mg; nicotinic acid 1.20 mg; ascorbic acid 22 mg; vitamin A 570 iu. The free amino acids present are: serine, aspartic acid, glycine, glutamic acid, alanine, tyrosine and all the essential amino acids except histidine and methionine.

Tubers—the tubers have a brown fibrous skin and white solid flesh, and have a high protein content (normally 12-15 per cent wet weight) compared with other tropical root crops, such as yams or cassava. The average composition of the dry roots has been widely reported as: moisture 9.0 per cent; protein 24.6 per cent; fat 1.0 per cent; carbohydrate 56.1 per cent; fibre 5.4 per cent; ash 3.9 per cent. A recent analysis of four genotypes grown at IITA gave the following results: dry matter 32.2–48.7 per cent; crude protein 17.0–20.0 per cent; fat 0.6–1.4 per cent; crude fibre 17.0–21.0 per cent; starch 48.5–54.0 per cent; ash 1.77–2.11 per cent; manganese 30.0–50.0 ppm; copper 6.0–68.0 ppm; iron 220–706 ppm; sulphur 0.16–0.36 per cent; S/N ratio 5.0–12.0 per cent.

Seeds—these consist of approximately 22 per cent skin and 88 per cent edible material. Protein contents ranging from 29.75 to 37.4 per cent and fat contents from 15.0–18.1 per cent have been reported. A sample of dry decorticated Ghanaian seeds which was analysed had the following composition: protein 37.3 per cent; fat 18.1 per cent; carbohydrate 25.2 per cent; moisture 9.7 per cent; fibre 5.4 per cent; ash 4.3 per cent; thiamine 13.9 ppm; riboflavin 1.8 ppm. The oil had the following characteristics $N_D{}^{30°C}$ 1.466; SG $^{30°C}$ 0.9284; sap. val. 175.6; iod. val. 82.1 and contained 125.9 mg tocopherol (γ + β)/100 g. The fatty acid composition (percentage by weight) was myristic 0.06; palmitic 8.9–9.7; palmitoleic 0.83; stearic 5.7–5.9; oleic 32.3–39.0; linoleic 27.2–27.8; linolenic 1.1–2.0; arachidic 2.0; parinaric 2.5; behenic 13.4–15.5. The amino acid composition of the seeds, expressed as a percentage of total protein content, has been reported as follows: cystine 1.6–2.6; lysine 7.4–8.0; histidine 2.7; arginine 6.5–6.6; aspartic acid 11.5–12.5; threonine 4.3–4.5; serine 4.9–5.2; glutamic acid 15.3–15.8; proline 6.9–7.6; glycine 4.3; alanine 4.3; valine 4.9–5.7; methionine 1.2; isoleucine 4.9–5.1; leucine 8.6–9.2; tyrosine 3.2; phenylalanine 4.8–5.8 tryptophan 1.0. The nutritive value of the winged bean is similar to that of the soyabean. It is deficient in the sulphur-containing amino acids.

Like many other legume seeds, the winged bean contains a trypsin inhibitor. Autoclaving for 10 minutes at 266°F (130°C), or boiling for 30 minutes after soaking in water for 10 hours, is reported to destroy the inhibitor. The application of dry heat at 347°F (175°C) for 10 minutes has been found to be unsatisfactory. The presence of a phytohaemagglutinin, which is apparently destroyed by cooking, has also been reported, but no urease activity has been detected. The presence of hydrocyanic acid has been reported in the plant stems, but not in the seeds.

Processing

Although currently the seeds are not processed, it has been suggested that the techniques for producing soyabean flour and protein concentrates could be applied to the winged bean.

Production and trade

No statistical data are available. The winged bean is traditionally a market garden crop of the New Guinea Papua highlands.

Major influences

With current world food shortages, the winged bean is of considerable interest as a high-protein, multi-purpose crop, particularly suitable for cultivation in the humid tropics, where the incidence of protein deficiency in human diets is often very difficult to remedy. All parts of the plant are edible, ie seeds, tubers, leaves and flowers. The seeds, which are very similar nutritionally to soyabeans, have the advantage that they have a pleasant sweet flavour in contrast to the rather bitter, beany flavour of the former. Like the soyabean, the winged bean could be utilized as a source of edible oil and has potential as a substitute if commercial production could be developed. Another interesting feature of the crop is the high protein content of the tubers which could help alleviate protein deficiency in local diets. In addition, the exceptional ability of the crop to fix atmospheric nitrogen by bacteria in the root nodules should not be overlooked, in view of the world shortage and rising prices of artificial nitrogenous fertilizers.

It is possible that in the future the winged bean could become as important as the soyabean in world agriculture, but considerable research is required, particularly on the agronomic aspects of the crop. There is a need to develop self-supporting types and to increase the availability of supplies of seed of improved strains.

Bibliography

AGCAOILI, F. 1929. Seguidillas bean. *Philipp. J. Sci.*, *40*, 513–514.

ANON. 1976. Nutritive value of some uncommon foods: Goa beans and root. *Natl. Inst. Nutr. Annu, Rep.* 1975. pp 21–22. Hyderabad, India, Indian Counc. Medic. Res., 155 pp.

AYANABA, A. AND NANGJU, D. 1973. Nodulation and nitrogen fixation in six grain legumes. *Proc. 1st Int. Inst. Trop. Agric., Grain legume improvement workshop*. pp. 198–204. Ibadan, Nigeria, Int. Inst. Trop. Agric., 325 pp.

BURKILL, I. H. 1906. Goa beans in India. *Agric. Ledger*, (4), pp. 51–64.

BURKILL, I. H. 1935. *Psophocarpus. A dictionary of the economic products of the Malay Peninsula.* Vol. 2. pp. 1818–1820. London: Crown Agents for the Colonies, 2402 pp.

CARANDANG, E. C. 1969. Sequidilla. *Culture of vegetables.* pp. 83–84. Manila, Repub. Philipp., Dep. Agric. Nat. Resource, Bur. Plant Ind., 150 pp.

ČERNÝ, K. AND ADDY, H. A. 1973. The winged bean (*Psophocarpus palustris* Desv.) in the treatment of kwashiorkor. *Br. J. Nutr.*, *29*, 105–111. [Although entitled *P. palustris*, the authors have subsequently pointed out that the winged bean referred to in this and the following paper was in fact *P. tetragonolobus*].

ČERNÝ, K., KORDYLAS, M., POSPÍSIL, F., ŠVÁBENSKÝ, O. AND ZAJIC, B. 1971. Nutritive value of the winged bean (*Psophocarpus palustris* Desv.). *Br. J. Nutr.*, *26*, 293–299.

Winged bean

CHANDRA, V. AND SRIVASTAVA, G. S. 1977. The winged bean. *Indian Farming*, *27* (4), 19; 29.

CLAYDON, A. 1975. The nutritive potential of the winged bean plant (*Psophocarpus tetragonolobus*). *Proc. Papua New Guinea food crop conf.* (Wilson, K. and Bourke, R. M. eds.). pp. 53–61. Port Moresby, Papua New Guinea, Dep. Primary Ind., 388 pp.

DEANON, J. R. (JR.) AND SORIANO, J. M. 1967. The legumes. *Vegetable production in South East Asia.* (Knott, J. E. and Deanon, J. R. (Jr.) eds.). pp. 66–96. Los Baños, Laguna: Univ. Philipp. Coll. Agric., 366 pp.

ELMES, R. P. T. 1976. Cross-inoculation relationships of *Psophocarpus tetragonolobus* and its rhizobium with other legumes and rhizobia. *Papua New Guinea Agric. J.*, *27*, 53–57.

ERSKINE, W. AND BALA, A. A. 1976. Crossing technique in winged bean. *Trop. Grain Legume Bull.*, (6), pp. 32–35.

HERKLOTS, G. A. C. 1972. Four-angled bean. *Vegetables in South-east Asia.* pp. 257–260. London: George Allen and Unwin Ltd, 525 pp.

HYMOWITZ, T. AND BOYD, J. 1977. Origin, ethnobotany and agricultural potential of the winged bean—*Psophocarpus tetragonolobus*. *Econ. Bot.*, *31*, 180–188.

INTERNATIONAL INSTITUTE OF TROPICAL AGRICULTURE. 1973. Grain legume improvement program, 1973 report. pp. 31–32. *Nigeria*, *Ibadan*, *Int. Inst. Trop. Agric.*, 78 pp.

JAFFÉ, W. G. AND KORTE, R. 1976. Nutritional characteristics of the winged bean in rats. *Nutr. Rep. Int.*, *14*, 449–455.

KARIKARI, S. K. 1972. Pollination requirements of winged beans (*Psophocarpus* spp. Neck.) in Ghana. *Ghana J. Agric. Sci.*, *5*, 235–239.

KHAN, T. N. 1976. Programs on winged bean in Papua New Guinea and their implications. *Trop. Grain Legume Bull.*, (5), pp. 24–26.

KHAN, T. N., BOHN, J. C. AND STEPHENSON, R. A. 1977. Winged beans: Cultivation in Papua New Guinea. *World Crops*, *29*, 208–209; 212–214.

KHAN, T. N. AND ERSKINE, W. 1976. Production and improvement of food legumes in Papua New Guinea and its implication to Malaysian self-sufficiency. *Proc. conf. Malaysian food self-sufficiency*, 1975. (Tan, B. T., Ch'ng, G. C., Cheam, S. T. and Wong, K. C. eds.). pp. 117–125. Kuala Lumpur, Malaysia, Univ. Malaya. Agric. Grad. Alumni, 353 pp.

KRISHNAMURTHI, A. (ed.). 1969. *Psophocarpus* D. C. *The wealth of India: Raw materials.* Vol. 8 (Ph-Re). pp. 294–295. New Delhi: Publications Inf. Dir., Indian Counc. Sci. Ind. Res., 394 pp.

MASEFIELD, G. B. 1952. The nodulation of annual legumes in England and Nigeria: Preliminary observations. *Emp. J. Exp. Agric., 20*, 175–186.

MASEFIELD, G. B. 1957. The nodulation of annual leguminous crops in Malaya. *Emp. J. Exp. Agric., 25*, 139–150.

MASEFIELD, G. B. 1961. Root nodulation and agricultural potential of the leguminous genus *Psophocarpus. Trop. Agric.*, (*Trinidad*), *38*, 225–229.

MASEFIELD, G. B. 1967. The intensive production of grain legumes in the tropics. *Proc. Soil Crop Sci., Florida., 27*, 338–346.

MASEFIELD, G. B. 1973. *Psophocarpus tetragonolobus*—a crop with a future. *Field Crop Abstr., 26*, 157–160.

OCHSE, J. J. AND BAKHUIZEN VAN DEN BRINK, R. C. 1931. *Psophocarpus tetragonolobus* (L.) A.DC. *Vegetables of the Dutch East Indies.* pp. 427–429. Buitenzorg, Java: Archipel Drukkerij, 1006 pp.

POSPÍSIL, F., KARIKARI, S. K. AND BOAMAH-MENSAH, E. 1971. Investigations of winged bean in Ghana. *World Crops, 23*, 260–264.

PURSEGLOVE, J. W. 1968. *Psophocarpus tetragonolobus* (L.) DC. *Tropical crops: Dicotyledons.* Vol. 1. pp. 315–318. London: Longmans, Green and Co Ltd, 332 pp.

RACHIE, K. O. AND ROBERTS, L. M. 1974. Grain legumes of the lowland tropics. *Adv. Agron., 26*, 1–132.

RUSKIN, F. R. (ed.). 1975. The winged bean a high-protein crop for the tropics. *Washington, Comm. Int. Relat., Natl. Acad. Sci., Advis. Stud. & Spec. Rep.* 17, 42 pp.

THOMPSTONE, E. AND SAWYER, A. M. 1914. *Psophocarpus tetragonolobus* DC. The peas and beans of Burma. pp. 80–84. *Rangoon, Burma, Dep. Agric. Bull.* 12, 107 pp.

WONG, K. C. 1976. The potential for four-angled bean (*Psophocarpus tetragonolobus* (L.) (DC.) in Malaysia to increase food supply. *Proc. conf. Malaysian food self-sufficiency*, 1975. (Tan, B. T., Ch'ng, G. C. Cheam, S. T. and Wong, K. C. eds.). pp. 103–115. Kuala Lumpur, Malaysia, Univ. Malaya. Agric. Grad. Alumni, 353 pp.

A Food legumes: distribution on a broad climatic basis

	Tropical rainforest	Tropical monsoon	Tropical savanna	Dry tropical steppe	Humid sub-tropical	Dry sub-tropical	Humid inter-mediate	Dry inter-mediate
Adzuki bean					X	X		
Asparagus bean	X	X	X					
Bambara groundnut			X	X				
Broad bean		X(CS, I)	X(CS, I)		X	X	X	X(I)
Chick pea		X(CS)	X(CS)	X(CS, I)	X	X		
Cluster bean			X(DS)	X		X		
Cowpea	X(MD, LD)	X(MD, LD)	X	X(SD)				
Grass pea		X(DS)	X(CS)	X(CS)	X	X		
Haricot bean	(X)	X	X	X(I)	X	X	X	X
Horse gram		X(DS)	X	X				
Hyacinth bean	X	X	X	X(I)				
Jack bean	(X)	X	X	X(I)				
Kersting's groundnut			X	X				
Lentil		X(H, CS)	X(H, CS)	X(H, I)	X	X		
Lima bean	X	X						
Lupin		X(H, CS)	X(H, CS)	X(H)		X	X	X
Moth bean			X	X				
Mung bean		X	X	X(I)				
Pea		X(H, CS)	X(H, CS)	X(H, CS, I)	X	X(I)	X	X
Pigeon pea		X	X	X	X			
Rice bean	X	X	X					
Runner bean		X(H)	X(H)				X	
Sword bean	X	X	X	X(I)				
Tepary bean			X(DS)	X				
Urd		X(DS)	X(DS)	X				
Velvet bean	X	X	(X)					
Winged bean	X	X	X(I)					

(X) Limited cultivation
(I) Supplementary irrigation may be required
(CS) Cool season crop
(H) Higher elevation
(DS) Dry season crop
(LD) Long-duration types
(MD) Medium-duration types
(SD) Short-duration types

B Grain legumes: Estimated world production[1]

Quantity thousand tonnes

Legume	Annual average 1965–69	Per-centage of total	Annual average 1970–74	Per-centage of total	1975	Per-centage of total	1976	Per-centage of total
Dry peas	10 149	24.2	11 326	25.3	10 221[3]	22.8	13 427	26.1
Dry beans[2]	10 674	25.4	11 551	25.8	12 745[3]	28.6	12 580	24.4
Chick peas	6 317	15.0	6 494	14.5	5 538[3]	12.4	7 466	14.5
Broad beans	4 931	11.7	5 214	11.7	6 201[3]	13.8	6 187	12.0
Pigeon peas	1 783	4.3	1 880	4.2	1 960	4.4	n/a	n/a
Cowpeas	898	2.1	1 100	2.5	1 097[3]	2.5	n/a	n/a
Lentils	1 018	2.4	1 094	2.4	1 120[3]	2.5	1 236	2.4
Lupins	820	2.0	730	1.6	577	1.3	n/a	n/a
Other grain legumes[4]	5 419	12.9	5 352	12.0	5 226	11.7	10 626	20.6
Total	42 009	100.0	44 741	100.0	44 793[3]	100.0	51 522	

[1] Source: *FAO Production yearbooks.*

[2] Principally the haricot bean, but also includes other beans such as mung, lima, moth, rice and adzuki.

[3] Revised figures.

[4] Includes various vetches.

n/a Not available.

C Index of botanical names

D Index of trivial names

(Other than the common name used for the title entries)

Abai	— Jack/Sword bean
Abangbang	— Lima bean
Aboboi	— Bambara groundnut
Aci bakla	— Lupin
Ackerbohne	— Broad bean
Aconite bean	— Cluster bean
Aconite leaved kidney bean	— Moth bean
Adad	— Urd
Adagora	— Haricot bean
Adagura (kwolla)	— Cowpea
Adanguare	— Haricot bean
Adas	— Chick pea/Lentil
Adasha tarbutit	— Lentil
Adashim	— Lentil
Adesi	— Lentil
Adhaki	— Pigeon pea
Adigura-tsada	— Haricot bean
Adonguari	— Cowpea
Ads masri	— Lentil
Ads sudani	— Pigeon pea
Adsuki bean	— Adzuki bean
Adungare	— Chick pea
Adzukibohne	— Adzuki bean
Afun tarbuti	— Pea
Afuna	— Pea
Afunat habakar	— Cowpea
Agaya	— Hyacinth bean
Agni guango ahrua	— Hyacinth bean
Agrâo de bico	— Chick pea
Agwa	— Cowpea
Ain-ater	— Pea
Akide enu	— Cowpea
Akidiani	— Cowpea
Akkerwt	— Pea
Akpaka (pakera)	— Lima bean
Akyii	— Bambara groundnut
Alberga	— Pigeon pea
Alhamos	— Chick pea
Almorta	— Grass pea
Alotoko	— Lima bean
Altramuz	— Lupin
Alverja	— Pea/Pigeon pea
Alverjas	— Grass pea
Alverjón	— Pea
Amali	— Winged bean
Amashaza	— Pea
Ambelophassula	— Cowpea
Ambérique (de Madagascar)	— Mung bean
Ambérique (haricot)	— Urd
Ambrevade	— Pigeon pea
Amora-guaya	— Hyacinth bean
Amuli	— Cowpea
Anataque	— Hyacinth bean
Angola pea	— Pigeon pea
Angolische erbse	— Pigeon pea
Anipay	— Rice bean
Anumulu	— Hyacinth bean
Apatram	— Lima bean

Apena — Pigeon pea
Apikak — Hyacinth bean
Arahar — Pigeon pea
Arhar — Pigeon pea
Arveja — Pea
Ashanguare — Haricot bean
Asparagus pea — Winged bean
Ataque — Hyacinth bean
Atari — Pea
Ater(o) — Pea
Atera argobba — Cowpea
Ater-bahari — Broad bean
Ater-bar-ativeri — Broad bean
Atera argobba — Cowpea
Ater cajeh — Chick pea
Atir saho — Chick pea
Atsuki bean — Adzuki bean
Attur — Pea
Atzuki bean — Adzuki bean
Australian pea — Hyacinth bean
Avarai — Hyacinth bean
Avarakai — Sword bean
Avare — Hyacinth bean
Avitas poroto — Lima bean
Awara — Jack bean
Awuje — Lima bean
Ayecote — Runner bean
Ayocote — Runner bean
Azuki bean — Adzuki bean

Bab — Haricot bean
Babricorn bean — Sword bean
Bagila — Broad bean
Bagoly borsó — Chick pea
Bakela — Broad bean
Bakla — Broad bean
Bakuchi — Cluster bean
Balatong — Mung bean/Urd
Baldenga — Broad bean
Baldunga — Broad bean
Ballar — Hyacinth bean
Bamboo bean — Rice bean
Banana stock pea — Velvet bean
Bannabees — Hyacinth bean
Bannette — Cowpea
Banor — Asparagus bean
Bara sem — Jack bean
Baran chaki — Jack bean
Barbata/i — Cowpea
Basilla — Pea
Batagadle — Pea
Batani — Pea
Batao — Hyacinth bean
Batau — Hyacinth bean
Batong — Cowpea
Batong-baimbing — Winged bean
Batong-hidjao — Mung bean
Batura — Grass pea
Bazilla — Pea
Bazzelah — Pea
Bean — Cowpea/Haricot/Lima bean
Beglau — Hyacinth bean
Beiqa — Chick pea
Bengal butter bean — Sword bean
Bengal gram — Chick pea

Bendi	— Kersting's groundnut
Benibanaigen	— Runner bean
Bersem	— Lentil
Bezelye	— Pea
Bindi	— Kersting's groundnut
Biri	— Urd
Birssin	— Lentil
Blabi	— Chick pea
Black gram	— Mung bean/Urd
Black-eye bean or pea	— Cowpea
Bob	— Broad bean
Bobik	— Broad bean
Boby kormouvyje	— Broad bean
Bodi bean	— Asparagus bean
Boerboon	— Broad bean
Bohne	— Haricot bean
Bolakadala	— Pea
Bonavis (pea)	— Hyacinth bean
Bonavist(a) bean	— Hyacinth bean
Bonchi (kai)	— Haricot bean
Bondböna	— Broad bean
Boo-ngor	— Cowpea
Boontje	— Haricot bean
Boontjis	— Haricot bean
Boot (kaley)	— Chick pea
Bo-sa-pè	— Haricot bean
Borsó	— Pea
Börülce	— Cowpea
Botol	— Runner bean
Boubour	— Mung bean
Boucouson	— Asparagus bean
Brazilian bean	— Haricot bean
Bsella	— Pea
Bunabis	— Hyacinth bean
Bunchu-kai	— Haricot bean
Bundo	— Mung bean
Burusa	— Pigeon pea
Burusu	— Pigeon pea
Burma bean	— Lima bean
Burma haricot	— Winged bean
Burssum	— Lentil
Bush bean	— Haricot bean
But (mah)	— Chick pea
Buthalai	— Chick pea
Butingi	— Haricot bean
Butter bean	— Hyacinth/Lima/ Runner bean
Cachito	— Pigeon pea
Cadios	— Pigeon pea
Café français	— Chick pea
Calamismis	— Winged bean
Calavance	— Cowpea
Callivance	— Cowpea
Caraota	— Haricot/Lima bean
Caraota chivata	— Hyacinth bean
Caraota Florida	— Runner bean
Caraota grande	— Sword bean
Carolina bean	— Lima bean
Carpet legume	— Hyacinth bean
Caseknife bean	— Runner bean
Catjangbohne	— Cowpea
Catjang	— Cowpea
Caupi	— Cowpea

Cece (bianco) — Chick pea
Ceci — Chick pea
Cecio — Chick pea
Cesari — Chick pea
Cesco — Chick pea
Ceseron — Chick pea
Chahna — Chick pea
Chala — Chick pea
Chania — Chick pea
Channa — Chick pea
Chapprada (avare) — Hyacinth bean
Chara-koni-sem — Winged bean
Charal — Grass pea
Chaucha japonese — Hyacinth bean
Chaudhaari-phali — Winged bean
Chaula — Cowpea
Chaunangi — Lentil
Chavdhari-ghevda — Winged bean
Chavli — Cowpea
Chela — Chick pea
Chemps — Chick pea
Cheong dau-kok — Asparagus bean
Cherupayaru — Mung bean
Ch'eung kong tau — Asparagus bean
Chevaux de frise bean — Winged bean
Chiapas — Runner bean
Chicaro — Chick pea
Chicharo — Pea
Chicharo de arbol — Pigeon pea
Chicharo de paloma — Pigeon pea
Chícharo de vaca — Cowpea
Chichaso — Hyacinth pea
Chicher — Chick pea
Chichipir — Winged bean
Chickasano — Mung bean
Chickasaw Lima — Jack bean
Chickasaw pea — Mung bean
Chickling pea or vetch — Grass pea
Chikkudu — Hyacinth bean
Chilemba — Haricot bean
Chimbamba — Haricot bean
Chimbera — Chick pea
Chimbolo verde — Hyacinth bean
Chimtza (tarbutit) — Chick pea
China pea — Cowpea
Chinchoncho — Pigeon pea
Chipir — Winged pea
Chirisanagalu — Lentil
Chiroko — Mung bean/Urd
Chocho — Lupin
Chola — Chick pea
Chono — Chick pea
Choroko — Urd
Chota but — Chick pea
Chowlee — Cowpea
Chowli — Cowpea
Chumbinho opaco — Haricot bean
Chunna — Chick pea
Chural — Grass pea
Cicererbis — Chick pea
Cicerchia coltivata — Grass pea
Cicérolé — Chick pea
Cicerewt — Chick pea
Ciche — Chick pea

Cigarillas	— Winged bean	Dâu-mwoi	— Urd
Citzaro	— Pea	Dâu ngu	— Lima bean
Civet bean	— Lima bean	Dâu trang	— Cowpea
Ckuku	— Lima bean	Dâu trieu	— Pigeon pea
Climbing mountain bean	— Rice bean	Dâu tua	— Cowpea
Coffee pea	— Chick pea	Dâu van	— Hyacinth bean
Common bean	— Haricot bean	Dâu xa	— Cowpea
Congo bean or pea	— Pigeon pea	Dâu xanh	— Mung bean
Congo goober	— Bambara groundnut	Dâu xay	— Pigeon pea
Cornfield pea	— Cowpea	Dew bean or gram	— Moth bean
Coupé	— Cowpea	Dhal	— Pigeon pea/Lentil
Cowgram	— Cowpea	Diéguem tenguéré	— Kersting's groundnut
Crab-eye pea	— Rice bean	Digir	— Cowpea
Cranberry bean	— Haricot bean	Dinawa	— Cowpea/Tepary bean
Csicseri borsó	— Chick pea	Dir-daguer	— Jack bean
Cumana tupi	— Hyacinth bean	Djelbane	— Chick pea
Cumandiata	— Hyacinth bean	Djokomaie	— Bambara groundnut
Curry bean	— Lima bean	Dog-toothed pea	— Grass pea
Cut-eye bean	— Jack/Sword bean	Dolic (d'Egypte)	— Hyacinth bean
Cyamopse à quatre ailes	— Cluster bean	Dolicho lablab	— Hyacinth bean
Cytise des Indes	— Pigeon pea	Dolichos (bean)	— Hyacinth bean
		Dolico	— Cowpea
Dafal	— Lima bean	Dolico cavallino	— Horse gram
Dagarti bean	— Cowpea	Dolico gigante	— Asparagus bean
Danguleh	— Pea	Dolique asperge	— Asparagus bean
Dao van	— Hyacinth bean	Dolique de Chine	— Cowpea
Dara-dhambala	— Winged bean	Dolique de Cuba	— Asparagus bean
Dâu bien	— Lima bean	Dolique d'Egypte	— Hyacinth bean
Dâu cau	— Winged bean	Dolique de Floride	— Velvet bean
Dâu chieu	— Pigeon pea	Dolique du Soudan	— Hyacinth bean
Dâu den	— Cowpea	Dolique geánt	— Asparagus bean

Dolique indigène — Cowpea
Dolique mongette — Cowpea
Dolique mougette — Cowpea
Doperwt — Pea
Dord — Mung bean
Double bean — Broad/Lima bean
Dougoufulo — Kersting's groundnut
Doyi — Kersting's groundnut
Dragon bean — Winged bean
Dridhabija — Cluster bean
Duffin bean — Lima bean
Duiveboon — Broad bean
Dungay — Rice bean
Dwergertjie — Chick pea

Earth pea — Bambara groundnut
Echte kicher — Chick pea
Echte linse — Lentil
Edihimba — Haricot bean
Egyptian (kidney) bean — Hyacinth bean
Egyptian pea — Chick pea
Eka-wohe — Cowpea
Embrevade — Pigeon pea
Endō — Pea
English bean — Broad bean
Enkoole — Cowpea
Enkoore — Cowpea
Epi roro — Bambara groundnut
Erbse — Pea
Ere(e) — Cowpea
Erebinthos — Chick pea
Erevinthos — Chick pea
Ertjie — Pea
Ervanços — Chick pea
Ervilha — Pea
Ervilha de Congo — Pigeon pea
Ervilha de vaca — Cowpea
Erwt(en) — Pea
European bean — Broad bean
Faba bean — Broad bean
Fagiola — Haricot bean
Fagiolini rampicanti — Runner bean
Fagiolino dall'occhio — Cowpea
Fagiolo asparagio — Asparagus bean
Fagiolo adzuki — Adzuki bean
Fagiolo di Lima — Lima bean
Fagiolo di Spagna — Runner bean
Fagiolo mungo — Mung bean
Fagiulo commune — Haricot bean
Fajola — Haricot bean
Faki — Lentil
Fasiolos — Haricot bean
Fasolea-dima — Cowpea
Fasolia — Haricot bean
Fasulia — Haricot bean
Fasûlya — Haricot bean
Fasûlyah nashef — Haricot bean
Fava de cavallo — Winged bean
Faveira — Broad bean
Feijão — Haricot bean
Feijão adzuki — Adzuki bean
Feijão brabham — Cowpea

Feijão de China — Cowpea
Feijão de corda — Cowpea
Feijão de porco — Jack bean
Feijão ervilha — Haricot bean
Feijão espadinho — Lima bean
Feijão fradinho — Cowpea
Feijão makunda — Cowpea
Feijoeiro — Haricot bean
Feldbohne — Broad bean
Feuerbohne — Runner bean
Fève — Broad bean
Fève creole — Lima bean
Fève d'Egypte — Hyacinth bean
Fève de kandela — Kersting's groundnut
Fève des marais — Broad bean
Fève Jack — Jack/Sword bean
Féverole — Broad bean
Févette — Broad bean
Field bean — Broad/Haricot/Hyacinth bean
Fiwi bean — Hyacinth bean
Flügelbohne — Winged bean
Fluweelboontjie — Velvet bean
Fontanellerbse — Chick pea
Four-angled bean — Winged bean
Four-cornered bean — Winged bean
Fovetta — Grass pea
Frash bean — Haricot bean
French bean — Haricot bean
Frijol(e) — Cowpea/Haricot bean
Frijol adzuki — Adzuki bean
Frijol arroz — Rice bean
Frijol ayocote — Runner bean
Frijol bocon — Hyacinth bean
Frijol cabellero — Hyacinth bean
Frijol carita — Cowpea
Frijol chileno — Hyacinth bean
Frijol chino — Pigeon pea
Frijol comba — Lima bean
Frijol de la tierra — Hyacinth bean
Frijol de ojo negro — Cowpea
Frijol de palo — Pigeon pea
Frijol diablito — Adzuki bean
Frijol gallinazo — Grass pea
Frijol gandul — Pigeon pea
Frijol japones — Pigeon pea
Frijol precioso — Cowpea
Frijol terciopelo — Velvet bean
Frijol trigo — Tepary bean
Frijolito de Cuba — Lima bean
Fuji-mame — Hyacinth bean
Ful — Broad bean
Ful abungawi — Bambara groundnut
Ful masri — Broad bean

Gai-kalai — Rice bean
Gairance — Chick pea
Gairoutte — Chick pea
Gaisa — Cowpea
Gallinazo blanco — Hyacinth bean
Gallinita — Hyacinth bean
Gandal — Pigeon pea
Gandul — Pigeon pea
Gandures — Pigeon pea

Garavance	—	Chick pea
Garbancillo bolando	—	Tepary bean
Garbanso	—	Winged bean
Garbanza	—	Chick pea
Garbanzo	—	Chick/Grass pea
Garbanzo falso	—	Pigeon pea
Garvance	—	Chick pea
Gavar	—	Cluster bean
Gawar	—	Cluster bean
Geocarpa (groundnut)	—	Kersting's groundnut
Gerenga	—	Hyacinth bean
Gertere	—	Bambara groundnut
Gesse blanche	—	Grass pea
Gesse chiche	—	Grass pea
Gesse commune	—	Grass pea
Gesette	—	Grass pea
Gewöhnliche Kichererbse	—	Chick pea
Ghurush	—	Rice bean
Gilban(eh)	—	Grass pea
Gishi-shato	—	Pea
Gisima	—	Jack bean
Goa bean	—	Jack/Winged bean
Gobbe	—	Bambara groundnut
Golden gram	—	Mung bean
Goodé	—	Pigeon pea
Goongo pea	—	Pigeon pea
Gorakshaphalini	—	Cluster bean
Gorani	—	Cluster bean
Gorchikuda	—	Cluster bean
Gori kayi	—	Cluster bean
Gotani bean	—	Jack/Sword bean
Goue mungboontjie	—	Mung bean
Gouree	—	Cluster bean
Govar	—	Cluster bean
Gowar(a)	—	Cluster bean
Grain de cheval	—	Horse gram
Gram	—	Chick pea
Grâo de bico	—	Chick pea
Grâo de pulha	—	Urd
Gravancos	—	Chick pea
Green gram	—	Mung bean
Green pea	—	Pigeon pea
Groote boon	—	Broad bean
Grosse fève	—	Broad bean
Ground bean	—	Bambara/Kersting's groundnut
Grudge pea	—	Jack bean
Guando/u	—	Pigeon pea
Guandu de fava larga	—	Pigeon pea
Guar	—	Cluster bean
Guar khurti	—	Cluster bean
Guaracaro	—	Lima bean
Guerte	—	Bambara groundnut
Gueshrangaig	—	Hyacinth bean
Guijiya	—	Bambara groundnut
Guisante	—	Pea
Guisante de paloma	—	Pigeon pea
Guisante enano	—	Pigeon pea
Gujuya	—	Bambara groundnut
Gungo pea	—	Pigeon pea

Guronsh — Rice bean
Gurush — Rice bean
Guwar — Cluster bean
Gwar — Cluster bean

Haba — Broad bean
Haba blanca — Jack bean
Haba caballar — Broad bean
Haba comun — Broad bean
Haba criolla — Jack bean
Haba de burro — Jack/Sword bean
Haba de terciopelo — Velvet bean
Haba lima — Lima bean
Habas — Jack/Sword bean
Habichuela — Haricot/Lima bean
Habichuela China — Asparagus bean
Hagoly-borsó — Chick pea
Halifax pea — Cowpea
Hamaz — Chick pea
Hamiça — Chick pea
Hammes — Chick pea
Ham(m)o(u)s — Chick pea
Hamtak — Asparagus bean
Hanasasage — Runner bean
Haricot adzuki — Adzuki bean
Haricot à feuilles angulaires — Adzuki bean
Haricot à fleurs rouges — Runner bean
Haricot à couper — Haricot bean
Haricot à oeil noir — Cowpea
Haricot commún — Haricot bean
Haricot cutelinho — Hyacinth bean
Haricot de behanzin — Kersting's groundnut
Haricot d'Espagne — Runner bean
Haricot de lima — Lima bean
Haricot dolique — Cowpea
Haricot doré — Mung bean
Haricot dragon — Winged bean
Haricot du cap — Lima bean
Haricot du kissi — Lima bean
Haricot (grain) de riz — Rice bean
Haricot indigène — Cowpea
Haricot kilomètre — Asparagus bean
Haricot konde — Adzuki bean
Haricot kunde — Cowpea
Haricot mungo — Mung bean/Urd
Haricot nain — Haricot bean
Haricot pain — Haricot bean
Haricot papillon — Moth bean
Haricot pistache — Bambara groundnut
Haricot princesse — Haricot bean
Haricot riz — Tepary bean
Haricot rouge — Sword bean
Haricot royal — Kersting's groundnut
Haricot sabre — Jack bean
Haricot Sudan — Tepary bean
Haricot velouté — Velvet bean
Haricot velu — Urd
Haricot vert — Haricot bean
Harimandha-kam — Chick pea
Hausa groundnut — Kersting's groundnut
Heerboontjie — Lima bean

Helmbohne — Hyacinth bean
Himmos (akhdar) — Chick pea
Hindu pea — Cowpea
Hiramame — Lentil
Hluba bean — Bambara groundnut
Hommes — Chick pea
Hommos malana — Chick pea
Horse bean — Broad bean/Jack/Sword bean
Horse grain — Horse gram
Hummous — Chick pea
Hurali — Horse gram
Hyokkomame — Chick pea

Icaraota — Haricot bean
Igiuhluba — Bambara groundnut
Ihora — Pigeon pea
Ihora parippu — Pigeon pea
Ikiker — Chick pea
Ilanda — Cowpea
Illundu — Urd
Imare — Cowpea
Imposo — Pigeon pea
Increase pea — Asparagus pea
Indian (butter) bean — Hyacinth bean
Indian pea — Cowpea/Grass pea
Indian vetch — Grass pea
Ingen (mame) — Haricot bean
Intongwe — Pea
Intoyo — Bambara groundnut
Itab — Hyacinth bean

Java bohne — Lima bean
Jerusalem pea — Mung bean
Judia (commun) — Haricot bean
Judia asparaga — Asparagus bean
Judia de arroz — Rice bean
Judia de mungo — Mung bean
Judia escarlata — Runner bean
Judia d'Espagna — Runner bean
Judia de lima — Lima bean
Judion — Lima bean
Juga/o bean — Bambara groundnut
Jumez — Chick pea
Juroku-sasage — Asparagus bean

Kabaro — Lima bean
Kabkaza — Chick pea
Kabuli mater — Pea
Kachang bêlimbing — Winged bean
Kachang bêlut — Asparagus bean
Kachang bogor — Bambara groundnut
Kachang bol — Cowpea
Kachang botol — Winged bean
Kachang botor — Winged bean
Kachang bunchis — Haricot bean
Kachang China — Lima bean
Kachang dal — Pigeon pea
Kachang embing — Winged bean
Kachang hijan — Mung bean
Kachang hijau — Mung bean/Urd
Kachang hiris — Pigeon pea
Kachang hitam — Urd
Kachang jawa — Lima bean

Kachang kara — Hyacinth bean
Kachang kayu — Pigeon pea
Kachang kélisah — Winged bean
Kachang kotor — Winged bean
Kachang kuda — Chick pea
Kachang menila — Bambara groundnut
Kachang panjang — Asparagus bean/ Cowpea
Kachang panjang — Cowpea
Kachang parang — Sword bean
Kachang parang puteh — Jack bean
Kachang pendek — Haricot bean
Kachang perut ayam — Asparagus bean
Kachang poi — Bambara groundnut
Kachang puteh — Pea
Kachang sepalit — Rice bean
Kachang serendeng — Lima bean
Kachang serinding — Lima bean
Kachang tanah — Bambara groundnut
Kadala(i) — Chick pea
Kadale — Chick pea
Kadios — Pigeon pea
Kadli — Chick pea
Kadyos — Pigeon pea
Kafferboon — Cowpea
Kaffir bean — Cowpea
Kaffir pea — Bambara groundnut/ Cowpea
Kajang kaokara — Lima bean
Kakunda bakishi — Pigeon pea
Kalai — Horse gram/Urd
Kalamismis — Winged bean
Kala-pè — Chick pea
Kalipan — Rice bean
Kallu — Horse gram
Kallupayaru — Moth bean
Kambulu — Urd
Kandalu — Pigeon pea
Kandela — Kersting's groundnut
Kandelabohne — Kersting's groundnut
Kandi — Pigeon pea
Kandulu — Pigeon pea
Kansari — Grass pea
Kanyensi — Mung bean
Kaos bakol — Sword bean
Kaos bebedogan — Sword bean
Kaos parasaman — Sword bean
Kaos pedang — Sword bean
Kapri — Pea
Kapucijners — Pea
Kara bedog — Sword bean
Kara we-dung — Sword bean
Karakala — Cowpea
Kara-kara — Hyacinth bean
Karas — Grass pea
Karbantos — Chick pea
Karikadale — Chick pea
Karil — Grass pea
Kashrengeig — Hyacinth bean
Kassar — Grass pea
Katchang botor — Winged bean
Katjang babi — Broad bean

Katjang djong — Mung bean
Katjang eedjo — Mung bean
Katjang ertjis — Pea
Katjang goode — Pigeon pea
Katjang (h)eeris — Pigeon pea
Katjang kadjoo — Pigeon pea
Katjang merah — Cowpea/Haricot bean
Katjang otji — Rice bean
Katjang padi — Mung bean
Katjang pandjang — Asparagus bean/Cowpea
Katjang toonggak — Cowpea
Katjeper — Winged bean
Katschang — Pigeon pea
Kebkabeik — Chick pea
Kêchipir — Winged bean
Kekara — Hyacinth bean
Kekara kratok — Lima bean
Keker — Chick pea
Kerara — Hyacinth bean
Kerstingiella groundnut — Kersting's groundnut
Kesari — Grass pea
Kêtjeepir (blinger) — Winged bean
Kêtjeper — Winged bean
Khasi kollu — Lima bean
Kheri — Moth bean
Khesari dhal — Grass pea
Khesari meh — Grass pea
Khesra — Grass pea
Khessary pea — Grass pea
Kibal — Cowpea
Kicher(en) — Chick pea
Kichererbse — Chick pea
Kidney bean — Haricot/Moth bean
Kifudu — Mung bean/Urd
Kiker — Chick pea
Kikuyu bean — Hyacinth bean
Kilkilang — Rice bean
Ki-mame — Pigeon pea
Kisari — Grass pea
Kokondo — Lima bean
Kollu — Horse gram
Konda kadala — Chick pea
Korkadala — Chick pea
Kotaranga — Cluster bean
Kothaverai — Cluster bean
Kothaveray — Cluster bean
Kottavarai — Cluster bean
Kouarourou — Kersting's groundnut
Krandang — Sword bean
Kulapia — Chick pea
Kulat — Horse gram
Kulatha — Horse gram
Kulthi (bean) — Horse gram
Kulti — Cluster bean
Kumkumapesalu — Moth bean
Kunde — Cowpea/Haricot bean
Kurtikalai — Horse gram
Kuru bakla — Broad bean
Kuru fasulya — Haricot bean
Kuwara — Cluster bean
Kwaruru — Kersting's groundnut

Kyamos — Broad bean

Laba major — Broad bean
Labboon — Broad bean
Labe-labe — Hyacinth bean
Lablab (bean) — Hyacinth bean
Lablabi — Chick pea
Lakar-sem — Winged bean
Lakh — Grass pea
Lakhodi — Grass pea
Lakhori — Grass pea
Lang — Grass pea
Laputu — Cowpea
Lathyrus pea — Grass pea
Latri — Grass pea
Lazy-man pea — Rice bean
Leblebi — Chick pea
Leca — Lentil
Lencse — Lentil
Lensie — Lentil
Lenteja — Lentil
Lenteja françesa — Pigeon pea
Lenticchia — Lentil
Lentilha — Lentil
Lentille — Lentil
Lentille d'Espagne — Grass pea
Lentille de terre — Kersting's groundnut
Liboshi — Cowpea
Likote — Cowpea
Likotini — Cowpea
Linay — Rice bean
Linse — Lentil
Linze — Lentil
Lóbab — Broad bean
Lobia — Cowpea/Lima bean
Lobiya — Lima bean
Long bean — Asparagus bean
Lopena — Pigeon pea
Loputa — Cowpea
Lou teou — Mung bean
Louvia — Hyacinth bean
Lubia — Cowpea
Lubia adassi — Pigeon pea
Lubia(h) bean — Hyacinth bean
Lubia beida — Cowpea
Lubia chiroko — Mung bean
Lubia elfil — Jack/Sword bean
Lubia helu — Cowpea
Lubia kordofani — Cowpea
Lubia tayiba — Cowpea
Lubya baladi — Cowpea
Lubya kusais — Runner bean
Lubya msallat — Cowpea
Lubya tarbutit — Cowpea
Lupina — Lupin
Lupine — Lupin
Lupino — Lupin

Macape — Hyacinth bean
Macululu — Hyacinth bean
Madagascar (butter) bean — Lima bean
Madagascar groundnut — Bambara groundnut
Madras gram — Horse gram

Mag	— Mung bean
Magtámbókan	— Jack bean
Magtambokau	— Sword bean
Maharage	— Haricot/Lima bean
Mahasha	— Urd
Makendal rouge	— Sword bean
Makhan s(h)im	— Jack bean
Makhmali sem	— Velvet bean
Maljoe	— Jack bean
Mambi bean	— Rice bean
Manawa	— Haricot bean
Mandubi d'Angola	— Bambara groundnut
Mangulasi	— Rice bean
Manila bean	— Bambara groundnut/ Winged bean
Mar	— Pea
Marble pea	— Cowpea
March	— Urd
Martar	— Pea
Marutong	— Jack bean
Mash	— Mung bean/Urd
Mashâza	— Pea
Mash kalai	— Urd
Massar	— Lentil
Masser	— Lentil
Masur	— Lentil
Mat (bean)	— Moth bean
Matar (mah)	— Pea
Matar mar	— Pea
Mate mah	— Urd
Math	— Moth bean
Matikalai	— Urd
Matki	— Moth bean
Matpe	— Moth bean
Mât-pè	— Urd
Matri	— Grass pea
Matur	— Grass pea
Mauritius bean	— Velvet/Winged bean
Mauritius pea	— Pea
Mbani	— Pigeon pea
Mbazi	— Pigeon pea
Mbwanda	— Jack bean
Mdengu	— Chick pea
Me-karal	— Cowpea
Mercimek	— Lentil
Mesh	— Mung bean
Messer	— Lentil
Meth	— Rice bean
Meth-kalai	— Moth bean
Mfini	— Lima bean
Mgwaru	— Cluster bean
Michigan pea bean	— Haricot bean
Minimulu	— Moth bean
Minumulu	— Urd
Misurpappu	— Lentil
Mittikelu	— Moth bean
Mochai	— Hyacinth bean
Mond-bohne	— Lima bean
Mongo	— Mung bean
Moong	— Mung bean
Moot	— Moth bean
Morisuavarai	— Winged bean
Moyashi-mame	— Mung bean

Mudgaha — Mung bean
Mug — Mung bean
Mukhudo — Chick pea
Mula — Haricot bean
Multiflora bean — Runner bean
Muneta — Mung bean
Muñggo — Mung bean/Urd
Mungobohne — Mung bean
Mungo-lising — Rice bean
Murukavarai — Winged bean
Musémana — Jack bean
Musimana — Jack bean
Musiripappu — Lentil
Muthera — Horse gram
Muthira — Horse gram
Muthiva — Horse gram
Mu-tsa — Rice bean
Mutter pea — Grass pea

Nachius — Chick pea
Nachunt — Chick pea
Nakhut — Chick pea
Nakud — Chick pea
Nakut — Chick pea
Naripayaru — Moth bean
Nata-mame — Sword bean
Navy bean — Haricot bean
Nela-kadalai — Bambara groundnut
Nga-choi — Mung bean
Ngeri — Cowpea
Nguno — Cowpea
Niébé — Cowpea
Nirikia — Adzuki bean
Njama — Bambara groundnut
Njegele — Pea
Njugo bean — Bambara groundnut
Njugu mawe — Bambara groundnut
Nkol — Pigeon pea
Nohot — Chick pea
Nohud — Chick pea
Nohut — Chick pea
Nokhut — Chick pea
No or Non-eye pea — Pigeon pea
Nori — Cowpea
Ntoyo — Bambara groundnut
Nut — Chick pea
Nyemba bean — Cowpea
Nyoari — Cowpea
Nzama — Bambara groundnut
Nzumbil — Bambara groundnut

Obushaza — Pea
O-cala — Hyacinth bean
Ohota-farengota — Pigeon pea
Ojo de venado — Velvet bean
Ojoo — Haricot bean
Okboli ede — Bambara groundnut
Omnos — Chick pea
Omugobe — Cowpea
One-eye bean — Jack/Sword bean
Ontjet — Broad bean

Oregon pea — Mung bean
Oror — Pigeon pea
Ossangue — Hyacinth bean
Osu — Cowpea
Otile — Pigeon pea
Otong — Cowpea
Overlook bean — Jack bean

Paardeboon — Broad bean
Paayap — Cowpea
Pacae — Broad bean
Pachapayaru — Mung bean
Pagapay — Rice bean
Pagsei — Rice bean
Pairu — Pea
Pai-si-gong — Pigeon pea
Paksai — Rice bean
Pallang — Winged bean
Panguita — Lima bean
Pani payeru — Moth bean
Papaya bean — Hyacinth bean
Pararu — Bambara groundnut/ Kersting's groundnut
Parda — Hyacinth bean
Paripu — Pigeon pea
Parupa-gulung — Winged bean
Pasalu — Mung bean
Pasipayeru — Mung bean
Passi payeru — Mung bean
Paszuly — Haricot bean
Patagonian bean — Jack bean
Patani — Lima bean/Pea
Pataning-espada — Jack bean
Pataning españa — Jack bean/ Sword bean
Patanlu — Pea
Patcha-payru — Mung bean
Pau maia — Rice bean
Pavta — Hyacinth bean
Pay-in-chong — Pigeon pea
Paythenkai — Cowpea
Pea bean — Asparagus/ Haricot bean
Pearson bean — Sword bean
Pè-bi-zât — Horse gram
Pè-bya-gale — Haricot bean
Pè-dalet — Jack bean
Pè-di-sein — Mung bean
Pè-di-wa — Mung bean
Pè-gin — Rice bean
Pè-gya(ni) — Haricot bean
Pè-gyi — Hyacinth bean
Pè-kyin-boung — Grass pea
Pè-lêt-ma — Broad bean
Pè-myît — Winged bean
Pè-nauk(sein) — Mung bean
Pè-sa-li — Grass pea
Pesalu — Mung bean
Pè-saung-sa — Winged bean
Pè-saung-ya — Winged bean
Pè-saung-za — Winged bean
Pèsigon — Pigeon pea
Pè-sin-gôn — Pigeon pea
Pè-walee — Cluster bean
Pè-yin — Rice bean

Pferdbohne	— Broad bean	Pois contour	— Hyacinth bean
Pferdekorn	— Horse gram	Pois coolis	— Hyacinth bean
Pharetta	— Grass pea	Pois cornu	— Chick pea
Phillipesara	— Moth bean	Pois d'Angola	— Pigeon pea
Pinto bean	— Haricot bean	Pois de Brazil	— Cowpea
Pisello	— Pea	Pois de brebis	— Chick pea
Pisello bretonne	— Grass pea	Pois de canne	— Cowpea
Pisello cece	— Chick pea	Pois de Congo	— Pigeon pea
Pisello cicerchia	— Grass pea	Pois de terre	— Bambara groundnut
Pisello cornuto	— Chick pea	Pois doux	— Lima bean
Pisson	— Pea	Pois du cap	— Lima bean
Pistache malagache	— Bambara groundnut	Pois d'un sou	— Hyacinth bean
Pithanga	— Lima bean	Pois en tout temps	— Hyacinth bean
Pizelli	— Pea	Pois gagne	— Jack bean
Pois	— Pea	Pois gagne rouge	— Sword bean
Pois ailé	— Winged bean	Pois gris	— Chick pea
Pois amer	— Lima bean	Pois indien	— Hyacinth bean
Pois arachide	— Bambara groundnut	Pois jaune	— Rice bean
Pois bambarra	— Bambara groundnut	Pois mascate	— Velvet bean
Pois bécu	— Chick pea	Pois nain	— Pigeon pea
Pois blanc	— Chick pea	Pois pigeon	— Pigeon pea/ Rice bean
Pois boucoussou	— Hyacinth bean	Pois pointu	— Chick pea
Pois breton	— Chick pea	Pois poona	— Cowpea
Pois café	— Chick pea	Pois sabre	— Jack bean/ Sword bean
Pois cajan	— Pigeon pea	Pois sabre rouge	— Sword bean
Pois carré	— Grass pea/ Winged bean	Pois souterrain	— Bambara groundnut
Pois chabot	— Chick pea	Pois tête de belier	— Chick pea
Pois chiche	— Chick pea	Pois vache	— Cowpea
Pois chouche	— Lima bean	Pois zombi	— Rice bean
Pois citron	— Chick pea	Pol hagina	— Broad bean

Pole bean — Haricot/Hyacinth/Lima/Runner bean
Polong — Pea
Polon-mé — Asparagus bean
Poncho — Cowpea
Poona pea — Cowpea
Poor man's bean — Hyacinth bean
Popat — Hyacinth bean
Popondo — Jack bean
Poponla — Jack bean
Porotillo — Haricot bean
Porotito del ojo — Cowpea
Poroto (comun) — Haricot bean
Poroto arroz — Adzuki bean
Poroto aterciopelado — Velvet bean
Poroto bombero — Hyacinth bean
Poroto de Egipto — Hyacinth bean
Poroto de España — Runner bean
Poroto de Lima — Lima bean
Poroto de manteca — Lima bean
Poroto japonese — Hyacinth bean
Poroto pallar — Runner bean
Pothudhambala — Lima bean
Princess bean — Haricot bean
Princess pea — Winged bean
Pronkboon — Runner bean
Puakani — Jack bean
Puerto Rican pea — Pigeon pea
Puffebohne — Broad bean
Pul — Broad bean
Purutu — Haricot bean
Puso-poroto — Pigeon pea

Quechua — Haricot bean
Quimbolillo — Pigeon pea
Quinchoncho — Pigeon pea

Rahar — Pigeon pea
Raima-mama — Lima bean
Raimame — Lima bean
Rajmah — Haricot bean
Rangoon bean — Lima bean
Rebinthia — Chick pea
Red bean — Adzuki/Rice bean
Red dhal — Lentil
Red gram — Pigeon pea
Reisbohne — Rice bean
Revithia — Chick pea
Rice haricot bean — Tepary bean
Roaj(galing) — Lima bean
Rounceval pea — Asparagus bean

Saar-sar — Jack bean
Saat-erbse — Pea
Saat platterbse — Grass pea
Sabberi — Grass pea
Sabidokong — Winged bean
Sabre bean — Jack/Sword bean
Sadaw-pè — Pea
Saeme — Hyacinth bean
Sai dau-kok — Cowpea
Salad bean — Haricot bean
Salboco bulluc — Haricot bean
Salbocoghed — Pigeon pea
Sangalu — Chick pea

San-du-si	— Broad bean	Sigró	— Chick pea
San-to-pè	— Broad bean	Sim	— Hyacinth/Rice/ Runner bean
Santal	— Grass pea	Sin-gaung-kala-pè	— Chick pea
Sasage	— Cowpea	Sisser erwt	— Chick pea
Saubohne	— Broad bean	Sita-mas	— Rice bean
Sawawa	— Pea	Sitao	— Asparagus bean
Scarlet runner bean	— Runner bean	Six weeks bean or pea	— Asparagus bean
Scimitar bean	— Sword bean	Slanutok	— Chick pea
See-kok-tau	— Winged bean	Snake bean	— Asparagus/Jack bean
Segidilla	— Winged bean	Snap bean	— Haricot bean
Seguidilla	— Winged bean	Sora-mame	— Broad bean
Seim bean	— Hyacinth bean	Southern pea	— Cowpea
Sem	— Hyacinth/Lima/ Runner bean	Spizole	— Chick pea
Sequidilla	— Winged bean	Split pea	— Lentil
Sererella	— Winged bean	Stick bean	— Runner bean
Sesquidilla	— Winged bean	Stizolobia	— Velvet bean
Sewee bean	— Lima bean	Stone groundnut	— Bambara groundnut
Shambe kayi	— Winged bean	Stragaliais	— Chick pea
Shembera	— Chick pea	Strauch erbse	— Pigeon pea
Sheuit aduma	— Runner bean	String bean	— Haricot/Runner bean
Sheuit hagina	— Haricot bean	Sundal kadalai	— Chick pea
Sheuit tzehuba	— Mung bean	Sunshine bean	— Jack bean
Shihu	— Chick pea	Sut(a)ri	— Rice bean
Shiltong	— Rice bean		
Shim(bi)	— Hyacinth bean	Tagalo patani	— Lima bean
Shimbera	— Chick pea	Takarmany borsó	— Pea
Shimbrah	— Chick pea	Take-azuki	— Rice bean
Shumbra	— Chick pea	Taklauo	— Rice bean
Siam bean	— Cluster bean	Tapilan	— Rice bean
Sibatse simaron	— Lima bean	Tau-kok	— Cowpea
Sieva bean	— Lima bean		
Sigarilya	— Winged bean		

Tayôk-pè — Broad bean
Teiko — Haricot bean
Teora — Grass pea
Texan bean — Tepary bean
Thattapayru — Cowpea
Thogari — Pigeon pea
Thovaroy — Pigeon pea
Thulukkappayar — Lentil
Thuvara(i) — Pigeon pea
Thuvaran — Pigeon pea
Tick bean — Broad bean
Tientsin green bean — Mung bean
Tikari — Urd
Timbolillo — Pigeon pea
Tiuri — Grass pea
Tjeepir (bee-bas or weloo) — Winged bean
Tjeetjeepir — Winged bean
Togare/i — Pigeon pea
Tonga bean — Hyacinth bean
Tonkin pea — Cowpea
Too-afuk yaou — Asparagus bean
Too-a kee-o — Mung bean
Too-a lund tow — Sword bean
Too-a-poo — Winged bean
Tovarai — Pigeon pea
Tremoçeiro — Lupin
Tremoço — Lupin
Tsuru-adsuki — Rice bean
Tua dam — Cowpea
Tua kack — Haricot bean
Tua kiew — Mung bean
Tua kio — Mung bean
Tua kok — Asparagus bean
Tua nang — Hyacinth bean
Tua pab — Hyacinth bean
Tua pep — Hyacinth bean
Tua phum — Haricot bean
Tua pra — Sword bean
Tua pu — Winged bean
Tua rachamat — Lima bean
Tua-re — Pigeon pea
Tua tawng — Mung bean
Tua tong — Mung bean
Tuibonen — Broad bean
Tulkayrai — Moth bean
Tukur-ater — Pea
Tum bekai — Sword pea
Tur(a) — Pigeon pea
Turkish gram — Moth bean
Turmas — Lupin
Turmus — Lupin
Tuvar(a)(i) — Pigeon pea
Tuvaray — Pigeon pea
Tuvarika — Pigeon pea
Tuver(a) — Pigeon pea
Twin flowered bean — Hyacinth bean

Udad — Urd
Uddulu — Urd
Udid — Urd
Uhunnu — Urd
Ulavalu — Horse gram
Ulundu — Urd
Undu — Urd
Urad — Urd

Urahi — Hyacinth/ Runner bean

Uri — Hyacinth/ Runner bean

Urid(kai) — Urd

Vah — Cluster bean

Vahki-phali — Cluster bean

Val — Hyacinth bean

Vanzon — Bambara groundnut

Valavarai — Sword bean

Vanlig blomsterböna — Runner bean

Vella tamma — Jack bean

Vellai tambattai — Jack bean

Vicos — Broad bean

Vigna einese — Cowpea

Vika bob — Broad bean

Vilaiti matar — Pea

Vilaiti sem — Haricot bean

Vio vio — Pigeon pea

Vlci — Lupin

Voamba — Cowpea

Voandzou — Bambara groundnut

Voanjobory — Bambara groundnut

Voehm — Cowpea

Voeme — Cowpea

Waalse boon — Broad bean

Waby bean — Hyacinth bean

Waken kurawa — Pigeon pea

Wal (papdi) — Hyacinth pea

Walee-pè — Cluster bean

Watani — Pea

Wierboon — Broad bean

Windsor bean — Broad bean

Winged pea — Winged bean

Wolfsbohne — Lupin

Wolfsboon — Lupin

Wonder bean — Sword bean

Wooly pyrol — Urd

Wuch — Cowpea

Yaenari — Mung bean

Yard (long) bean — Aspargus bean

Yayenari — Mung bean

Yeguas — Lima bean

Yellow dhal — Pigeon pea

Yellow gram — Chick pea

Yeshil bakla — Broad bean

Yeshil fasulya — Haricot bean

Yewof-ater — Pigeon pea

Zabache — Lima bean

Zada-adagonna — Haricot bean

Ziesererbsen — Chick pea

Ziserbohne — Chick pea

0543695/3M/7.79/LOM